Maarif Global Education Series

Series Editors

Birol Akgün,
Ankara Yıldırım Beyazıt University
Ankara, Türkiye

Yusuf Alpaydın,
Marmara University
Istanbul, Türkiye

Maarif Global Education Series addresses the transformations happening in information science, economy, politics, technology, people, and society and discusses how these affect education policies from preschool to higher education on a global scale, with a macro perspective. The series offers practical, realistic, sustainable policy recommendations and strategies.

The concept of Maarif, included in the name of the series, represents self-knowledge and respect for one's self and others, excitement in learning and teaching, and humility towards truth. Therefore, the works published in the series in the form of edited books and monographs discuss the theory and philosophy of education with a global perspective, including both western and eastern educational experiences.

The books published under the series would be beneficial for decision-makers, educators, academicians, and higher education students studying in education fields to learn about educational experiences of different cultures and to understand contemporary transformation in the field. The books in the series will be prepared in academic style contributed by authors from different countries.

Rıdvan Elmas • Murat Akarsu
Editors

STEM Education

Shaping Future Learning Practices in the Age of AI

palgrave
macmillan

Editors
Rıdvan Elmas
Atatürk Faculty of Education
Marmara University
Istanbul, Türkiye

Murat Akarsu
Faculty of Education
Ağrı İbrahim Çeçen University
Ağrı, Türkiye

ISSN 2730-7646 ISSN 2730-7654 (electronic)
Maarif Global Education Series
ISBN 978-981-95-9022-3 ISBN 978-981-95-9023-0 (eBook)
https://doi.org/10.1007/978-981-95-9023-0

This book is an open access publication.

Turkish Maarif FoundationThis work was supported by Turkish Maarif Foundation.

Cover Credit: @ Getty image: Issarawat Tattong

This Palgrave Macmillan imprint is published by the registered company Springer Nature Singapore Pte Ltd.
The registered company address is: 152 Beach Road, #21-01/04 Gateway East, Singapore 189721, Singapore

Contents

Contributors

Amber Adgerson University of North Dakota, Grand Forks, ND, USA

Murat Akarsu Faculty of Education, Ağrı İbrahim Çeçen University, Ağrı, Türkiye

Ayşe Akkır Turkish Maarif Foundation, Istanbul, Türkiye

Muhammet Arıcan Boğaziçi University, Istanbul, Türkiye

Marta Romero-Ariza University of Jaén, Jaén, Spain

Martin Bilek Charles University, Prague, Czechia

Ismail Çuhadar Ministry of Education, Istanbul, Türkiye

Mutlu Cukurova University College London, London, UK

Kosta Dolenc University of Maribor, Maribor, Slovenia

Rıdvan Elmas Atatürk Faculty of Education, Marmara University, Istanbul, Türkiye

Kristof Fenyvesi University of Jyväskylä, Finnish Institute for Educational Research, Jyväskylä, Finland
Global Research Institute for Finnish Education, The Education University of Hong Kong, Ting Kok, Hong Kong

Douglas K. Hartman Department of Teacher Education, College of Education, Michigan State University, East Lansing, MI, USA

Anna-Leena Kähkönen University of Jyväskylä, Jyväskylä, Finland

Anssi Lindell University of Jyväskylä, Jyväskylä, Finland

Jan Lundell University of Helsinki, Helsinki, Finland

Brooke Max Purdue University, West Lafayette, IN, USA

Christine McGrail University of North Dakota, Grand Forks, ND, USA

Monika Pelikanova Charles University, Prague, Czechia

Şenay Purzer Purdue University, West Lafayette, IN, USA

Antonio Quesada University of Jaén, Jaén, Spain

Elizabeth Suazo-Flores University of North Dakota, Grand Forks, ND, USA

Pierre Sutherland Sparx Learning Ltd., Exeter, UK

Merve Adıguzel Ulutaş Gazi University, Ankara, Türkiye

Mateja Ploj Virtic University of Maribor, Maribor, Slovenia

Peter Wulff Ludwigsburg University of Education, Ludwigsburg, Germany

Beyzanur Yalvaç Boğaziçi University, Istanbul, Türkiye

Tuğba Yüksel Recep Tayyip Erdoğan University, Çayeli/Rize, Türkiye

List of Figures

List of Tables

CHAPTER 1

What Can a Century of Scholarship on Teacher Professional Development Teach Us About Equipping Today's STEM Teachers to Use GenAI Effectively and Ethically?

Douglas K. Hartman

Introduction

Generative AI offers hitherto unimagined possibilities to improve student learning in STEM, yet evidence-based guidance on how to prepare teachers for this new frontier is scarce.

Educators broadly agree that many teachers would benefit from professional development (PD) to use Generative AI (GenAI) effectively (e.g., Miao et al., 2021). Recommendations abound, with optimism and funding flowing from many corners (e.g., Exec. Order No. 14277, 2025; Tomaškinová & Tomaškin, 2024; The White House, 2025). Promises of

D. K. Hartman (✉)
Department of Teacher Education, College of Education, Michigan State University, East Lansing, MI, USA
e-mail: dhartman@msu.edu

R. Elmas, M. Akarsu (eds.), *STEM Education*, Maarif Global Education Series,
https://doi.org/10.1007/978-981-95-9023-0_1

widely accessible, better optimized, and more intuitive AI PD are frequently made (e.g., Gârdan et al., 2025). Framework, roadmap, competency, and policy documents for GenAI PD are paraded on social media (e.g., Chan, 2023; Fadel et al., 2024). And AI Literacy courses and workshops for teachers abound (e.g., AI for Education, 2025). While only a few small-scale, high-intensity studies are underway, the fact remains that there is a "notable gap in empirical research" on effectively preparing teachers to utilize generative AI in ways that measurably improve student learning in STEM (Zhai & Krajcik, 2025, p. 13).

This research gap is understandable. Demonstrating a causal connection between PD and student learning is notoriously difficult (Kennedy, 2016). In the last 20 years there have been a few studies that do so (e.g., Atlay et al., 2019; Grönqvist, Öckert & Rosenqvist (in press); Gustafsson & Nilsen, 2016; Harris & Sass, 2011; Johnson & Fargo, 2010 Wallace, 2009). The largest, most sophisticated, and most ambitious of these studies was published recently. A team of Swedish researchers tackled a fundamental question: What effect does typical STEM PD have on student learning? They analyzed data from more than 5000 grade 4 and 8 students in more than 2000 STEM classrooms from over two dozen countries across a 16-year period (Kirsten et al., 2023).

Results from the study were clear: typical STEM PD has a slightly negative effect on student learning in mathematics and science. Put plainly, this result means that grade 4 students in a US urban classroom understand slightly less about subtracting fractions with like denominators after their teacher participates in PD that focuses on helping students deepen their understanding of number and operations with fractions. Or that grade 8 students in a Turkish village classroom learned slightly less about energy transformations after their teacher participated in PD that focused on using hands-on lab-work for students to investigate heat from friction or build a rubber band car.

Even more concerning than the overall decrease was the finding that higher-performing students were disproportionately negatively affected by typical STEM PD. Simply put, higher-performing students were affected more negatively when their teachers participated in typical STEM PD. In practical terms, this result meant that high-achieving grade 4 students in a US suburban primary classroom learned less about solving multi-digit multiplication problems when their teacher had more PD. Or that more able grade 8 students in a Turkish public middle school classroom learned less about photosynthesis and plant cell functions when their teacher had more PD.

Why would typical STEM PD have a negative effect on student learning, especially among higher-performing students? Indirect evidence indicates a number of reasons.

1. *Generic Content*—Teachers experience PD featuring generic STEM content disconnected from their actual classroom, curriculum, teaching, goals, and student needs (Garet et al., 2001). For example, an elementary PD session in mathematics (Grade 4 fractions) that broadly covers mathematics pedagogy but does not specifically address teaching challenges related to fraction equivalence.
2. *Passive Learning*—STEM PD primarily involves passive, lecture-based presentations rather than engaging teachers in active participation, interaction, or hands-on practice (Desimone et al., 2013). For instance, a middle-school science PD (Grade 7 ecosystems) where teachers simply watch a PowerPoint presentation on ecosystem interactions without actively participating in collaborative problem-solving or inquiry-based methods.
3. *Limited Collaboration*—STEM teachers often engage in PD individually, with limited or no collaboration with peers, preventing mutual learning or the exchange of ideas (Penuel et al., 2012). For example, a PD session in Grade 9 biology (cellular processes) that provides online self-paced modules where teachers have no opportunities for peer collaboration or group discussions to deepen their understanding.
4. *Insufficient Modeling*—STEM Teachers experience PD that lacks explicit modeling of instructional techniques or practical examples, making it difficult to apply learned strategies in the classroom (Opfer & Pedder, 2011). A PD workshop in Grade 8 physics (forces and motion) that discusses conceptual frameworks but does not provide demonstrations or modeling of inquiry-based activities involving force diagrams.
5. *Inadequate Coaching*—STEM PD does not provide access to qualified mentors or experts who can offer personalized feedback, guidance, or support (Kraft et al., 2018). A PD in high school chemistry (Grade 11 stoichiometry) in which teachers receive instructional materials without expert coaching or feedback to address misconceptions or implementation challenges.

6. *Minimal Reflection*—STEM teachers are given minimal or no structured opportunities to reflect on their instructional practices, thus limiting their professional growth and self-assessment (Jay & Johnson, 2002). A PD in Grade 10 geometry where teachers learn new technology tools for visualization but have no structured reflection sessions afterward to evaluate their implementation effectiveness.
7. *Standalone Sessions*—STEM teachers engage primarily in short, episodic PD sessions without ongoing support or sustained professional learning opportunities, limiting deep changes in instructional practice (Yoon et al., 2007). A single half-day PD session in middle school technology (Grade 8 coding/programming) introducing teachers to Python basics, with no subsequent follow-up or continued learning opportunities.

These patterns are not new; they echo centuries of critiques of PD, which we examine in the next section.

Background

These reasons for PD's negative effects are not new. There have been long-standing efforts by educators to grapple with these issues, as far back as the fifth- and fourth-century BCE. Evidence indicates that the Socratic and Philosophical Schools had discussions about how best to equip those who teach (Jaeger, 1986). The Confucian Education System about the same time prepared teachers through intensive personal apprenticeship, and later the Sui/Tang Dynasty's Exam System required mastery of Confucian texts and pedagogy by tutors (Lee, 2000). Rome's Rhetorical Schools offered PD with a flourish: training in oratory, persuasion, and pedagogy, with Quintilian's first-century textbooks *Institutio Oratoria* offering guidance to teachers on what we'd recognize today as educational theory, classroom management, and instructional methods for that time (Quintilian, 1920–1922).

The Islamic Madrasas of the eighth through fifteenth centuries provided structured apprenticeships for teachers, emphasizing methodologies for textual analysis, public speaking, debate, and classroom interaction, with the Endowment-Funded (Waqf) Madrasas supporting and developing instructional standards (Makdisi, 1981). And the Ottoman Empire's Madrasses and Palace Schools offered PD to religious teachers and leaders with explicit pedagogical methods and formal certification processes (Ergin, 1977; Flerischer, 1986; Şahin, 2013).

During Europe's medieval era (ninth through sixteenth centuries), the Monastic Schools structured teacher development through mentoring, manuals, and formal pedagogical practices. And the Cathedral Schools provided practical PD for clergy-teachers and employed a curriculum that included instructional methods (Leclercq, 1982). European efforts to expand teacher PD were considerable during the sixteenth through nineteenth centuries.

- The *Jesuit Schools* established professional training for teachers, emphasizing instructional methods and classroom discipline (Grendler, 2019; O'Malley et al., 1999).
- The *Christian Brothers* in France established teacher-training institutions that also carried out teacher PD on practical classroom management, curriculum instruction in reading, writing, and arithmetic, and the use of pedagogical methods (Koch et al., 2004; Pauly, 2016). Their instructional manuals like *The Conduct of Schools* (1720) were widely used across Europe and beyond.
- The *Teacher Seminaries* of the Pietist Movement in Germany promoted teacher PD through seminaries for primary teachers, with the first Prussian teacher seminary in Berlin established in 1748 to focus on pedagogy and curriculum (Barnard, 1876; Gawthrop, 1993).
- The *Monitorial System* of Joseph Lancaster and Andrew Bell in England and Scotland focused on PD that included training sessions and observation of demonstration lessons in Model Schools (Kaestle & Vinovskis, 1980; Oswald, 1987).
- The post-Gutenberg *Texts and Teachers Manuals* provided PD advice to teachers on practical pedagogy, curriculum design, and classroom management (e.g., Comenius, 1657/1957; de La Salle, 1720/1996; Pestalozzi, 1801/1977), and were often the basis for teacher study groups, seminars, and workshops across Europe (Cubberley, 1920).
- The *Teacher Seminars* that were state-sponsored in Prussia began in the late 1700s, offered PD to teachers in state-supported primary schools that focused on pedagogy, psychology, curriculum, and classroom management (Melton, 2003; Schleunes, 1989). These institutions profoundly influenced educational leaders worldwide, including Horace Mann in America (Messerli, 1972).

In North America, concerns about the systematic design of teacher PD and its impact on student learning appeared in the early 1800s (Frazier, et al., 1933). Prior to this, PD on the continent was largely informal and apprenticeship-based, as practiced in dame schools (Cremin, 1970), apprenticeship to a schoolmaster (Urban & Wagoner, 2009), and ministerial training (Kaestle, 1983). Those concerns were addressed in a re-shaped way when the *Lexington Normal School* was established in 1839 (Cubberley, 1920). While the Massachusetts school initially prepared new teachers, its foundational principles and models came to influence PD practices in the decades that followed (Labaree, 2005). This influence spotlighted a number of elements viewed as key PD practices during that era.

Above all, the *School* put pedagogical content knowledge front and center, embedding it into every strand of PD. Mann advocated for teacher learning that was explicitly connected to the subject areas they would teach, emphasizing reading, arithmetic, geography, and moral education. He voiced this content focus by writing, "Teachers must possess an accurate knowledge of the branches which they are to teach," underscoring content specificity as foundational (Messerli, 1972, p. 306).

Active, experiential learning formed the cornerstone of the *School's* approach to teacher learning. Teachers practiced instructional methods actively through demonstration lessons and peer teaching experiences. Mann stressed practical learning, asserting, "Knowledge, until it has been digested and assimilated by the learner's own mind, is not his own" (Fraser, 2001, p. 70). To his way of thinking, the experience of actively learning equipped teachers to directly apply instructional methods more effectively.

The *School's* influence on PD explicitly encouraged collaborative learning among practicing teachers. Mann emphasized the professional community, fostering regular dialogue among teachers for mutual improvement. Historical records indicate that the PD shaped by Lexington and other normal schools facilitated ongoing interaction among teachers, establishing a precedent for peer-based learning communities (Cubberley, 1919). Mann's approach illustrates how early models of teacher education recognized peer collaboration as integral to professional growth.

The *School's* incorporation of demonstration or model classrooms exemplified a commitment to modeling. Teachers regularly observed master teachers modeling effective instructional practices in real classrooms as part of PD. Mann's reports noted the effectiveness of model schools and modeling, observing, teachers learn best by observing the "exemplification and embodiment of [excellent teaching principles], in practice" (Mann, 1845, p. 71). These demonstrations provided opportunities for teachers to model for colleagues, as well as to observe fellow teachers modeling their practice.

Mann's approach included direct mentorship and feedback from experienced educators, akin to contemporary expert coaching. At Lexington, skilled mentors provided personalized guidance, observing teachers during lessons and offering immediate feedback. This early form of coaching was highlighted by Mann himself, who praised individualized attention from experienced teachers as crucial for professional growth (Peterson, 2010).

The PD shaped by the *Lexington Normal School* also encouraged teachers to continuously reflect on their practices as a way to refine their methods. It was a key element of the School's philosophy. Mann articulated the importance of reflective practice, stating teachers must "reflect critically upon their own experiences" to ensure effective teaching (Messerli, 1972, p. 310). This ongoing commitment to reflection established a mindset that was integral to the *School's* PD.

The *School's* approach to teacher learning signified a commitment to long-term, sustained PD rather than brief training. Mann envisioned the education of teachers as an enduring professional endeavor, emphasizing ongoing improvement rather than short-term interventions. He advocated for regular teacher institutes and ongoing support over extended periods, a principle reflected in the establishment of annual institutes and continuous learning cycles (Groen, 2008).

Through the start of the twentieth century, PD continued to focus on content (Frazier, et al., 1933).The idea of "professionalized subject matter" (p. 428) had gained traction as evidenced in Gray's (1919) article, "Methods of Improving the Technique of Teaching." Gray's emphasis on the scientific analysis of teaching practice marked a pivotal moment, shifting the focus of PD toward systematic, classroom-based inquiry. This perspective broadened in the mid-twentieth century, a period characterized by the growth of structured "in-service education" designed to keep practicing teachers updated on new curricula and pedagogical theories (Henry, 1957; Mattock, 1989/2012). Influenced by the curriculum reform movement of the post-Sputnik era, this approach often involved experts disseminating knowledge to teachers in workshops, reinforcing a model where best practices were developed externally and delivered to educators for implementation (Tyack & Cuban, 1995). While this expanded access to formal training, the emphasis remained largely on the technical aspects of teaching, with less focus on teachers' own decision-making processes or the specific contexts of their schools and students.

The latter half of the twentieth century ushered in a significant cognitive shift in the understanding of teaching and, consequently, in the design of PD. Moving away from a purely behavioral focus on teaching "techniques," researchers began to investigate the complex mental lives of teachers—how they planned, made instructional decisions, and reflected on their practice (Clark & Peterson, 1986). This cognitive revolution gave rise to influential concepts such as Shulman's (1986) development of pedagogical content knowledge (PCK), which highlighted the critical importance of teachers' deep, specialized knowledge when blending content and pedagogy. This new focus suggested that effective PD could not simply transmit skills; it had to cultivate teachers' professional judgment and their unique knowledge of how to represent subject matter in ways that were comprehensible to students.

Internationally, efforts to design and implement effective PD have been unfolding. In the United Kingdom, for instance, policy-driven reforms from the 1980s onward led to a focus on standards-based accountability, which heavily influenced the nature of teacher training and in-service support (Department for Education, 2011). More recently, collaborative, inquiry-based models like Lesson Study, adapted from Japan, have gained significant traction as a powerful form of classroom-embedded professional learning (Dudley, 2014). Across Europe, large-scale comparative studies like the OECD's Teaching and Learning International Survey (TALIS) have provided a panoramic view of PD, revealing a common emphasis on collaborative learning while also highlighting variations in duration and intensity across nations (OECD, 2019). Singapore has scaled up Japan's lesson study (Lewis et al., 2006) into a system-wide cycle of collaborative lesson design and analysis (Hairon, 2020). Finland emphasizes sustained teacher autonomy, making reflection and long-term growth the backbone of PD (Sahlberg, 2021). Shanghai's research lesson tradition builds on open-class peer critique (Yang et al., 2022), while Chile and Kenya are experimenting with teacher networks and mobile-based coaching (Miranda & Chapman, 2024; RTI International, n.d.). In Türkiye, national education reforms have repeatedly targeted teacher development as a central pillar for system-wide improvement, with a growing body of scholarship examining the challenges of implementing sustainable, high-quality PD within a centralized education system (e.g., Bümen, 2009).

This historical and global evolution has gradually led to a broad, research-based consensus about what makes PD effective. The early-twentieth-century focus on specific content and technique, the mid-century emphasis on curriculum implementation, and the later cognitive turn toward teacher knowledge have not been discarded but rather integrated into a more holistic framework. Scholars began synthesizing decades of findings to identify the core features of high-quality PD that were consistently linked to changes in teacher practice and improvements in student learning (e.g., Garet et al., 2001).

These efforts have culminated in the contemporary understanding that effective professional learning is not a singular event but a complex, multifaceted process. It is this rich, century-long legacy of research and practice, a kind of collective *marifet*—from Gray's early calls for a "science of teaching" to the global pursuit of collaborative, sustained, and context-specific models—that informs the core elements of effective PD detailed in the next section. These elements represent the distillation of what has been learned about supporting and improving the intricate work of teacher learning.

Elements

A number of researchers have identified elements that are associated with effective PD in education (e.g., Cordingley et al., 2015; Darling-Hammond et al., 2017; Desimone, 2009; Hill & Papay, 2022; Sykes & Wilson, 2016; Timperley, 2015). While there is some overlap among elements, there is no consensus among scholars (Sims & Fletcher-Wood, 2021). Seven of the most common elements are described, explained, and illustrated below.

Seven long-standing elements of effective PD stand out in the literature. While each of these elements is informed by evidence and the "wisdom of practice" from the century preceding the advent of pre-GenAI. Waiting for sufficient evidence can lead to systematic neglect of certain equity-oriented practices that could benefit teachers and students. And waiting also presents an inherent paradox: focusing too much on gathering evidence before implementing GenAI-embedded PD can paradoxically make it harder to collect the necessary information to gather evidence to understand and improve Gen-AI embedded PD efforts. This chapter, modest in its scope, proceeds with extrapolating findings from the pre-GenAI era to PD that has GenAI embedded in it. A description, explanation, and depiction of six key findings follow.

Content Focus

The literature indicates that effective PD focuses on a specific curricular area, connecting teacher learning with the subject she teaches (Darling-Hammond et al., 2017). For example, a grade 8 science teacher benefits when PD connects to her subject area. While this connection is necessary, it can still fall short if the connection is too loose or broad (Mundry, 2005). PD becomes less effective when it offers generic practices without meaningfully integrating science-specific content. Instead, high-impact PD drills deeply into specific concepts, processes, and practices central to both the teacher's curriculum and the day-to-day experiences of her grade 8 students. By zooming in on particular science topics—such as the difference between *contact* and *non-contact* forces within Newton's three laws of motion—teachers deepen their understanding of how a student learns a concept, better anticipate potential misconceptions, clarify difficult ideas, and tailor the progression of content in a lesson so it addresses students' learning needs more directly.

GenAI can serve as a partner for teachers in this science-focused PD. For instance, a grade 8 science teacher teaching Newtonian forces could use GenAI to locate and curate customized resources, such as classroom-ready demonstrations or explanations tailored to common student misconceptions, precisely when they are needed. In Japan, PD often zeroes in on dissecting a single math problem in exquisite detail, showing how depth—rather than breadth—sharpens both teacher insight and student understanding (Nakamura, 2019; Takahashi, 2017) By drawing on datasets of curricular materials and student learning progressions, PD that is science-specific focused can use GenAI tools to suggest sequences of learning activities closely aligned with key scientific content, thereby helping teachers plan lessons with greater fidelity to diverse learning needs. Furthermore, by integrating GenAI into the PD process, a teacher receives feedback on how well lesson plans or assessments match specific science standards and objectives, ensuring that PD keeps the focus firmly on the teacher's actual curriculum and the distinct needs of her students.

To effectively use GenAI in STEM PD, teachers benefit from moving beyond broad, one-off AI interactions. Instead of simply learning generic "best practices" for using GenAI, they could learn how to craft a prompt for the learning of a fine-grained science concept or how to build a series of linked ("daisy-chained") prompts that iteratively scaffold the design of a customized lesson. With this focus, GenAI can assist with recommendations or resources that directly address specific content—like *contact* and *non-contact* forces—rather than offer one-size-fits-all suggestions. When STEM PD aligns GenAI use with specific science content, feedback becomes precise rather than generic.

Active Learning

Effective PD engages teachers in actively learning how to prepare a lesson, analyze a student's work, and solve an instructional problem. For instance, a grade 11 calculus teacher participates in PD where she analyzes the homework of students. Yet active learning is more than just having a teacher be active in her learning; it becomes less effective when the tasks are generic or disconnected from the authentic work of teaching. Instead, high-impact PD engages teachers in the specific practices of their discipline, moving beyond abstract discussions to the concrete analysis of classroom artifacts (Garet et al., 2001). By examining a student's specific algebraic error in applying the chain rule, for example, a calculus teacher can practice diagnosing the misconception, anticipating similar errors in other students, planning a targeted re-teaching strategy, and rehearsing that strategy in a simulation session.

GenAI can function as a dynamic simulation partner to deepen this active learning. The same grade 11 calculus teacher could use a GenAI tool to generate multiple, varied examples of student work that contain common but distinct errors in applying the chain rule. In Brazil, inquiry-based PD programs encourage teachers to rehearse hands-on science experiments as if they were their own students (de Castro et al., 2018). By drawing on data about student misconceptions, GenAI can generate realistic practice scenarios for teachers, permitting the teacher to hone her diagnostic skills. Furthermore, the teacher could engage the GenAI in a role-play, prompting it to act as a struggling student, so she can actively practice her questioning and explanation techniques in a low-stakes setting.

To leverage GenAI for active learning, STEM PD is more useful when it guides teachers beyond using it as a simple information-retrieval engine. Instead of just asking GenAI to *list* common student errors, teachers can be supported in learning to prompt it to *simulate* authentic classroom artifacts or *embody* a student persona or role-playing experience. With this focus, GenAI is transformed from a passive knowledge repository into an interactive sparring partner that facilitates hands-on practice. By integrating GenAI's simulative capabilities into active learning PD, STEM teachers will be better equipped to analyze student thinking and refine their instructional responses in ways that directly shape their words and actions in the classroom.

Peer Collaboration

Effective PD brings teachers together to work with grade level or subject area colleagues at their school, actively sharing ideas about how to solve problems of practice from their teaching of specific STEM concepts, processes, and practices, and in turn working to improve their in situ practice collectively. For example, a grade 3 teacher participates in PD that regularly supports the co-planning of integrated science lessons with other grade 3 teachers in his school. This collaboration can be ineffective, however, if it remains at the surface level of sharing materials without a shared inquiry into teaching and learning (Penuel et al., 2012). High-impact PD, in contrast, structures collaboration around specific classroom artifacts. For instance, deep collaboration among a grade 3 teaching team would move beyond just co-planning into collaboratively analyzing student work from a shared science lesson on plant life cycles, then using that direct evidence to identify common misconceptions and collectively refine their instructional approach for the next day.

GenAI can serve as an analytical partner to streamline and deepen this collaborative work. The grade 3 teaching team, for example, could upload anonymized samples of their students' written explanations of the plant life cycle to a GenAI tool. Singapore's adaptation of Japan's Lesson Study illustrates this well: teachers co-design, observe, and revise lessons together, showing how structured collaboration can become a professional 'laboratory of practice' (Hairon, 2020). By using natural language processing to identify patterns across the dataset of student responses, the GenAI could generate a summary of the most prevalent misconceptions, providing the team with a data-informed starting point for their discussion. Furthermore, after analyzing the student work, the teachers could use the GenAI to co-author differentiated follow-up activities, ensuring that their collaborative planning is responsive to the specific learning needs they have collectively identified.

To effectively use GenAI in collaborative PD, STEM teachers may be better served by learning to use it as a shared engine for inquiry rather than an individual resource-gathering tool. Instead of teachers individually prompting an AI for lesson ideas, the team learns to interactively craft shared prompts that ask the GenAI to analyze their collective student data or act as a "critical friend" to interrogate a proposed lesson plan. With this focus, GenAI is transformed from a digital filing cabinet into a facilitator that can help provide substance for deep, evidence-based conversations among colleagues. By integrating GenAI's analytical capabilities into peer collaboration, PD can equip STEM teaching teams to more fully and efficiently identify student needs and co-design materials and instruction, fostering a professional learning community whose responses are more targeted and timelier.

Modeled Practice

Effective PD incorporates real-world demonstrations of exemplary teaching to illustrate specific instructional practices clearly. For instance, a grade 11 physics teacher can participate in PD that involves analyzing video demonstrations of inquiry-based lab investigations by a master physics teacher. However, this approach can be ineffective if the models presented are disconnected from the specific content of a teacher's own classroom context or fail to show the nuances of student interaction. High-impact PD moves beyond simply showing a polished final performance. It instead models the entire instructional cycle, including the messy parts of teaching (Opfer & Pedder, 2011). This could involve the physics teacher collaboratively watching a video of a master teacher, not just conducting a flawless demonstration, but also struggling with a student's misconception about kinetic energy and then modeling, in real-time, how to diagnose and address that confusion through targeted questioning.

GenAI can serve as a repository and a curation partner, providing teachers with on-demand access to a wide array of instructional models. The grade 11 physics teacher could use a GenAI tool to find and filter a library of video-based teaching models, searching specifically for clips that show experienced teachers addressing the exact misconception about kinetic energy that his students are exhibiting. In Shanghai, the tradition of open classes permits teachers to publicly demonstrate lessons for peer critique (Yang et al., 2022). Drawing on vast datasets of educational videos and instructional materials, the GenAI can locate hyper-specific examples of practice, saving the teacher valuable time and providing targeted support. Furthermore, the teacher could upload a transcript of his own teaching and ask the GenAI to compare his questioning patterns to those in the expert models, providing private, personalized feedback on his instructional technique.

To leverage GenAI effectively for modeling, STEM PD can guide teachers to use it as an analytical tool, not just as a video search engine. Instead of simply watching a video, teachers can learn how to prompt GenAI to create an annotated transcript of the model lesson, highlighting key instructional moves, decision points, and pedagogical reasoning. With this focus, GenAI is transformed from a passive library into an interactive seminar that deconstructs expert practice. By integrating GenAI's teacher-initiated analytical capabilities with a rich library of teaching models, STEM PD can offer teachers a customized way to observe, analyze, and ultimately incorporate exemplary models of instructional practices into their own teaching.

Expert Coaching

Effective PD employs on-the-ground assistance from an expert in the teacher's subject area to provide direct guidance, individual observation, personalized feedback, and on-the-spot demonstrations. For example, a grade 6 middle school mathematics teacher participates in PD that provides a coach who conducts in-class observations and feedback for her as she implements a new curricular approach. This coaching is less effective, however, when it consists of infrequent, "drive-by" observations with generic checklist feedback that feels more evaluative than supportive. In contrast, high-impact coaching is cyclical and collaborative (Kraft et al., 2018). It involves a pre-conference where the grade 6 math teacher and her coach set a specific instructional goal for specific content—such as improving questioning to elicit student thinking about proportional reasoning in next week's lesson—followed by the observation and a post-conference where they analyze lesson evidence together to plan concrete next steps.

GenAI can function as an analytical assistant, augmenting the coaching cycle by providing data for the teacher and coach to discuss. The grade 6 math teacher could record her lesson on proportional reasoning, and a

GenAI tool could then generate an anonymized transcript and teacher-initiated analysis of the classroom discourse. In Türkiye, reforms in the early 2000s emphasized school-based coaching to strengthen teacher self-efficacy (Bümen, 2009; Karagöz & Ror, 2020). By using natural language processing, GenAI could quantify the ratio of teacher-to-student talk, categorize the cognitive demand of the teacher's questions, and map patterns of student participation, providing specific data points to ground the coaching conversation. Furthermore, between coaching sessions, the teacher could use the GenAI as a simulator, prompting it to role-play a student with a specific misconception about proportions, allowing her to practice the new questioning techniques recommended by her coach in a safe, repeatable environment.

To leverage GenAI effectively within a coaching framework, STEM PD can help equip teachers and coaches to use it as a tool for generating actionable insights, not just descriptive reports. Instead of merely reviewing a dashboard of metrics, the teacher and coach can work toward co-authoring prompts that recursively ask the GenAI for deeper analysis, such as, "Identify three moments in the transcript where a student's partial understanding of ratios was evident and suggest an alternative question I could have asked." This focus transforms the GenAI from a passive data collector into an active contributor to the coaching dialogue. By integrating GenAI as an analytical partner, expert STEM coaching can become more precise, evidence-based, and efficient, allowing the teacher and coach to devote their valuable time to the pedagogical conversations that inform instructional improvement.

Ongoing Reflection

Effective PD provides regular opportunities for a teacher to critically evaluate and reflect on their teaching, so they continuously improve and refine their instruction. For instance, a grade 9 biology teacher participates in PD that supports her regularly reflecting on lab activities with peer and coach feedback. This process becomes less effective when reflection is unstructured or superficial, amounting to little more than a brief, intuitive summary of how a lesson felt (Jay & Johnson, 2002; Korthagen & Vasalos, 2005). High-impact PD, by contrast, structures reflection around specific evidence and a clear focus (Little, 2002). This would involve the biology teacher using a protocol to analyze student lab notebooks from a lesson on cellular respiration, specifically looking for evidence of how students connected data to their scientific claims. This transforms reflection from a casual practice into a rigorous, data-informed analysis of teaching and learning.

GenAI can function as a "thought partner," creating a private, non-judgmental space for teachers to engage in deep, evidence-based reflection. The grade 9 biology teacher could upload an anonymized transcript of a student discussion from her cellular respiration lab and prompt the GenAI tool to analyze it using specific criteria or pedagogical frameworks. Finland's emphasis on teacher autonomy underscores the essential role reflection plays in teacher PD (Chung, 2023; Niemi, 2015). The GenAI could identify patterns in the discourse, highlighting moments where student thinking was made visible and posing reflective questions like, "At this point in the discussion, what evidence did the students use to support their claim, and how might you have prompted them to consider alternative explanations?" Furthermore, the teacher could engage the GenAI in a conversational dialogue about her instructional dilemmas, with the AI acting as a Socratic partner that asks probing questions to help her clarify her thinking and explore potential solutions.

To effectively leverage GenAI for reflection, STEM PD can guide teachers to use it as an interactive analytical tool, not just a digital diary. Instead of simply recording their thoughts, teachers can learn how to prompt the GenAI with specific analytical tasks, such as a starter prompt like, "Analyze this lesson transcript and identify all student questions. Based on the types of questions they asked, what can you infer about their understanding versus their confusion about cellular respiration?" This approach transforms GenAI from a passive repository for thoughts into a cognitive partner that actively scaffolds the reflective process. By integrating GenAI's analytical capabilities into the practice of reflection, STEM PD can help teachers cultivate a systematic, evidence-based approach to continuously improve their instruction.

Sustained Duration

Effective PD occurs over extended periods, providing teachers with multiple touchpoints across months or years to acquire, practice, and adapt the teaching of specific concepts, processes, and practices meaningfully. For example, a grade 4 teacher might participate in a focused two-year PD initiative focused on integrating the computational-thinking topic of decomposition and pattern recognition across the science curriculum.

This approach stands in contrast to the short-term, episodic sessions that limit deep changes in practice (Yoon et al., 2007). High-impact PD is not a single event but a coherent, long-term program (Garet et al., 2001). For the grade 4 teacher, this means the two-year initiative would include initial workshops, structured time for classroom implementation, regular peer collaboration to troubleshoot challenges, and ongoing coaching to refine her practice as she sequentially integrates concepts like decomposition and pattern recognition into her science units.

GenAI can serve as a just-in-time support tool and a continuity partner, bridging the gaps between formal PD and coaching sessions and sustaining momentum over the long term. During her multi-year initiative, the grade 4 teacher might face a specific challenge while trying to design a lesson on algorithms. Between scheduled meetings with her coach, she could consult a GenAI tool for immediate, customized support, asking it to suggest grade-appropriate activities that connect algorithmic thinking to her current math curriculum. Chile's teacher networks, which connect educators across schools for multi-year collaborative inquiry, illustrate how sustained PD is woven into the fabric of everyday teaching (Miranda & Chapman, 2024). By drawing on a repository of educational materials and learning standards, the GenAI can provide curated, on-demand resources that keep her professional learning moving forward. Furthermore, she could use a GenAI tool to build a personal portfolio over the 2 years, periodically uploading lesson plans and reflections into a personal GPT assistant to track her growth and identify areas for continued focus.

To leverage GenAI effectively within long-term PD, STEM teachers can gain more from using it as a continuous learning companion rather than for isolated, one-off queries. Instead of using discrete prompts for quick ideas, the teacher can begin building a sustained, threaded conversation with a GenAI tool dedicated to her computational thinking journey, allowing her to revisit past ideas and build upon them. For example, she might prompt, "Building on the decomposition lesson we planned for my ecosystems unit last semester, how can I now introduce pattern recognition in my upcoming unit on weather?" This transforms the GenAI from a simple search engine into a personalized learning partner that supports an ongoing developmental arc. By embedding GenAI as a continuous support system, STEM PD models with sustained duration can provide the personalized, just-in-time assistance needed to translate new pedagogical knowledge into lasting instructional change.

Implementation

The elements associated with effective PD in education are most useful when implemented in thoughtful ways. While the scholarship on implementation in educational contexts has grown in recent years (e.g., Century & Cassata, 2016; Cook et al., 2019; Hill et al., 2023; Lyon, 2017), the full body of research on implementation is vast, covering the fields like social services, health, organizational management, and public policy, to name a few (e.g., Ashcraft et al., 2024; Fixsen et al., 2005; Moir, 2018; Wang et al., 2023). Five of the features from this literature have bearing on the implementation of PD that uses GenAI effectively and ethically to equip today's STEM teachers.

Before exploring these five features, it is worth noting that implementation is not only a matter of structure but also of culture. Implementation is more likely to take hold in schools that cultivate a climate of trust where experimentation with GenAI tools is encouraged, and missteps are treated as opportunities for collective learning rather than individual failure.

Organizational Backing

Organizational backing means leaders provide the time, resources, and policies needed to sustain teachers' adoption of new practices. For example, a school system or district that provides its math and science teachers with access to a GenAI-powered tutoring platform—along with PD with features outlined in the previous section—demonstrates organizational support in action. This support appears foundational for equipping today's teachers, since without leadership allocating resources for PD, meaningful integration of GenAI into daily STEM instruction is not likely to happen. Unlike past technologies, GenAI requires us to think about organizational support not just as provisioning hardware, apps or certificates, but as building an ecosystem aimed at equipping teachers to focus exclusively on student learning.

Professional Engagement

Active commitment means that teachers buy into the value of a new practice and are motivated to use it. For instance, a physics teacher experimenting with GenAI to generate multiple solution strategies for a problem of practice when teaching the topic of waves and optics can build confidence in students and colleagues once the benefits are seen in practice. For GenAI adoption, buy-in appears particularly important because many educators are cautious or skeptical about the role of AI in teaching and learning. For GenAI, buy-in depends not only on evidence of effectiveness but also on addressing teachers' concerns about integrity, bias, and replacement.

Pedagogical Catalysts

Pedagogical catalysts are teachers who lead by example, champion the new practice, and support colleagues during implementation. For example, a STEM biology teacher who pilots GenAI lesson-planning tools in her biology class, shares results with colleagues, and coaches others in adapting prompts and threading dialogues effectively serves as a champion. Such champions can play a pivotal role when implementing GenAI, as teachers are more likely to trust and act with peers who have firsthand experience with integrating GenAI tools into their pedagogy. In the GenAI era, champions can strengthen implementation by both demonstrating effective classroom uses and modeling critical reflection about when *not* to use GenAI, highlighting that discernment and balance are as important as enthusiasm.

System Synchrony

System synchrony means ensuring that new practices fit within existing policies, curricula, assessments, and broader educational goals. A good example is when a regional education authority integrates GenAI-driven formative assessment tools in math classes, aligning them with state standards and high-stakes testing requirements. This alignment can help teachers see that GenAI use supports, rather than conflicts, with their mandated goals. Unlike other tools, GenAI makes systems alignment more complex, as it requires reconciling fast-evolving technologies with slow-moving policy frameworks while ensuring that curricular integrity, equity, and accountability remain intact.

Deictic Adaptation

Finally, deictic adaptation refers to the process of modifying new practices and evolving technologies to fit local contexts while preserving their essential features. For instance, a chemistry teacher might adapt GenAI's lab-simulation outputs to match the equipment and resources actually available in their school lab. Adaptation becomes especially important for equipping teachers today, since GenAI tools are not static but learn and improve over time, creating shifting affordances and risks. This means teachers and educational systems may need to reconceptualize adaptation as an ongoing process of co-evolution with the technology, not a one-time adjustment, ensuring that GenAI remains pedagogically meaningful and contextually appropriate.

Final Thoughts

Drawing upon a century of scholarship, this chapter has sketched an outline for PD that can equip STEM teachers to use Generative AI (GenAI) effectively and ethically. The rationale for this framework begins with a difficult truth: typical STEM PD has often been ineffective—sometimes even negatively affecting student learning—because of persistent flaws such as generic content, passive learning formats, and lack of sustained support.

History reminds us that questions about how best to develop teachers are not unique to the GenAI era. Ancient rhetorical schools, medieval madrasas, and early modern seminaries all grappled with how to connect knowledge, pedagogy, and practice. In the United States, Horace Mann's normal schools crystallized many principles still seen as essential today: content-focused instruction, collaborative learning, modeling of practice, mentorship, reflection, and long-term professional growth. What has changed is not the core challenge but the tools and contexts in which it unfolds. By situating GenAI within this longer trajectory, we see that today's dilemmas echo enduring concerns, though they demand reinterpretation in light of technologies that evolve at a pace unimaginable to earlier reformers.

A century of PD scholarship points to design elements and implementation features consistently tied to quality teacher learning. This chapter draws upon that scholarship by highlighting seven elements that improve teaching: content focus, active learning, peer collaboration, modeled practice, expert coaching, ongoing reflection, and sustained duration. Each was illustrated with examples showing how GenAI can extend or reshape its role in practice.

Alongside these, five implementation features were highlighted: organizational backing, professional engagement, pedagogical catalysts, system synchrony, and deictic adaptation. Without these enabling conditions, the effects of the PD elements are unlikely to be realized.

The examples offered here are suggestive rather than prescriptive, meant to provide direction at a time when empirical evidence on STEM GenAI-PD is scarce (Zhai & Krajcik, 2025). Looking ahead to the next century of teacher education, this framework is perhaps best understood as a starting point rather than an endpoint. This is crucial because the velocity of GenAI's development demands an intellectual agility that questions beyond the current horizon. To cultivate this agility, this discussion invites a number of foundational questions, *sonsuza dek*.

1. *What professional identities might STEM PD cultivate?* Rather than stopping at technical proficiency, the formation of professional identities could center on cultivating *irfan*—a deep, discerning form of knowing that would help teachers to partner with GenAI while simultaneously distinguishing their own expertise from an algorithm's output (Mutahhari, 1986–1987; 2002).
2. *What structures will optimize continuity with adaptability in the design of STEM PD?* Rather than settling for simple trade-offs, structural designs could be guided by *mizan*—a carefully calibrated equilibrium between enduring principles and agile responses to the evolving GenAI landscape (Chittick, 1989).
3. *How do we design STEM PD approaches that are as iterative and responsive as GenAI itself?* Instead of periodic updates, approaches to PD could lean toward *tekâmül*—an intentional, ongoing evolution of PD experiences that tracks with GenAI's developmental arc (Bergson & Tunç, 1947).
4. *How can PD integrate human-AI co-agency?* Beyond treating GenAI as a support tool, integration could cultivate *müşterek*—a type of agency in which teachers co-create with AI while sustaining human-in-the-loop decision-making authority (Dardot & Laval, 2018).
5. *What accountability frameworks are needed?* Rather than stopping at compliance, accountability frameworks could cultivate *basiret*—the foresight and prudence needed to navigate GenAI's complex ethical terrain, anticipating downstream effects and choosing wise courses of action (Uludağ, 1992).

6. *What is the long-term trajectory of teacher expertise in the GenAI era?* Rather than continuously charting new GenAI competencies, trajectories could embody *feraset*—seeing past the moment, farsightedly, toward the wider and deeper horizon of human capacities teachers can and will need to cultivate (Ghersetti, 2018).

These and other questions are challenges to ensure that this new frontier of PD for STEM teachers does not widen pedagogical inequities but design more personalized, engaging, and effective learning opportunities for all teachers … and in turn their students.

References

AI for Education. (2025). *AI literacy training*. AI for Education. Accessed March 25, 2025 at https://www.aiforeducation.io/ai-literacy-training

Ashcraft, L. E., Goodrich, D. E., Hero, J., Phares, A., Bachrach, R. L., Quinn, D. A., Qureshi, N., Ernecoff, N. C., Lederer, L. G., Scheunemann, L. P., Rogal, S. S., & Chinman, J. J. (2024). A systematic review of experimentally tested implementation strategies across health and human service settings: Evidence from 2010-2022. *Implementation Science, 19*(43). https://doi.org/10.1186/s13012-024-01369-5

Atlay, C., Tieben, N., Hillmert, S., & Fauth, B. (2019). Instructional quality and achievement inequality: How effective is teaching in closing the social achievement gap? *Learning and Instruction, 63*, 101211. https://doi.org/10.1016/j.learninstruc.2019.05.008

Barnard, H. (1876). German pedagogy: Education, the school, and the teacher. In *German literature* (2nd ed.). Brown & Gross.

Bergson, H., & Tunç, M. S. (1947). *Yaratıcı tekâmül [Creative evolution]*. Millî Eğitim Basımevi. Accessed June 24, 2025 at https://archive.org/details/henri-bergson-yaratici-tekamul/page/n27/mode/2up

Bümen, N. T. (2009). Possible effects of professional development on Turkish teachers' self-efficacy and classroom practice. *Professional Development in Education, 35*(2), 261–278. https://doi.org/10.1080/13674580802568385

Chan, C. K. Y. (2023). A comprehensive AI policy education framework for university teaching and learning. *International Journal of Educational Technology in Higher Education, 20*(38), 1–25. https://doi.org/10.1186/s41239-023-00408-3

Chung, J. (2023). Research-informed teacher education, teacher autonomy and teacher agency: The example of Finland. *London Review of Education, 21*(1), 13. https://doi.org/10.14324/LRE.21.1.13

Century, J., & Cassata, A. (2016). Implementation research: Finding common ground on what, how, why, where, and who. *Review of Research in Education, 40*(1), 169–215. https://doi.org/10.3102/0091732X16665332

Chittick, W. C. (1989). *The Sufi path of knowledge: Ibn al-'Arabi's metaphysics of imagination.* State University of New York Press. ISBN: 978-0-88706-885-0.

Clark, C. M., & Peterson, P. L. (1986). Teachers' thought processes. In M. C. Wittrock (Ed.), *Handbook of research on teaching* (3rd ed., pp. 255–296). Macmillan. ISBN: 978-0029003107.

Comenius, J. A. (1957). *Opera didactica omnia* (Photolithographic reprint). Academiae Scientiarum Bohemoslovenicae. (Original work published in 1657).

Cook, C. R., Lyon, A. R., Locke, J., Waltz, T., & Powell, B. J. (2019). Adapting a compilation of implementation strategies to advance school-based implementation research and practice. *Prevention Science, 20*, 914–935. https://doi.org/10.1007/s11121-019-01017-1

Cordingley, P., Higgins, S., Greany, T., Buckler, N., Coles-Jordan, D., Crisp, B., Saunders, L., & Coe, R. (2015). *Developing great teaching: Lessons from the international reviews into effective professional development.* Teacher Development Trust. https://chartered.college/wp-content/uploads/2026/03/DGT-Full-report.pdf

Cremin, L. A. (1970). *American education: The colonial experience, 1607–1783.* Harper & Row. ISBN: 978-0061316708.

Cubberley, E. P. (1919). *Public education in the United States: A study and interpretation of American educational history.* Houghton Mifflin.

Cubberley, E. P. (1920). *The history of education: Educational practice and progress considered as a phase of the development and spread of western civilization.* Houghton Mifflin.

Darling-Hammond, L., Hyler, M. E., & Gardner, M. (2017). *Effective teacher professional development.* Palo Alto, CA: Learning Policy Institute.

Dardot, P., & Laval, C. (2018). *Müşterek: 21. yüzyılda devrim üzerine deneme* [Common: Essay on revolution in the 21st century] (E. Sarıkartal & F. Taylan, Trans.). İstanbul Bilgi Üniversitesi Yayınları. ISBN: 978-605-399-523-4.

de Castro, R. M., Siqueira, S. W. M., Bastos, C. A. R., & Pfeiffer Fernandes, M. C. (2018). Active learning in practice: Techniques and experiences in information systems courses in Brazil. In A. Misseyanni, M. D. Lytras, P. Papadopoulou, & C. Marouli (Eds.), *Active learning strategies in higher education* (pp. 273–292). Emerald. https://doi.org/10.1108/978-1-78714-487-320181012

Department for Education. (2011). *Teachers' standards: Guidance for school leaders, school staff and governing bodies.* Crown. Accessed July 11, 2025 at https://www.gov.uk/government/publications/teachers-standards

Desimone, L. M. (2009). Improving impact studies of teachers' professional development: Toward better conceptualizations and measures. *Educational Researcher, 38*(3), 181–199. https://doi.org/10.3102/0013189X08331140

Desimone, L. M., Smith, T. M., & Phillips, K. J. R. (2013). Linking student achievement growth to professional development participation and changes in instruction: A longitudinal study of elementary students and teachers in Title I schools. *Teachers College Record, 115*(5), 1–46. https://doi.org/10.1177/016146811311500508

Dudley, P. (2014). *Lesson study: A handbook.* Lesson Study UK. Accessed August 15, 2025 at https://lessonstudy.co.uk/wp-content/uploads/2012/03/new-handbook-revisedMay14.pdf

Ergin, O. N. (1977). *Türk maarif tarihi* (Vols. 1–5). Eser Neşriyat. Accessed June 12, 2015 at https://archive.org/details/turkiye-maarif-tarihi-i-ii.-cilt-osman-nuri-ergin-1977/T%C3%BCrkiye%20Maarif%20Tarihi%20I-II.%20Cilt%20%2D%2D%20Osman%20Nuri%20Ergin%20%2D%2D%201977%20/page/XVI/mode/2up

Exec. Order No. 14277 (2025). *Advancing artificial intelligence education for American youth.* 90 FR 17519. Accessed May 29, 2025 at https://www.federalregister.gov/documents/2025/04/28/2025-07368/advancing-artificial-intelligence-education-for-american-youth#:~:text=U.S.C.%201503%20&%201507.-,Published%20Document:%202025%2D07368%20(-90%20FR%2017519),14277%20of%20April%2023%2C%202025

Fadel, C., Black, A., Taylor, R., Slesinski, J., & Dunn, K. (2024). *Education for the age of AI: Why, what and how should students learn for the age of artificial intelligence?* Independently published. ISBN: 979-8871151150.

Fixsen, D. L., Naoom, S. F., Blase, K. A., Friedman, R. M., & Wallace, F. (2005). *Implementation research: A synthesis of the literature [monograph].* University of South Florida, Louis de la Parte Florida Mental Health Institute, The National Implementation Research Network (FMHI Publication #231). Accessed July 2, 2025 at https://fpg.unc.edu/publications/implementation-research-synthesis-literature

Flerischer, C. H. (1986). *Bureaucrat and intellectual in the Ottoman Empire: The historian Mustafa Ali (1541–1600).* Princeton University Press. ISBN: 978-0691054643.

Frazier, B.W., Betts, G.L., Greenleaf, W. J., Waples, D., Dearborn, N. H., Carney, M., & Alexander, T. (1933). *National survey of the education of teachers: Vol. 5. Special survey studies* (Bulletin No. 10). U. S. Department of the Interior, Office of Education; U.S. Government Printing Office.

Fraser, J. W. (2001). *The school in the United States: A documentary history* (1st ed.). McGraw-Hill. ISBN: 0072324481.

Gârdan, I. P., Manu, M. B., Gârdan, D. A., Negoiță, L. D. L., Paştiu, C. A., Ghiță, E., & Zaharia, A. (2025). Adopting AI in education: Optimizing human resource management considering teacher perceptions. *Frontiers in Education*, 1–24. https://doi.org/10.3389/feduc.2025.1488147

Garet, M. S., Porter, A. C., Desimone, L., Birman, B. F., & Yoon, K. S. (2001). What makes professional development effective? Results from a national sample of teachers. *American Educational Research Journal, 38*(4), 915–945. https://doi.org/10.3102/00028312038004915

Gawthrop, R. L. (1993). *Pietism and the making of eighteenth-century Prussia.* Cambridge University Press. https://doi.org/10.1017/CBO9780511523311

Ghersetti, A. (2018). Physiognomy. In K. Fleet, G. Krämer, D. Matringe, J. Nawas, & D. J. Stewart (Eds.), *Encyclopaedia of Islam three online.* Brill. https://doi.org/10.1163/1573-3912_ei3_COM_46183

Gray, W. S. (1919). Methods of improving the technique of teaching. *Elementary School Journal, 20*(4), 263–275. https://www-journals-uchicago-edu.proxy2.cl.msu.edu/doi/10.1086/454742

Grendler, P. F. (2019). *Jesuit schools and universities in Europe, 1548–1773.* Brill. https://doi.org/10.1163/25897454-12340001

Groen, M. (2008). The Whig party and the rise of common schools, 1837–1854: Party and policy reexamined. *American Educational History Journal, 35*(1–2), 251–260. Accessed May 19, 2025 at https://eric.ed.gov/?id=EJ818633

Grönqvist, E., Öckert, B., & Rosenqvist, O. (in press). Does the 'Boost for Mathematics' boat mathematics? *American Economic Journal: Economic Policy.* https://www.aeaweb.org/articles?id=10.1257/pol.20220009&&from=f

Gustafsson, J.-E., & Nilsen, T. (2016). The impact of school climate and teacher quality on mathematics achievement: A difference-in-differences approach. In T. Nilsen & J.-E. Gustafsson (Eds.), *Teacher quality, instructional quality and student outcomes: Relationships across countries, cohorts and time* (pp. 81–95). Springer. https://doi.org/10.1007/978-3-319-41252-8_4

Hairon, S. (2020). Back to the future: Professional learning communities in Singapore. *Asia Pacific Journal of Education, 40*(4), 501–515. https://doi.org/10.1080/02188791.2020.1838880

Harris, D. N., & Sass, T. R. (2011). Teacher training, teacher quality and student achievement. *Journal of Public Economics, 95*(7), 798–812. https://doi.org/10.1016/j.jpubeco.2010.11.009

Henry, N. B. (1957). In-service education for teachers, supervisors, and administrators. In *Fifth-sixth yearbook of the National Society for the Study of Education, part I.* University of Chicago Press.

Hill, C. J., Scher, L., Haimson, J., & Granito, K. (2023). *Conducting implementation research in impact studies of education interventions: A guide for researchers.* (NCEE 2023-005). U.S. Department of Education, Institute of Education Sciences, National Center for Education Evaluation and Regional Assistance. Retrieved September 1, 2025 from https://ies.ed.gov/ies/2025/01/conducting-implementation-research-impact-studies-education-interventions-guide-researchers

Hill, H. C., & Papay, J. P. (2022). *Building better PL: How to strengthen teacher learning*. Research Partnership for Professional Learning. https://rpplpartnership.org/external-resource/building-better-pl/

Jaeger, W. (1986). *Paideia: The ideals of Greek culture: Volume I. Archaic Greece: The mind of Athens* (2nd ed.). Oxford University Press. ISBN: 978-0195004250.

Jay, J. K., & Johnson, K. L. (2002). Capturing complexity: A typology of reflective practice for teacher education. *Teaching and Teacher Education, 18*(1), 73–85. https://doi.org/10.1016/S0742-051X(01)00051-8

Johnson, C. C., & Fargo, J. D. (2010). Urban school reform enabled by transformative professional development: Impact on teacher change and student learning of science. *Urban Education, 45*(1), 4–29. https://doi.org/10.1177/0042085909352073

Kaestle, C. F. (1983). *Pillars of the Republic: Common schools and American Society, 1780–1860*. Hill and Wang. ISBN: 978-0-809001-54-5.

Kaestle, C. F., & Vinovskis, M. A. (1980). *Education and social change in nineteenth-century Massachusetts*. Cambridge University Press. https://doi.org/10.1017/CBO9780511759833

Karagöz, S., & Ror, Z. (2020). The effects of professional development on Turkish teachers' self-efficacy and job satisfaction. *Australian Journal of Teacher Education, 45*(3). https://doi.org/10.14221/ajte.2020v45n3.4

Kennedy, M. M. (2016). How does professional development improve teaching? *Review of Educational Research, 86*(4), 945–980. https://doi.org/10.3102/0034654315626800

Kirsten, N., Lindvall, J., Ryve, A., & Gustafsson, J.-E. (2023). How effective is the professional development in which teachers typically participate? Quasi-experimental analyses of effects on student achievement based on TIMSS 2003-2019. *Teaching and Teacher Education, 132*(104242), 1–13. https://doi.org/10.1016/j.tate.2023.104242

Koch, C., Calligan, J., & Gros, J. (Eds.). (2004). John Baptist de La Salle: The spirituality of Christian education. Paulist Press. ISBN: 9780809141623.

Korthagen, F., & Vasalos, A. (2005). Levels in reflection: Core reflection as a means to enhance professional growth. *Teachers and Teaching: Theory and Practice, 11*(1), 47–71. https://doi.org/10.1080/1354060042000337093

Kraft, M. A., Blazar, D., & Hogan, D. (2018). The effect of teacher coaching on instruction and achievement: A meta-analysis of the causal evidence. *Review of Educational Research, 88*(4), 547–588. https://doi.org/10.3102/0034654318759268

Labaree, D. F. (2005). An uneasy relationship: The history of teacher education in the university. In M. Cochran-Smith & K. M. Zeichner (Eds.), *Studying teacher education: The report of the AERA Panel on Research and Teacher Education* (pp. 290–325). Lawrence Erlbaum. ISBN: 978-0-805855-92-0.

de La Salle, J. B. (1996). The conduct of the Christian schools (F. de La Fontainerie & R. Arnandez, Trans.; W. Mann, Ed.). Lasallian Publications. (Original work published 1720). https://lasallian.info/wp-content/uploads/2012/12/Conduct-2007-reprint.pdf

Leclercq, J. (1982). *The love of learning and the desire for god: A study of monastic culture* (C. Misrahi, Trans.). Fordham University Press. ISBN: 9780823204076.

Lee, T. H. C. (2000). *Education in traditional China: A history*. Brill. https://doi.org/10.1163/9789004389557

Lewis, C., Perry, R., & Murata, A. (2006). How should research contribute to instructional improvement? The case of lesson study. *Educational Researcher, 35*(3), 3–14. https://doi.org/10.3102/0013189X035003003

Little, J. W. (2002). Locating learning in teachers' communities of practice: Opening up problems of analysis in records of everyday work. *Teaching and Teacher Education, 18*(8), 917–946. https://doi.org/10.1016/S0742-051X(02)00052-5

Lyon, A. R. (2017). *Implementation science and practice in the education sector* [Research Brief]. Project AWARE: SAMHSA. Accessed August 27, 2025 at https://education.uw.edu/sites/default/files/Implementation%20Science%20Issue%20Brief%20072617.pdf

Makdisi, G. (1981). *The rise of colleges: Institutions of learning in Islam and the West*. Edinburgh University Press. ISBN: 0852243758.

Mann, H. (1845). Eighth annual report of the Board of Education, together with the eighth annual report of the secretary of the board. Massachusetts Board of Education. https://archive.org/stream/eighthann00mass/eighthann00mass_djvu.txt?utm_source=chatgpt.com

Mattock, J. G. (1989/2012). The in-service education and training of teachers. In V. McClelland & V. Varma (Eds.), *Advances in teacher education* (pp. 179–208). Routledge. https://doi.org/10.4324/9780203125977

Melton, J. V. H. (2003). *Absolutism and the eighteenth-century origins of compulsory schooling in Prussia and Austria*. Cambridge University Press. ISBN: 9780521528566.

Messerli, J. (1972). *Horace Mann: A biography*. Alfred A. Knopf. ISBN: 9780394429205.

Miao, F., Holmes, W., Huang, R., & Zhang, H. (2021). AI and education: Guidance for policy-makers. *UNESCO*. https://doi.org/10.54675/PCSP7350

Miranda, R. M., & Chapman, C. (2024). Towards a network learning system: Reflections on a university initial teacher education and school-based collaborative initiative in Chile. *Professional Development in Education, 50*(4), 760–774. https://doi.org/10.1080/19415257.2021.1902840

Moir, T. (2018). Why is implementation science important for intervention design and evaluation within educational settings? *Frontiers in Education, 3*. https://doi.org/10.3389/feduc.2018.00061

Mundry, S. (2005). Changing perspectives in professional development. *Science Education, 14*(1), 9–15. https://files.eric.ed.gov/fulltext/EJ740949.pdf.

Mutahhari, M. M. (1986–1987). An introduction to İrfan. *Al-Tawhid: A Journal of Islamic Thought and Culture, 4*(1), 40–94. Accessed July 17, 2025 at https://www.academia.edu/2920265/Al_Tawhid_A_Journal_Of_Islamic_Thought_And_Culture_

Mutahhari, M. (2002). *Understanding Islamic sciences*. Saqi Books. ISBN: 978-1904063032.

Nakamura, K. (2019). How lesson study helps student teachers learn how to teach mathematics through problem-solving: Case study of a student teacher in Japan. In R. Huang, A. Takahashi, & J. P. da Ponte (Eds.), *Theory and practice of lesson study in mathematics* (pp. 507–525). Springer. https://doi.org/10.1007/978-3-030-04031-4_25

Niemi, H. (2015). Teacher professional development in Finland: Towards a more holistic approach. *Psychology, Society & Education, 7*(3), 279–294. https://doi.org/10.25115/psye.v7i3.519

OECD. (2019). *TALIS 2018 results (Volume I): Teachers and school leaders as life-long learners*. OECD. https://doi.org/10.1787/1d0bc92a-en

O'Malley, J. W., Bailey, G. A., Harris, S. J., & Kennedy, T. F. (Eds.). (1999). The Jesuits: Cultures, sciences, and the arts, 1540–1773. University of Toronto Press. ISBN: 9780802042873.

Opfer, V. D., & Pedder, D. (2011). Conceptualizing teacher professional learning. *Review of Educational Research, 81*(3), 376–407. https://doi.org/10.3102/0034654311413609

Oswald, J. M. (1987). *The Lancastrian monitorial system of instruction* (ED285838). ERIC. Accessed May 15, 2025 at https://eric.ed.gov/?q=ed285838&id=ED285838

Pauly, E. L. (2016). Jean-Baptiste de La Salle's contributions for the formation of the modern school. *Revista Brasileira de Educação, 21*(66), 32–63. https://doi.org/10.4025/rbhe.v16i4.699pt

Penuel, W. R., Sun, M., Frank, K. A., & Gallagher, H. A. (2012). Using social network analysis to study how collegial interactions can augment teacher learning from external professional development. *American Journal of Education, 119*(1), 103–136. https://doi.org/10.1086/667756

Pestalozzi, J.H. (1977). *How Gertrude teaches her children* (L. E. Holland & F. C. Turner, Trans.; E. T. Cooke, Eds.). University Publications of America. (Original work published in 1801). ISBN: 0890931631.

Peterson, P. E. (2010). *Saving schools: From Horace Mann to virtual learning*. Harvard University Press. ISBN: ISBN 9780674062153.

Quintilian. (1920–1922). *The Institutio oratoria of Quintilian* (H. E. Butler, Trans.; 4 vols.). Loeb Classical Library. Harvard University Press/William

Heinemann. ISBNs: Vol. 1 9780674991385; Vol. 2 9780674991392; Vol. 3 9780674991408; Vol. 4 9780674991415.

RTI International. (n.d.). The National Rollout of coaching with tangerine in Kenya. *SharEd*. Accessed September 5, 2025 at https://shared.rti.org/content/national-rollout-coaching-tangerine-kenya

Şahin, K. (2013). *Empire and power in the reign of Süleyman: Narrating the sixteenth-century Ottoman world*. Cambridge University Press.

Sahlberg, P. (2021). *Finnish lessons 3.0: What can the world learn from educational change in Finland?* Teachers College Press. ISBN: 978-0807765169.

Schleunes, K. A. (1989). *Schooling and society: The politics of education in Prussia and Bavaria, 1750–1900*. Berg Publishers/St. Martin's Press. ISBN: 0854962670.

Shulman, L. S. (1986). Those who understand: Knowledge growth in teaching. *Educational Researcher, 15*(2), 4–14. https://doi.org/10.3102/0013189X015002004

Sims, S., & Fletcher-Wood, H. (2021). Identifying the characteristics of effective teacher professional development: A critical review. *School Effectiveness and School Improvement, 32*(1), 47–63. https://doi.org/10.1080/09243453.2020.1772841

Sykes, G., & Wilson, S. M. (2016). Can policy (re)form instruction? In D. H. Gitomer & C. A. Bell (Eds.), *Handbook of research on teaching* (5th ed., pp. 851–916). American Educational Research Association. https://doi.org/10.3102/978-0-935302-48-6_13

Takahashi, A. (2017). Lesson study: The fundamental driver for mathematics teacher development in Japan. In B. Kaur, O. N. Kwon, & Y. H. Leong (Eds.), *Professional development of mathematics teachers* (pp. 47–61). Springer. https://doi.org/10.1007/978-981-10-2598-3_4

The White House. (2025). *Pledge to America's youth: Investing in AI education* [Press Release]. Accessed September 15, 2025 at https://www.whitehouse.gov/edai/

Timperley, H. (2015). Continuing professional development. In J. D. Wright (Ed.), *International encyclopedia of the social and behavioral sciences* (2nd ed., pp. 796–802). Elsevier. https://doi.org/10.1016/B978-0-08-097086-8.92134-2

Tomaškinová, J., & Tomaškin, J. (2024). Unlocking the future of education: Empowering educators with AI by overcoming professional development challenges. In L. G. Chova, C. G. Martínez, & J. Lees (Eds.), *17th International Conference of Education, Research and Innovation Proceedings* (pp. 10633–10642). https://doi.org/10.21125/iceri.2024.2759

Tyack, D., & Cuban, L. (1995). *Tinkering toward utopia: A century of public school reform*. Harvard University Press. ISBN: 978-067489283-5.

Uludağ, S. (1992). Basiret. In *İslâm Ansiklopedisi* [*Encyclopedia of Islam*] (Vol. 5, p. 103). Türkiye Diyanet Vakfı. https://cdn2.islamansiklopedisi.org.tr/dosya/5/C05001845.pdf

Urban, W. J., & Wagoner, J. L. (2009). *American education: A history*. Routledge. https://doi.org/10.4324/9780203872246

Wallace, M. R. (2009). Making sense of the links: Professional development, teacher practices, and student achievement. *Teachers College Record, 111*(2), 573–596. https://doi.org/10.1177/016146810911100205

Wang, Y., Wong, E. L.-Y., Nilsen, P., Chung, V. C., Tian, Y., & Yeoh, E.-K. (2023). A scoping review of implementation science theories, models, and frameworks–an appraisal of purpose, characteristics, usability, applicability, and testability. *Implementation Science, 18*(43). https://doi.org/10.1186/s13012-023-01296-x

Yang, X., Ran, H., & Zhang, M. (2022). The Shanghai model: An innovative approach to promote teacher professional development through teaching-research system. *Educational Philosophy and Theory, 54*(10), 1581–1592. https://doi.org/10.1080/00131857.2020.1794155

Yoon, K. S., Duncan, T., Lee, S. W. Y., Scarloss, B., & Shapley, K. (2007). *Reviewing the evidence on how teacher professional development affects student achievement*. Issues & Answers Report, REL 2007–No. 033. U.S. Department of Education (ED498548). Accessed March 1, 2025 at https://files.eric.ed.gov/fulltext/ED498548.pdf

Zhai, X., & Krajcik, J. (2025). Artificial intelligence-based STEM education. In X. Zhai & J. Krajcik (Eds.), *Uses of artificial intelligence in STEM education* (pp. 3–14). Oxford University Press. ISBN: 978-0198882077.

CHAPTER 2

A Conceptual Framework for Teacher Agency in STEM Education in the AI Era

Mutlu Cukurova

Introduction

In the twenty-first-century STEM classroom, the growing presence of artificial intelligence technologies presents a paradox of empowerment and concern. On one hand, AI promises to relieve teachers of burdensome administrative tasks (Roy et al., 2024), provide personalized learning insights ((VanLEHN, 2011), and complement instructional capabilities including, (i) for lesson planning (e.g., by defining students' needs and familiarizing teachers with such needs (van den Berg & du Plessis, 2023)); (ii) classroom implementations (e.g., through immediate feedback to students (Di Mitri et al., 2022) or to teachers about their practice (Demszky et al., 2025) or about their interventions (Aslan et al., 2019)); (iii) assessment (e.g., through modeling students' mastery (Minn, 2022) or assessment item generation (Chan et al., 2025) or supported essay scoring (Seßler et al., 2025) or AI-supported feedback generation (Zhang et al., 2025)) aspects of their teaching). On the other hand, there are significant concerns regarding potential threats, including the erosion of teacher

M. Cukurova (✉)
University College London, London, UK
e-mail: m.cukurova@ucl.ac.uk

R. Elmas, M. Akarsu (eds.), *STEM Education*, Maarif Global Education Series,
https://doi.org/10.1007/978-981-95-9023-0_2

autonomy (Miao & Cukurova, 2024), de-professionalization of teaching (Berendt et al., 2020), and ethical pitfalls if AI is misapplied (Holmes et al., 2022). Many education experts are also concerned that over-reliance on AI to perform teachers' tasks, such as marking, feedback generation, and lesson planning could risk skill atrophy for teachers (Felix & Webb, 2024). Therefore, the notion of teacher agency (i.e., teachers' ability to exercise professional judgment, innovate in practice, and maintain control over pedagogical decisions) has come to the forefront of debates on AI in education. This chapter emphasizes the need and urgency for moving beyond polarized narratives, rejecting both the dystopian fear of AI teachers replacing humans and the utopian hype that AI alone will solve all educational problems. Instead, it calls for evidence-informed strategies that harness AI's potential while safeguarding teachers' agency, rights, and professional integrity. As AI increasingly takes on tasks traditionally viewed as distinctively human, we must engage in profound and forward-looking reflection about their essential roles and identities. What are the enduring and uniquely human qualities at the heart of STEM education? How can we safeguard, elevate, and flourish them as technology transforms lives and the broader teaching and learning ecosystem? What is the role of a teacher in STEM education where technological advancements allow the generation and sharing of information effectively and efficiently?

To frame the inquiry into these difficult questions, the chapter outlines a continuum of AI integration in teaching, distinguishing between *replacement*, *complementarity*, and *augmentation* of teacher competence with AI in educational practice. Replacement through automation refers to AI systems taking over tasks independently. Complementarity requires a careful consideration of what teacher competence means for a given task and a specific AI tool's design, development, and deployment features that can complement specific aspects of a teacher's competence at the given task with the help of AI. Augmentation suggests a deeper assimilation of AI models into teachers' cognitive and pedagogical processes to be able to improve teachers' competence at a given task so that augmented teachers can achieve the task better than AI or teacher alone (i.e., competence augmentation requires teacher-AI teams to perform better than the best of teacher or AI alone). Understanding these conceptual models is crucial for identifying when AI use supports teacher agency, when it might erode it, and when teachers' agentic interactions can lead to competence augmentation for teachers.

Benefits of AI in Education and Early Concerns

A brief review of the literature reveals that while AI applications in education are burgeoning, our understanding of how exactly STEM teachers are integrating AI into their practices at a scale remains nascent and continually evolving. Additionally, the ways in which teachers engage with AI vary significantly across countries and jurisdictions, making it challenging to comprehensively capture the full range of practices in this chapter. However, for instance, based on the open call for evidence of the use of GenAI in education by the UK Department for Education, which received responses from 567 participants, the majority of whom were teachers, the public generally believed AI to offer various opportunities (Use Cases for Generative AI in Education User Research Report, 2024). These include freeing up teachers' time, improving teaching and educational materials, providing additional support, particularly for students with special educational needs and disabilities (SEND) and those for whom English is an additional language (EAL), and enhancing subject-specific applications (e.g., STEM). Overall, these perceived benefits outweigh concerns (students' overreliance on AI, academic misconduct, fear of AI replacing f2f teaching, and 'digital divide') about AI in education. Most use cases observed in self-declared surveys and interviews of teachers also indicate that teachers use AI to develop lesson materials, ensuring alignment with curricular objectives while saving time on content preparation. In assessment, teachers tend to use it to support their marking and provide personalized formative feedback for students. Beyond the classroom, teachers appear to use AI in drafting statutory policies, streamlining administrative tasks, and aiming to reduce bureaucratic burdens.

While self-reported survey data offer some insights into teachers' usage of AI, this method faces inherent limitations, notably biases arising from external pressures or social desirability that may prevent teachers from accurately reporting their AI practices. Recent research from Anthropic.ai (Handa et al., 2025) analyses over four million conversations with AI and shows high reliance on AI for some professions, using it for more than 75% of their professional tasks, among certain professional groups, including language and literature teachers. While understanding the extent of AI usage among teachers is crucial, examining precisely how teachers employ AI, whether primarily for augmentation or automation, is equally significant. Although it is a task-dependent discussion, evidence suggests a tendency towards automating routine tasks (Handa et al., 2025), with AI directly executing responsibilities requiring minimal teacher involvement. Such automated applications, although efficient, raise critical questions regarding the broader implications for teacher agency and educational quality. It is imperative to further investigate the balance between automation and augmentation, prioritizing AI implementations that enhance, rather than diminish, teachers' professional roles, their agency, capacities, and human engagement in education.

Beyond the recent use cases of generative AI, the past few decades have witnessed an explosion of interest in applying AI to education, from intelligent tutoring systems and automated grading tools to AI-driven decision-making tools and adaptive learning platforms. There is substantial amount of evidence that shows the positive impact of using these AI applications to support students' academic performance (Li et al., 2019; VanLehn et al., 2020), their affective engagement (Bosch et al., 2016; Calvo & D'Mello, 2010), and metacognitive development (Azevedo et al., 2004; Laru & Järvelä, 2015) in controlled experimental evaluations. Although these small-scale empirical studies provide valuable insights, their outcomes typically reflect carefully designed academic AI tools evaluated under controlled conditions in which teacher implementation is guided by researchers in high fidelity. However, alongside these promises, early scholarly and practitioner commentary raised flags about possible pitfalls. These concerns have been amplified by the rapid rise of generative AI (e.g., large language models) in recent years, which brought AI's capabilities, along with its risks, into mainstream awareness. Most educational stakeholders recently began grappling with scenarios that once seemed futuristic or of interest to a small group of scholars.

One core concern in these discussions is the potential erosion of teacher agency in AI integrations in education. As AI systems begin to handle not just typical monotonous administrative tasks of teachers but also complex pedagogical and instructional decisions, such as selecting content, assessing student work, or providing feedback, teachers might find their professional judgment marginalized by algorithmic outputs. Recent empirical research is starting to document these dynamics. For instance, Zawacki-Richter et al. (2019) reviewed AI applications in higher education and noted that teacher involvement is often lacking in the research and development of AI in education. Similarly, a more recent systematic review by Tan et al. (2024) found that while numerous studies focus on how AI can support teaching, far fewer examine how AI influences teacher professional development or autonomy. In their work, only about one-third of the studies reviewed addressed teacher learning or empowerment, indicating that the teacher capacity-building aspect often lags behind technological innovation. Without deliberately centering teachers in AI integration efforts, we risk a technocentric approach that may not meet the real needs of educators or could undermine their professional roles and agency.

A Working Definition of Teacher Agency

In this chapter, teacher agency refers to teachers' active capacity to make choices and exert influence in their professional practice. It encompasses the autonomy to make instructional decisions, the ability to adapt and innovate pedagogy, and the power to shape the educational environment in accordance with their professional values and their students' needs. In essence, teacher agency is about teachers "acting rather than being acted upon" in the educational process. It is rooted in professional competence and confidence, and is often enabled or constrained by the tools and technology teachers use in their practice as well as the broader institutional and policy context in which they are situated. Teacher agency is not an all-or-nothing attribute; it exists in degrees and forms. Educational sociologists have described multiple forms of teacher autonomy, for example, autonomy over curriculum content, over pedagogy, over student assessment, and over professional development pathways (Frostenson, 2015). A supportive school culture and policy framework can expand these autonomies, whereas top-down mandates, high-stakes accountability regimes, or technology systems with a higher degree of machine agency to make autonomous decisions on behalf of teachers (e.g. AI-based educational technologies) can compress them.

Teacher agency is a key concept in STEM education since teachers' sense of agency is linked to their motivation, job satisfaction, and willingness to embrace new pedagogies or tools. When teachers feel empowered to make decisions, they are more likely to take initiative in improving their teaching and respond creatively to challenges in the classroom. In STEM fields, which are rapidly evolving, teacher agency enables teachers to update curricula, incorporate emerging real-world problems, and contextualize learning experiences to social, emotional, and relational needs of their students in ways that AI algorithms cannot. Moreover, teacher agency is tightly connected to teacher identity, that is, the sense of oneself as a professional with a meaningful mission. The introduction of AI into the classroom can perturb that identity as some teachers may fear being displaced or judged by AI systems, while others might see AI as an opportunity to enhance their effectiveness. For instance, a recent study by Guan et al. (2025) on pre-service teachers indicated that exposure to AI in education prompted reflections on their evolving role and anxiety about role change, highlighting a need to prepare teachers for new hybrid roles working alongside AI.

From a learner's perspective, teacher agency translates to richer educational interactions. A teacher with high agency will actively interpret AI-generated insights or recommendations and adapt them to the context of their students. For example, if an AI system flags certain students as at-risk in math, a teacher with professional agency will use that information as one data point, combine it with her own observations, and decide on a tailored intervention (perhaps a targeted small-group session or a new way of explaining the concept) that her professional judgment deems appropriate. In contrast, a low-agency scenario might involve a teacher mechanically following the AI's prescription (e.g., assigning remedial modules) without contextualization, or conversely, ignoring useful AI suggestions due to mistrust. Both extremes are suboptimal, and the goal is a balanced partnership where the teacher remains the orchestrator of instruction, using AI tools as informative assistants. Although most researchers and practitioners would agree with the proposed need for balance, there is little understanding regarding where this balance stands for a given teacher task and how it can be achieved. In the following sections, the chapter dives deeper into how a theoretical conceptualization of AI's role vis-à-vis teacher agency, introducing the framework of automation, complementarity, and augmentation, and then explores the potential and evidenced implications of each conceptualization for education.

A Theoretical Framework for AI in Education and Implications on Teacher Agency

This section presents three conceptual modes of integration as a framework to differentiate AI's roles and their implications for teacher agency. These modes can be considered as points along a spectrum from AI operating independently of teachers to AI becoming deeply embedded in teachers' cognitive routines to augment their competence. By delineating these, we aim to clarify which approaches threaten teacher agency and which can potentially bolster it.

1. Competence Atrophy through Replacement refers to AI systems executing tasks with minimal or no teacher intervention. This means replacing what a teacher or student would typically do with an AI-driven process. Classic examples include automated grading of exams or essays, asking an AI model to create lesson plans, questions, materials, algorithmic scheduling of student practice, such as homework, or AI tutors delivering content directly to learners, automating the pedagogical practice of teachers. The main appeal of automation is efficiency and scalability. Indeed, certain routine or labor-intensive tasks in teaching (e.g., grading multiple-choice quizzes, drafting lesson plans, generating practice problems) can be reliably automated, potentially freeing teachers' time for more complex work (e.g., Leiker et al., 2023; Pea et al., 2023). However, how exactly the saved time is used by teachers is not always a decision teachers have freedom to take, but rather is influenced by the political, managerial, or leadership level ecosystemic issues of their practice. In addition, teachers often engage in complex decision-making processes in their everyday practice, which are not possible to be fully automated and replaced by AI.

Regardless, even for relatively routine teaching tasks, full automation raises significant concerns about the loss of teacher agency. When tasks are automated, teachers lose direct engagement with that aspect of instruction and, importantly, the insight that comes from performing the task. For example, while grading essays is time-consuming, teachers often learn a great deal about their students' understanding, misconceptions, and writing skills by reading their work. An automated essay scorer might save time, but the teacher might miss the nuanced window into student thinking that manual grading provides. Furthermore, if important decisions, such as diagnosing a student's learning gap or deciding the next topic to teach, are always handed over to AI, the teacher's role can be reduced to that of a proctor or technician, merely overseeing the functioning of the software for this task. This scenario directly conflicts with the teacher agency, and there are multiple social, ethical, and pedagogical issues associated with this automation conceptualization. The following subsections cover some of these issues with a particular focus on those that have direct implications for teacher agency.

Risk of Dehumanization in Education

Dehumanization in the context of this chapter refers to the erosion of the human elements that are fundamental to teaching and learning. Education is an interactive process of human development. When AI is used as a substitute for teachers or peers, there is a danger that learning becomes overly mechanistic, losing the empathy and social dynamics that characterize many effective pedagogies and educational practices.

One aspect of dehumanization is the vision of "teacherless schools" of the future. This practice is particularly considered appropriate in STEM education due to misconceptions regarding the dogmatic and factual knowledge acquisition-driven nature of STEM subjects. Proponents often highlight potential benefits like personalization at scale or cost savings for schools. First of all, it is important to highlight that for such a technology to replace human teachers completely, one has to imagine a future AI with emotional intelligence capabilities. Yet, even if such technology were to mature, the potential loss of such a replacement would be immeasurable. Human teachers contribute numerous intangible qualities (e.g. moral judgment, inspiration, role-modeling, the ability to build trust with other humans and mentor them) that no algorithm can replicate. The dehumanization risk is that AI, no matter how advanced, lacks human understanding (i.e., the same way humans understand other humans), and human care for students.

Dehumanization can also occur even with teachers still present, but AI is used in ways that sideline human interaction. For instance, many AI learning companion arguments made in today's technology discourse imagine a class where each student sits at a computer, working through an AI-personalized curriculum, with the teacher relegated to tech support. In such a scenario, students might progress academically, but miss out on collaborative learning, dialogue, and the social construction of knowledge with other humans. Such isolated learning can also increase feelings of loneliness and reduce future engagement. When AI substitutes for human interaction (e.g., chatbots replacing teacher or tutor communication), it can negatively affect students' sense of belonging, their motivation, and retention. STEM education often thrives on inquiry-based instruction, collaborative lab-work, group discussions, and debate, which are facets that require human presence and guidance.

Crucially, dehumanization also threatens the inculcation of values and wisdom as educational outcomes. STEM education in the twenty-first century is increasingly called upon to include ethics, social implications of technology, and interdisciplinary thinking. Teachers are the ones who contextualize scientific knowledge within real-world and ethical frameworks. If AI systems deliver content without those human conversations, students may not develop the same level of moral reasoning or social conscience regarding the use of their STEM knowledge. A dehumanized AI-centric model of education could produce scientists and technicians who lack the ethical grounding or societal understanding that human teachers impart as part of their mentorship.

Dehumanization can also manifest in assessment and feedback. If students come to see that all their feedback and evaluation is from impersonal algorithms, they might feel less seen or valued as individuals. Recent research indicates that students have lower trust and acceptance of AI-generated feedback compared to human feedback (Nazaretsky et al., 2024; Zhang et al., 2025). An entirely automated feedback regime might thus diminish students' motivation or sense of accountability (e.g., it's easier to ignore a computer's judgment than a respected teacher's guidance).

Finally, it's worth noting that teachers themselves can feel dehumanized if AI is implemented poorly. If teachers are asked to work under algorithmic scripts or performance dashboards that dictate their every move (e.g., teaching by dashboard and performance orientation), their professional identity can erode. They might feel more like assembly line workers than educators. Protecting teacher agency is thus directly tied to keeping education human-centered. Teachers need the freedom to exercise compassion and adapt to the unpredictable, beautifully messy nature of human learning, which no AI can fully capture, not at the moment, and unlikely in the near future.

Cognitive Atrophy

The integration of generative AI into educational environments in ways that do not allow teacher agency also raises significant questions about its impact on educational stakeholders' various higher-order thinking skills, particularly through its influence on critical and reflective thinking practices. In this sense, generative AI represents the latest in a historical sequence of innovations that prompt concerns regarding their influence on human cognitive abilities. Throughout history, various technologies have each faced skepticism due to fears of their potential detrimental impacts on intellectual capacities. For instance, writing was objected to by Socrates, who feared writing would weaken memory since we wouldn't have to engage with knowledge through dialogue, printing was objected to by Trithemius, who argued that printed books would devalue the work of scribes and hand-written manuscripts to be more spiritually significant, calculators were objected to by many math and arithmetic teachers, who argued calculators would impair students' ability to perform mental arithmetic and understand basic mathematical concepts, and the Internet is objected to by many teachers, who fear easy access to information can lead to superficial learning and a decline in deep knowledge.

Such concerns are indeed grounded in evidence, as improper utilization of technological tools for automation can weaken the cognitive functions that society aims to preserve. The point about the automation of routine tasks inadvertently depriving individuals of opportunities to cultivate their judgment and cognitive resilience, leaving them ill-prepared for situations that require exceptional thinking, is periodically reminded by scholars (e.g., Bainbridge, 1983; Cukurova, 2025). Consequently, there are multiple research studies increasingly focusing on the cognitive impacts of generative AI, examining issues of cognitive offloading, memory retention, and creative thought processes. For instance, recent experiments focusing on the cognitive cost of using an LLM in the educational context of writing an essay, show that students who write without an LLM support exhibit the strongest, widest-ranging networks whereas those using a search engine showing intermediate engagement, and those using an LLM eliciting the weakest overall coupling indicating the limited cognitive engagement of students while writing an essay using GenAI (Kosmyna et al., 2025).

However, the evidence on such a negative impact on the cognitive capabilities of stakeholders is just emerging and often stems from a limited number of specific studies conducted in particular contexts. Multiple studies indicate that generative AI, as a generic technology, has the potential to both enhance and diminish students' and teachers' thinking practices for creativity, criticality, and reflection. Observed outcomes depend critically on the AI system's design and deployment, the supported task, user competence, and broader ecosystem enablers for effective AI integration in education. Thus, addressing potential cognitive atrophy among education stakeholders necessitates a nuanced, critical assessment of educational generative AI, considering design, deployment, and contextual factors specific to each use case. Professional development should also integrate emerging evidence on the detrimental effects of general-purpose generative AI on teacher and student thinking practices. This knowledge empowers teachers to engage cautiously and guide their students effectively when interacting with these tools.

There are also many lessons to be learned from other domains in which specific strategies have been developed to keep humans actively engaged in their practice. For instance, in the aviation industry, a classic domain where automation saves lives but also can dull pilot skills, pilots undergo regular simulation training on manual flying and handling without autopilot, to ensure they can take over when needed. In STEM education, this could translate to purposefully having "unplugged" activities embedded into STEM activities, not as an anti-technology stance, but to ensure that students and teachers can still perform and think without AI crutches and that teachers can still do organic assessments of their students' learning when needed. It can also mean designing AI that, paradoxically, encourages human engagement rather than replacing it. For example, an AI tutor might intentionally and irregularly ask the student to explain their reasoning, a task the AI could potentially infer itself based on the student's interaction logs but delegates to the human to keep them active. In essence, certain design choices to keep the human-in-the-loop in meaningful ways should be integrated into AI applications in STEM education.

Accountability

Another important aspect associated with the teacher agency is the concept of accountability. If an AI system used in education makes a harmful recommendation (e.g., misidentifies a student as low-ability, affecting their educational trajectory), who is responsible? Is it the company that made the algorithm, the school that implemented it, or the teacher who acted on it? These are complex questions. To maintain teacher agency, teachers should not be put in positions where they are asked to blindly implement AI outputs without recourse or where they might be blamed for AI errors beyond their understanding. Imagine a scenario where an AI scheduling system assigns students to certain tracks and a teacher is expected to go along with it; if it later emerges that the algorithm was biased, the teacher might be unfairly implicated.

Multiple accountability frameworks are beginning to emerge. The European Union's proposed AI Act, for instance, classifies AI in education as "high risk" and would require rigorous transparency, risk assessment, and human oversight for such systems. This means developers must document how the AI works, test for bias, and enable human intervention. For teachers, these regulatory moves can be empowering because they aim to provide teachers with the information and authority to challenge or opt out of AI suggestions that seem problematic. In practice, this might take the form of an AI system providing a clear explanation of its recommendation *and* an easy way for teachers to say "no, override" or "provide feedback" to improve it next time. The STEM education culture should be that teachers are accountable to students and society, not to the AI. If an AI says pass a student and the teacher disagrees based on evidence, the system must allow the teacher's judgment to prevail and record that decision (and perhaps learn from it, see the synergy conceptualization below). Only with such transparency and accountability can teachers confidently integrate AI, knowing it's a tool under their purview, not an opaque mandate.

AI Bias

AI tools carry biases (e.g., due to their training data, algorithms they are built on, and fine-tuning processes); if they are automating these tasks without caution, this can perpetuate existing cultural, linguistic, and socioeconomic biases, reinforce systemic inequalities, and exacerbate exclusion. The paradox is further pronounced in contexts where communities contribute through low-paid labor in data labeling, annotation, and other forms of digital labor required to train AI systems, yet these same communities rarely reap meaningful benefits from the resultant technological advances. Such inequities constitute a profound ethical concern amounting to a "double curse" of exploitation and exclusion. To confront and address this growing digital divide effectively, it is imperative that global policymakers, educational leaders, and technology developers commit explicitly to inclusive, equitable, and socially responsible frameworks for the deployment of AI in education, ensuring that all learners and educators, irrespective of their geographic, economic, or social circumstances, have genuine opportunities to engage with and benefit from emerging AI technologies. Further amplifying these concerns, Benjamin (2023) critically conceptualizes AI technologies as systems potentially propagating "engineered inequality." Rooted in societies characterized by deeply embedded structural disparities, AI products often unintentionally perpetuate, or even exacerbate, existing inequities through biased outcomes.

In education, such engineered inequalities could manifest through unequal access to resources, differential quality of educational opportunities, or perpetuation of stereotypes that disadvantage historically marginalized groups. AI's potential to reproduce stereotypes, amplify biased information, and facilitate the spread of mis- and disinformation must be explicitly acknowledged. Recognizing AI's potential to produce oppressive outcomes underscores the urgency of embedding ethical considerations and robust regulatory frameworks into AI design and implementation, ensuring the benefits of technology accrue equitably across diverse populations.

Due to the vast size of training datasets and extensive efforts by their producers to fine-tune them, generative AI models might appear as if they are not influenced as much by the impact of training data on their performance. However, recent research reveals troubling implicit biases inherent within large language models (LLMs) when applied in educational settings and demonstrates that AI-generated content consistently mirrors and reinforces existing social biases and stereotypes (Du et al., 2025). For instance, as evidenced in reference letters written by teachers using LLMs (Wan et al., 2023), male students are frequently characterized using formal and professionally favorable descriptors, whereas female students are portrayed in more communal, emotional, and less agentic terms. Men tend to be presented as proactive agents, competent and professional, whereas women are often described as approachable but less professionally competent, receiving disproportionately negative evaluations. Furthermore, LLMs tend to have implicit biases in academic recommendations, notably showing tendencies to assign higher-status roles and opportunities, such as STEM majors or PhD roles, disproportionately to males over females and entirely excluding non-binary individuals (Zheng, 2024). These biases reflect and risk perpetuating entrenched societal inequalities, underscoring an urgent need for vigilant and ongoing evaluation of AI tools' bias to prevent the reinforcement of harmful stereotypes in educational practices.

Furthermore, while AI is frequently advocated as a promising means of enabling teachers to make "objective" decisions, existing evidence from systematic reviews remains limited regarding its effectiveness in mitigating teacher biases or fostering equitable instructional practices. A systematic literature review by Gauthier et al. (2022) on data-driven educational technologies highlighted a notable gap that few studies have rigorously evaluated whether these technologies genuinely reduce teacher bias or promote equitable teaching practices. Consequently, optimism regarding AI's transformative potential to mitigate bias in education must be tempered by caution and a commitment to further empirical research. Future studies should prioritize robust evaluations of AI's role in addressing equity, exploring deeply how AI tools influence teacher behaviors and practices, with particular attention to whether they exacerbate or reduce systemic educational disparities when AI is implemented to replace human competence through automation.

Trust

Recent empirical research further underscores the complexity of integrating AI within education, highlighting the preference of students for teacher connections and credibility. For instance, Nazaretsky et al. (2024) demonstrated that students exhibit considerable caution towards AI-generated feedback, demonstrating a clear preference for personalized, human-generated responses from human teachers. Within a higher education context based on data from 457 students across diverse academic programs and levels, the experimental quantitative study revealed that students' perceptions of feedback quality diminish significantly upon discovering its AI-generated origin, reflecting deeper issues of trust and credibility inherent within automated systems. These results are further replicated in other studies, such as Zhang et al. (2025). This potential lack of trust and credibility of AI content underscores the vital importance of human agency and relational dynamics within educational interactions, emphasizing that certain educational interventions require not merely accuracy or relevance, but also the essential qualities of human empathy, relational understanding, and pedagogical intuition, which can only be delivered by human teachers. Although domain-specific explainability and particular framing of AI can indeed help improve user trust (e.g., Feldman-Maggor et al., 2025), such framing interventions should be carefully balanced. For instance, research by Hou and Jung (2021) demonstrated that how an AI is framed (e.g., as an "expert" system) can increase users' trust in it to the point of over-reliance. Without careful checks and balances, automated educational tools could thus induce teachers to abandon their active, reflective practice and instead become passive implementers of AI outputs.

2. Competence Amplification Through Complementarity

Amplification refers to AI systems functioning as complementary, supportive tools that magnify a teacher's capabilities, while the teacher remains actively involved. In the complementarity paradigm, AI and teacher work in tandem, each contributing what they do best without necessarily interacting with each other for augmentation of each other's competence. The underlying philosophy of complementarity is that leveraging AI's strengths (data processing, pattern recognition, speed, scale, no exhaustion, efficiency and time-saving opportunities, etc.) to complement human strengths (relational interpretation, empathy, moral discernment, contextual judgment, etc.). Rather than taking over tasks entirely, or competence augmentation in AI or humans, complementary systems can share tasks with teachers or handle sub-tasks under teacher direction.

For instance, regarding the complementarity paradigm, Holstein et al. (2020) provide a useful framework for thinking about what specific aspects of teacher competence can be complemented with the help of AI. The authors identify four dimensions where AI can meaningfully complement teacher competence without replacing the teacher's agency:

- *Complementary goal setting*: AI helps teachers set, monitor, and evaluate learning goals. For example, an AI system might track student progress in real-time and alert the teacher if the class is veering off from curriculum objectives or suggest intermediate milestones for a project. This supports strategic planning, enabling the teacher to adjust pacing or focus based on data, while the teacher still defines the ultimate goals and interprets the relevance of AI's feedback.
- *Complementary perceptual input*: AI expands what teachers can perceive about student learning processes. For instance, advanced analytics might detect subtle patterns, like a student's increased pause time before answering questions, or social network analysis showing a student's peripheral engagement in group work, that a teacher might miss in the moment or simple doesn't have the time to attend to all students and student groups simultaneously. Wearable or sensor data can alert a teacher to student confusion or disengagement, for instance, based on their gaze or posture type of nonverbal interactions. By surfacing these insights, AI complements the teacher's situational awareness, helping them better understand learners' needs. The teacher then uses these perceptions to inform their

actions (e.g., intervening with a timely question or encouragement of their students).

- *Complementary actions*: AI can increase the scale or scope of teacher actions. A common example is intelligent tutoring systems that can provide one-on-one practice for students on certain skills (e.g., algebra exercises), essentially allowing a single teacher to support many students simultaneously through a combination of direct and AI-mediated instruction. Another example is tools that automate the first draft of a lesson plan or a quiz, which the teacher can then integrate into their lesson delivery, taking charge of the implementation of what AI produces. Here, AI does the heavy lifting of generating materials or handling routine interactions to help teachers save time, enabling the teacher to manage larger classes or provide individualized attention that would otherwise be impossible. Importantly, the teacher is still orchestrating these actions, choosing when and how to deploy the AI-driven activities.
- *Complementary decisions*: AI offers data-informed recommendations to assist teacher decision-making. For instance, an AI might analyze a student's performance and recommend a personalized learning activity or flag which students could use extra help on a topic. These recommendations can improve the effectiveness and efficiency of a teacher's pedagogical reasoning by bringing evidence or options they might not have considered. Crucially, decision complementarity protects teacher agency in the sense that the AI only suggests, but the teacher actually takes the final decisions. In practice, this also has significant implications regarding AI's design features, since the AI's suggestions need to be transparent, explainable, and optional. A teacher with their professional knowledge might accept, modify, or reject an AI recommendation. The aim here is to improve the final decision quality by giving teachers better information, not to dictate decisions, or not to necessarily improve the decision-making competence of teachers in an augmentative sense.

If implemented appropriately, complementarity conceptualization further reinforces the teacher's agency. For example, in the context of STEM education, an AI system can evaluate student lab reports for common errors in data analysis and provide the teacher with a synthesized list of misconceptions along with suggested feedback for each student. The teacher reviews this, perhaps noticing patterns (e.g., many students misunderstood a control variable concept), and then the teacher both uses the AI-suggested comments to save time rather than writing the same comment 30 times, and plans a quick remedial discussion with the students on the concept. Here, the AI has saved the teacher time and pointed out an issue, but the teacher remains the one delivering the actual feedback and deciding how to address the misconceptions in the next lesson. Misconceptions detected by AI are likely to be the ones the teacher would be able to detect on their own if they were to go through all the student reports themselves, so the AI does not necessarily aim to augment a teachers' competence of misconception detection, but it complements the teachers' competence of making interventional decisions by taking charge of misconception diagnostics aspects of their work. It's important to acknowledge that AI systems, particularly those trained on large-scale datasets, can themselves reproduce or amplify existing societal or scientific misconceptions, as shown in previous literature on STEM education (e.g., Elmas et al., 2024). Therefore, such diagnostics would necessitate careful oversight and/or bespoke generative AI models that have been tested to show that they can reliably and accurately diagnose student misconceptions. Regardless, in this conceptualization, the teacher's agency, which involves deciding what to teach next or how to communicate feedback to their students, considering the whole socio-emotional complexities of a given context, is intact or potentially strengthened by having more time and better intel with the help of AI.

It's worth noting that complementarity is not achieved by just introducing any AI tool to STEM education, yet it requires intentional design. If an AI tool is built without considering an appropriate cognitive model of teacher competence for a given task (e.g. their cognitive, affective, situation-specific skills, and performance metrics) as well as teachers' workflow and decision points, it is likely to end up either being ignored or inadvertently undermining teacher agency. For example, a myriad of learning analytics dashboards are considered too complicated or provide insights without considering what teachers' competence is being complemented for a given task, have been shown to have little to no impact on educational practice. These tools are likely to confuse teachers and not be adopted by them, rather than help them. On the other hand, a well-designed complementary AI tool involves teachers in co-design, offers controls to customize AI behaviors, and provides clear explanations for its outputs regarding what aspects of teachers' competence they can complement in practice. Such complementary interactions with AI can have profound implications for STEM education as they can lead to transformative changes in teacher agency and identity. Complementary AI means the teacher has fully adapted to wield the AI in a thoughtful manner, much like a proficient craftsperson incorporates a new tool into their craft. The teacher's agency is exercised through the tool seamlessly, and they have a feel for AI just as most good drivers have a feel for their cars while they are driving them. For example, an educator might develop an internal radar for when to trust AI suggestions versus when to rely on their own expertise, akin to a seasoned pilot knowing when to rely on autopilot and when to take manual control. The teacher might also develop new reflective practices, such as regularly reviewing AI-generated analytics after each class as a feedback loop for their teaching. When such practices become second nature, the AI is effectively a part of the teacher's practice model, which exemplifies the complementarity conceptualization described here.

3. Competence Augmentation Through Synergy
   The concept of augmentation in this chapter refers to the process by which AI tools, and the new practices they enable, become woven into the internal repertoire of teachers and students in ways that these human–AI interactions do not lead to a decrease in human competence. The main condition for such augmentation is synergistic human–AI interaction, where human–AI combination yields emergent competence exceeding the maximum of individual AI or human competence at a given task. This idea, informed by theories of distributed and extended cognition, faces significant empirical hurdles. A recent comprehensive meta-analysis of 106 experimental studies, with 370 effect sizes reported, reveals that human–AI combinations underperform the better performer, human or AI alone, in 58% of the cases reviewed (Vaccaro et al., 2024), demonstrating that synergy is rare and context-dependent.

Critically, synergy emerges only under specific conditions. For instance, (i) task type matters, in a way that synergy is more likely to occur in content creation tasks (e.g., open-ended science education questions) but fails in complex decision-making tasks (e.g., classification of a student's emotional state given their relationship with other students); (ii) when humans outperform AI alone, synergy is more likely to arise, yet when AI outperforms humans, human–AI teaming tends to degrade the performance. This indicates that synergy depends on human competencies to meta-cognitively assess when to trust AI input and when to ignore it which emphasizes the importance of improving teachers and students' AI competencies to be able to protect their own agency and have a critical and informed trust in AI (see, for instance, UNESCO AI competency framework for teachers (Miao & Cukurova, 2024); (iii) instructional design in which AI is implemented is very important. The evidence from the metareview indicates that only 3 of 106 experiments tested predetermined subtask delegation between humans and AI. These showed non-significant synergy gains, underscoring that effective augmentation requires systematic scaffolding of interactions with AI as part of a carefully thought instructional design structure, rather than unstructured AI integration into STEM education.

These findings affirm that synergistic augmentation is very difficult to achieve in STEM education and many other domains (e.g., healthcare, law). They show that future AI implementations in STEM education aiming to achieve augmentation through synergistic interactions must prioritize task-specific AI design, development, and deployment; scaffolding robust human–AI synergistic interactions; and providing relevant AI competency training for teachers and students to educate them about how to unlock synergistic emergent competence in their interaction with AI. It is also important to have appropriate evaluation metrics for success regarding what aspects of competence is exactly augmented in humans and AI for a given task. Without these, human–AI combinations are unlikely to achieve synergy and risk diminishing returns compared to standalone human or AI performance at a given task.

In many senses, the concept of hybrid intelligence also captures this competence augmentation through synergy conceptualization presented here. For instance, Cukurova (2025) describes a vision for hybrid intelligence in education where human and AI intelligence are tightly coupled to produce better outcomes than either alone. In a hybrid system, tasks are not rigidly partitioned between human and machine; rather, there is a fluid interaction. For instance, in a writing exercise, a student might alternate between drafting text and consulting an AI for suggestions, learning from the AI's inputs, and the AI adjusting its model parameters and weights to the student's input as part of the co-regulation of the writing process. If teachers carefully observe such patterns, their role to a certain extent also shifts to teaching how to learn with AI, which can be considered a meta-level of instruction. In STEM education, particularly, this might involve guiding students on how to critically evaluate AI-provided answers or how to use AI tools to explore "what if" scenarios with counterfactuals. The teacher often takes the stance of being a mentor for AI-augmented learners, which is a different mode of teaching but one that retains teacher agency in guiding the overall learning process. Similarly, Molenaar (2022) advocates for hybrid intelligence approaches, where teachers and AI continuously co-adapt to each other. In such hybrid systems, the AI is expected to solicit teacher feedback on its recommendations and learn from it over time, as well as the learning occurring in teachers as part of the synergistic interactions. The goal is a synergy where the combined human–AI system is more competent and more effective than the best of either alone.

The concept of augmentation also highlights the need for developing new competencies for teachers and students. As AI becomes internal to practice, teachers need more than superficial AI tool use skills; they require deeper AI literacy and metacognitive skills about AI. They must understand AI's limitations, biases, and failure modes to avoid blindly accepting flawed AI recommendations and practices. For example, if a teacher starts to rely on an AI for grammar feedback on student essays, they should know that the AI might over-correct certain vernacular expressions or unique voices, and thus they need to routinely check for such biases. Professional learning for STEM educators, therefore, is of utmost importance. For instance, UNESCO's AI competency framework for teachers (Miao & Cukurova, 2024) and similar initiatives aim to equip educators with not just technical know-how, but also ethical awareness, pedagogical strategies for AI use, and most importantly a human-centered AI mindset to be able to protect their agency in their interactions with AI.

It is also important to acknowledge that since synergy requires mutual interactions of two-way information flow between AI and humans, one could also speak of the internalization of human values into AI systems as part of this process. As teachers work closely with AI, ideally, they would influence the design and tuning of these systems (e.g., through feedback, pushback on recommendations and reasons provided for such pushback, usage choices, participatory design, etc.). Over time, AI used in classrooms should learn the priorities of good teaching for a given teacher dependent upon their pedagogical philosophy, needs and requirements (promoting inquiry, not giving away answers too easily, respecting diverse solutions, prioritizing certain instructional choices over others, etc.) because teachers enforce these in their interactions with the AI. In a sense, the AI system learns some of the teacher's pedagogical wisdom through these synergistic interactions. Thus, augmentation would also require appropriate spaces for the reflection of humans and space for model updates for AI. That is, teachers reflecting on how AI is affecting their practice and their students' learning, AI using logged interaction data to update its model parameters and weighs to learn from its interactions with teachers. This two-way internalization is at the heart of human–AI co-evolution in education for synergistic augmentation; teachers shape AI just as AI shapes teaching practices synergistically.

This synergistic augmentation paradigm represents the deepest form of teacher-AI interaction, where AI is no longer a distinct and complementary "add-on" to their competence, but part of the fabric of teaching and learning. This holds the promise of truly hybrid pedagogies that leverage AI's capabilities while being guided by human wisdom in synergistic interactions that lead to an augmented competence that is greater than the maximum of an individual human or AI alone. It demands high levels of teacher agency to negotiate the human–AI relationship. In addition, teachers must remain self-aware, continually asserting human values and expertise in the loop even as they embrace new AI-empowered methods. Achieving success at this augmentation paradigm is extremely challenging in education, but it aligns with the vision of education where AI serves as a competence augmentation tool for both teachers and students, ultimately enriching the teaching and learning experiences.

Conclusion

One of the most salient points of this chapter is that AI integration in STEM education is not a purely technical matter; it is a profoundly socio-technical issue with significant ethical and pedagogical considerations one has to take on board. By differentiating three conceptualizations of AI including replacement through automation, amplification though complementarity, and augmentation through synergy, the chapter aims to underscore that the highest goal of education is not to see how much of it can be turned over to machines to save time and gain productivity through automation, but how AI can be used in education in ways that are complementary to human competence, and potentially augment human competence by protecting and promoting teacher agency and competence.

STEM education should neither unconditionally embrace every new AI gadget nor reject all AI on principle; instead, it should pilot innovations in safe and controlled, research-informed ways with rapid cycle evaluations in authentic settings. And when something works (like a workload-reducing tool that doesn't harm quality in other aspects), scale it thoughtfully, and when something causes harm or diminishes competence, dare to halt or redesign it. For these rapid cycle evaluations to be impactful, listening to teachers' and students' experiences is vital since they are the ones living the integration of AI in authentic educational contexts.

The risks of automation through replacement conceptualization discussed in this chapter must be proactively addressed. But they are not insurmountable. History has shown that societies can adapt to powerful new technologies by developing new norms, regulations, and practices that align those technologies with human values. The task before us with AI is to harness it in the service of STEM education as a human-centered endeavour. That means emphasizing qualities like critical thinking, creativity, ethical reasoning, collaboration, and well-being in any AI-enhanced curriculum. It means using AI to free up time for what machines cannot do, including having a heart-to-heart talk with a student, the spontaneous exploration of a question that sparks in class, and the mentorship that inspires a young person's life trajectory.

Human flourishing, perhaps the broadest goal of education, entails that students emerge from STEM education not only with robust knowledge and skills but also as well-rounded individuals ready to contribute positively to society. AI's encroachment into every aspect of life makes it even more crucial that education develops human traits. STEM education should thus integrate social sciences and ethics alongside technical content, perhaps inspired by the synergistic augmentation paradigm presented here in which the best outcomes arise when human and AI strengths are combined in ways that are superior to the best of human or AI. Reclaiming teacher agency is about reclaiming education as a practice of freedom, creativity, and collective betterment, now empowered by AI, but always led by humans with vision and virtue.

Acknowledgments No generative AI tools were used in the creation or development of the ideas presented in this chapter. This chapter's content is based on the keynote talk given by the author at the ACM International Conference on Learning Analytics & Knowledge (LAK) in March 2024 in Kyoto, Japan, and the UNESCO International Task Force on Teachers for Education 2030 position paper on the protection and promotion of teacher agency in the age of AI he authored. The author takes full responsibility for the content and its intellectual integrity. The author acknowledges Dr. Qi Zhou for his assistance with references and proofreading of this chapter, as well as the book editors, Drs. Ridvan Elmas and Murat Akarsu, for their insightful feedback during the review process and their patience and professionalism throughout the preparation of this chapter.

References

Aslan, S., Alyuz, N., Tanriover, C., Mete, S. E., Okur, E., D'Mello, S. K., & Arslan Esme, A. (2019). Investigating the impact of a real-time, multimodal student engagement analytics technology in authentic classrooms. In *Proceedings of the 2019 CHI Conference on Human Factors in Computing Systems* (pp. 1–12). https://doi.org/10.1145/3290605.3300534

Azevedo, R., Cromley, J. G., & Seibert, D. (2004). Does adaptive scaffolding facilitate students' ability to regulate their learning with hypermedia? *Contemporary Educational Psychology, 29*(3), 344–370.

Bainbridge, L. (1983). Ironies of automation. In *Analysis, design and evaluation of man–machine systems* (pp. 129–135). Elsevier. https://www.sciencedirect.com/science/article/pii/B9780080293486500269

Benjamin, R. (2023). Race after technology. In *Social Theory Re-Wired* (pp. 405–415). Routledge. https://www.taylorfrancis.com/chapters/edit/10.4324/9781003320609-52/race-technology-ruha-benjamin

Berendt, B., Littlejohn, A., & Blakemore, M. (2020). AI in education: Learner choice and fundamental rights. *Learning, Media and Technology, 45*(3), 312–324. https://doi.org/10.1080/17439884.2020.1786399

Bosch, N., D'Mello, S. K., Baker, R. S., Ocumpaugh, J., Shute, V., Ventura, M., Wang, L., & Zhao, W. (2016). Detecting student emotions in computer-enabled classrooms. *IJCAI, 16*, 4125–4129. https://pnigel.com/papers/bosch-pp-ijcai16-camera.pdf

Calvo, R. A., & D'Mello, S. (2010). Affect detection: An interdisciplinary review of models, methods, and their applications. *IEEE Transactions on Affective Computing, 1*(1), 18–37.

Chan, K. W., Ali, F., Park, J., Sham, K. S. B., Tan, E. Y. T., Chong, F. W. C., Qian, K., & Sze, G. K. (2025). Automatic item generation in various STEM subjects using large language model prompting. *Computers and Education: Artificial Intelligence, 8*, 100344.

Cukurova, M. (2025). The interplay of learning, analytics and artificial intelligence in education: A vision for hybrid intelligence. *British Journal of Educational Technology, 56*(2), 469–488. https://doi.org/10.1111/bjet.13514

Demszky, D., Liu, J., Hill, H. C., Sanghi, S., & Chung, A. (2025). Automated feedback improves teachers' questioning quality in brick-and-mortar classrooms: Opportunities for further enhancement. *Computers & Education, 227*, 105183.

Di Mitri, D., Schneider, J., & Drachsler, H. (2022). Keep me in the loop: Real-time feedback with multimodal data. *International Journal of Artificial Intelligence in Education, 32*(4), 1093–1118. https://doi.org/10.1007/s40593-021-00281-z

Du, Y., Borchers, C., & Cukurova, M. (2025). Benchmarking educational llms with analytics: A case study ongender bias in feedback. arXiv preprint arXiv:2511.08225.

Elmas, R., Adiguzel-Ulutas, M., & Yılmaz, M. (2024). Examining ChatGPT's validity as a source for scientific inquiry and its misconceptions regarding cell energy metabolism. *Education and Information Technologies, 29*(18), 25427–25456. https://doi.org/10.1007/s10639-024-12749-1

Feldman-Maggor, Y., Cukurova, M., Kent, C., & Alexandron, G. (2025). The impact of explainable AI on teachers' trust and acceptance of AI EdTech recommendations: The power of domain-specific explanations. *International Journal of Artificial Intelligence in Education, 35*, 1–34.

Felix, J., & Webb, L. (2024). Use of artificial intelligence in education delivery and assessment. *Parliamentary Office of Science and Technology*. https://aiadvisoryboards.wordpress.com/wp-content/uploads/2024/01/use-of-artificial-intelligence.pdf

Frostenson, M. (2015). Three forms of professional autonomy: De-professionalisation of teachers in a new light. *Nordic Journal of Studies in Educational Policy, 2015*(2), 28464. https://doi.org/10.3402/nstep.v1.28464

Gauthier, A., Rizvi, S., Cukurova, M., & Mavrikis, M. (2022). Is it time we get real? A systematic review of the potential of data-driven technologies to address teachers' implicit biases. *Frontiers in Artificial Intelligence, 5*, 994967.

Government Social Research. (2024). *Use cases for generative AI in education: User research report*. Department for Education, United Kingdom.

Guan, L., Zhang, Y., & Gu, M. M. (2025). Pre-service teachers preparedness for AI-integrated education: An investigation from perceptions, capabilities, and teachers' identity changes. *Computers and Education: Artificial Intelligence, 8*, 100341.

Handa, K., Tamkin, A., McCain, M., Huang, S., Durmus, E., Heck, S., Mueller, J., Hong, J., Ritchie, S., Belonax, T., Troy, K. K., Amodei, D., Kaplan, J., Clark, J., & Ganguli, D. (2025). Which economic tasks are performed with AI? evidence from millions of claude conversations. *arXiv*. https://doi.org/10.48550/arXiv.2503.04761

Holmes, W., Porayska-Pomsta, K., Holstein, K., Sutherland, E., Baker, T., Shum, S. B., Santos, O. C., Rodrigo, M. T., Cukurova, M., Bittencourt, I. I., & Koedinger, K. R. (2022). Ethics of AI in education: Towards a community-wide framework. *International Journal of Artificial Intelligence in Education, 32*(3), 504–526. https://doi.org/10.1007/s40593-021-00239-1

Holstein, K., Aleven, V., Rummel, N. (2020). A Conceptual Framework for Human–AI Hybrid Adaptivity in Education. In: Bittencourt, I., Cukurova, M., Muldner, K., Luckin, R., Millán, E. (eds) Artificial Intelligence in Education. AIED 2020. *Lecture Notes in Computer Science*(), vol 12163. Springer, Cham. https://doi.org/10.1007/978-3-030-52237-7_20

Hou, Y. T.-Y., & Jung, M. F. (2021). Who is the expert? reconciling algorithm aversion and algorithm appreciation in AI-supported decision making. *Proc ACM Hum-Comput Interact,* *5*(CSCW2), 1–25. https://doi.org/10.1145/3479864

Kosmyna, N., Hauptmann, E., Yuan, Y. T., Situ, J., Liao, X.-H., Beresnitzky, A. V., Braunstein, I., & Maes, P. (2025). Your brain on ChatGPT: Accumulation of cognitive debt when using an AI assistant for essay writing task. *arXiv.* https://doi.org/10.48550/arXiv.2506.08872

Laru, J., & Järvelä, S. (2015). Integrated use of multiple social software tools and face-to-face activities to support self-regulated learning: A case study in a higher education context. In L.-H. Wong, M. Milrad, & M. Specht (Eds.), *Seamless learning in the age of mobile connectivity* (pp. 471–484). Springer Singapore. https://doi.org/10.1007/978-981-287-113-8_24

Leiker, D., Finnigan, S., Gyllen, A. R., & Cukurova, M. (2023). Prototyping the use of Large Language Models (LLMs) for adult learning content creation at scale. *arXiv.* https://doi.org/10.48550/arXiv.2306.01815

Li, H., Gobert, J., & Dickler, R. (2019). Evaluating the transfer of Scaffolded inquiry: What sticks and does it last? In S. Isotani, E. Millán, A. Ogan, P. Hastings, B. McLaren, & R. Luckin (Eds.), *Artificial intelligence in education* (Vol. 11626, pp. 163–168). Springer International Publishing. https://doi.org/10.1007/978-3-030-23207-8_31

Miao, F., & Cukurova, M. (2024). *AI competency framework for teachers.* UNESCO Publishing. https://unesdoc.unesco.org/ark:/48223/pf0000391104

Minn, S. (2022). AI-assisted knowledge assessment techniques for adaptive learning environments. *Computers and Education: Artificial Intelligence, 3,* 100050.

Molenaar, I. (2022). Towards hybrid human-AI learning technologies. *European Journal of Education, 57*(4), 632–645. https://doi.org/10.1111/ejed.12527

Nazaretsky, T., Mejia-Domenzain, P., Swamy, V., Frej, J., & Käser, T. (2024). AI or human? evaluating student feedback perceptions in higher education. In R. Ferreira Mello, N. Rummel, I. Jivet, G. Pishtari, & J. A. Ruipérez Valiente (Eds.), *Technology enhanced learning for inclusive and equitable quality education* (Vol. 15159, pp. 284–298). Springer Nature Switzerland. https://doi.org/10.1007/978-3-031-72315-5_20

Pea, R. D., Biernacki, P., Bigman, M., Boles, K., Coelho, R., Docherty, V., Garcia, J., Lin, V., Nguyen, J., & Pimentel, D. (2023). Four surveillance technologies creating challenges for education. *Learning: Designing the Future, 317.*

Roy, P., Poet, H., Staunton, R., Aston, K., & Thomas, D. (2024). *ChatGPT in lesson preparation: A teacher choices trial.* http://www.mmiweb.org.uk/gaiandre/literature/EEF.pdf

Seßler, K., Fürstenberg, M., Bühler, B., & Kasneci, E. (2025). Can AI grade your essays? A comparative analysis of large language models and teacher ratings in multidimensional essay scoring. In *Proceedings of the 15th International Learning Analytics and Knowledge Conference* (pp. 462–472). https://doi.org/10.1145/3706468.3706527

Tan, X., Cheng, G., & Ling, M. H. (2024). Artificial intelligence in teaching and teacher professional development: A systematic review. *Computers and Education: Artificial Intelligence*, 100355.

Vaccaro, M., Almaatouq, A., & Malone, T. (2024). When combinations of humans and AI are useful: A systematic review and meta-analysis. *Nature Human Behaviour*, 1–11.

van den Berg, G., & du Plessis, E. (2023). ChatGPT and generative AI: Possibilities for its contribution to lesson planning, critical thinking and openness in teacher education. *Education Sciences, 13*(10), 998.

VanLEHN, K. (2011). The relative effectiveness of human tutoring, intelligent tutoring systems, and other tutoring systems. *Educational Psychologist, 46*(4), 197–221. https://doi.org/10.1080/00461520.2011.611369

VanLehn, K., Banerjee, C., Milner, F., & Wetzel, J. (2020). Teaching Algebraic model construction: A tutoring system, lessons learned and an evaluation. *International Journal of Artificial Intelligence in Education, 30*(3), 459–480.

Wan, Y., Pu, G., Sun, J., Garimella, A., Chang, K.-W., & Peng, N. (2023). 'Kelly is a Warm Person, Joseph is a Role Model': gender biases in LLM-Generated Reference Letters. *arXiv*. https://doi.org/10.48550/arXiv.2310.09219

Zawacki-Richter, O., Marín, V. I., Bond, M., & Gouverneur, F. (2019). Systematic review of research on artificial intelligence applications in higher education—where are the educators? *International Journal of Educational Technology in Higher Education, 16*(1), 39. https://doi.org/10.1186/s41239-019-0171-0

Zhang, A., Gao, Y., Suraworachet, W., Nazaretsky, T., & Cukurova, M. (2025). Evaluating trust in AI, human, and co-produced feedback among undergraduate students. *arXiv*. https://doi.org/10.48550/arXiv.2504.10961

Zheng, A. (2024). Dissecting bias of ChatGPT in college major recommendations. *Information Technology and Management*. https://doi.org/10.1007/s10799-024-00430-5

CHAPTER 3

Redefining STEM Education Approach for an AI-Driven Future

Rıdvan Elmas and Merve Adıguzel Ulutaş

Introduction

STEM education has gained significant popularity over the past few decades. These fields have gained considerable importance, especially in recent years, due to the crucial role of STEM in developing and advancing nations. The current global situation, marked by wars, crises, and similar challenges, encourages a defensive mindset, leading to increased investments in the core areas of STEM: science, technology, engineering, and mathematics (Li et al., 2020; Wang et al., 2025a, b). The increasing complexity of global issues has often led to a recognition that no single discipline can provide sufficient solutions; as a result, there is a growing shift toward integrating disciplines and collectively addressing these problems (Nugraha et al., 2024). As the intricate nature of these challenges underscores the need for multidisciplinary approaches, educational methods

R. Elmas (✉)
Atatürk Faculty of Education, Marmara University, Istanbul, Türkiye
e-mail: relmas@marmara.edu.tr

M. A. Ulutaş
Gazi University, Ankara, Türkiye
e-mail: merveadiguzel@gazi.edu.tr

R. Elmas, M. Akarsu (eds.), *STEM Education*, Maarif Global Education Series,
https://doi.org/10.1007/978-981-95-9023-0_3

that take a more holistic view, such as the STEM education approach, are becoming increasingly prominent in education (National Academies of Sciences, Engineering, and Medicine [NASEM], 2025).

The STEM education approach, derived from the STEM disciplines, has multiple definitions and implementations and is used as a broad concept—characterized and described in various ways (Minsky, 2007; Cambria et al., 2012; Akarsu et al., 2020). There is no single, universally accepted definition of the STEM Education Approach among researchers and practitioners (Hasanah, 2020). Most teachers and researchers implement the approach in their classrooms based on their understanding and previous experiences. Since the STEM education approach remains open to development and interpretation in various ways, it is influenced by technological advancements. The rise of sophisticated digital technologies, from artificial intelligence (AI), machine learning, robotics, to hyper-scale cloud infrastructure, has revolutionized research across various disciplines, reducing development cycles from years to mere minutes. Today, AI, once thought to introduce new paradigms to education in multiple ways, has evolved into an agent, not just a tool, such as virtual reality (VR), augmented reality (AR), television (TV), or a computer (Jiang et al., 2024a, b; Kamalov et al., 2025; Yan, 2025). This section was written to reconsider, explain, and define STEM education in the context of an AI-driven future (Zhai & Krajcik, 2024a; Acemoglu, 2025).

The Need to Redefine STEM Education

The Science and Engineering Indicators report published by the National Science Board (NSB) (2018) drew attention to US students' lack of interest and inadequacy in STEM disciplines. The decline in interest in STEM disciplines has alarmed not only the United States but several countries (Wang et al., 2011). STEM disciplines are crucial for a country's economic development (Kayan-Fadlelmula et al., 2022; Kennedy & Odell, 2014). Therefore, in many countries, studies have been conducted to investigate students' perceptions, attitudes, and motivations toward STEM disciplines and to enhance their interest (Hackman et al., 2021; Ogegbo & Aina, 2024; Mohd Shahali et al., 2019; Tsai et al., 2021). Funding for STEM education has begun to increase (Sanders, 2009; Li et al., 2020; Taylor et al., 2023). Therefore, the STEM Education approach has been adopted to direct students toward STEM careers, revitalize these sectors, and enhance students' skills in these disciplines. According to Nadelson and Seifert (2017), two important functions of integrated STEM education are to train a qualified workforce and to create an effective teaching environment. The STEM Education approach has been defined in the literature by different researchers (Kennedy & Odell, 2014; Moore et al., 2014; Hasanah, 2020; Akarsu et al., 2020).

> STEM education is an integration of those four disciplines with two possibilities: as an input and output in education (Hasanah, 2020).

> STEM education has evolved into a meta-discipline, an integrated effort that removes the traditional barriers between these subjects, and instead focuses on innovation and the applied process of designing solution to complex contextual problems using current tools and technologies. Engaging students in high quality STEM education requires programs to include rigorous curriculum, instruction, and assessment, integrate technology and engineering into the science and mathematics curriculum, and also promotes scientific inquiry and the engineering design process (Kennedy & Odell, 2014 p. 246).

> It is an effort to combine some or all of the four disciplines of science, technology, engineering, and mathematics into one class, unit, or lesson that is based on connections between the subjects and real-world problems (Moore et al. 2014 p. 38).

> It is an educational approach that integrates the disciplines of science, technology, engineering and mathematics to solve real-life problems, facilitates the understanding of real-life problems with engaging and motivating experiences, and is not only product-oriented but also process and skill-oriented (Akarsu et al., 2020).

Although numerous definitions exist in the literature, a globally agreed-upon definition does not currently exist (English, 2016; Madani, 2020). Moreover, there are unclear conditions for the effective implementation of STEM education (Breiner et al. 2012). Researchers shape STEM activities around their understanding of the STEM education approach. This situation has led to different, inconsistent practices under the title of STEM education globally. The study by Adiguzel-Ulutas et al. (2023) reviewed existing research on STEM education. It was revealed that various activities were carried out within the context of multiple different theoretical engineering frameworks in STEM education. For example, while the engineering design process, the theoretical framework frequently preferred by researchers in the STEM education approach, is followed by some researchers, others have combined it with different teaching methods such as argumentation, inquiry-based learning, context-based learning, and project-based learning (Akarsu et al., 2020). According to Kelley and Knowles (2016), using an engineering design approach in STEM education is vital for effectively integrating STEM disciplines.

On the other hand, teachers have some difficulties in applying the STEM education approach in their classes. For example, some teachers lack the competence to implement STEM module designs (Dare et al., 2018). According to Nadelson et al. (2012), teachers can implement STEM education effectively with sufficient content knowledge. Some teachers believe there is an insufficient amount of time and resources to implement these student-centered designs (Margot & Kettler, 2019; Sellami et al., 2024). There are problems in integrating STEM education into the curriculum (Heba et al., 2017; Nadelson & Seifert, 2017; Park et al., 2016). Current curricula are not designed to be flexible enough to include and implement STEM education practices. Discipline-based learning makes it challenging to implement STEM education, which is an interdisciplinary and problem-based approach. Teachers also worry about straying from the curriculum (Bagiati & Evangelou, 2015).

Apart from these implementation difficulties, the STEM education approach is grounded in real-life problems (Ashgar et al., 2012; Struyf et al., 2019) in context (Elmas, 2020a, b; McDonnell et al., 2025) and incorporates interdisciplinary perspectives to address these issues. It is student-centered (Guzey et al., 2016; Mathis et al., 2017; Stohlmann et al., 2012), with the teacher as a guide on the side. Students learn from their mistakes and are encouraged to design solutions to problems. They follow the engineering design process (Bryan et al., 2015; Elmas & Gül, 2020; Guzey et al., 2016; Moore & Smith, 2014) and strive to solve engineering problems. Collaboration with peers and group work (Burrows et al., 2014; Roehrig et al., 2012; Stump et al., 2016) are key components of this approach. These characteristics define STEM education. Thus, the STEM education approach combines many good features. This teaching environment prepares students for real-life problems and increases their motivation to pursue degrees in STEM fields. Additionally, STEM education fosters the development of twenty-first-century skills, including cooperation, collaboration, problem-solving, and creativity (Arlinwibowo et al., 2021; Tytler, 2020; Widya et al., 2019). As a result, it has been suggested that the STEM education approach is crucial in producing a skilled labor force, given the complexity of real-life problems. Because in real-life problems, disciplines cannot be isolated from one another (Martín-Páez et al., 2019).

For instance, global warming is a worldwide issue, not a local problem (Shepherd & Sobel, 2020). Many countries have taken action to prevent global warming. However, this issue cannot be resolved with a one-discipline perspective (Schipper et al., 2021). Besides, no one can solve this global problem just through their own efforts (Twomey et al., 2025). Similarly, waste management and recycling are other essential problems currently. Traditional and one-sided approaches are no longer effective in solving these complex problems. Creativity and innovative approaches are necessary to address these kinds of global problems. Apart from such examples, creative, sustainable, and engineering-oriented approaches are necessary for addressing various issues, including the extinction of endemic species, water scarcity, declining biodiversity, and the production of sustainable energy (Ibrahim et al., 2024; Lea-Smith et al., 2025). Policies, government interventions, resource allocation, social movements, scientific research, innovations, and technological developments all directly and/or indirectly affect global issues. According to Mark and Semaan (2008), technology plays a crucial role in the development and transformation of various areas. For example, the discovery of electricity in the nineteenth century led to significant changes in industry and daily life.

With the common advent of information and communication technologies, significant changes have occurred in various areas, including professional services and the entertainment industry (OECD, 1999). Smartphones have affected sectors such as photography and banking (Zhu & Wang, 2022; Awinzeligo et al., 2024). The education sector has also been affected by all these developments (Johnson et al., 2016). It has been exposed to change and transformation processes. E-learning and online learning platforms have become widespread in education (McBrien et al., 2009; Wallace, 2003). Then, mobile technologies, virtual and augmented reality (VR/AR), and robotics and coding subjects have become part of classroom instruction and extracurricular activities (Benitti, 2012; Martin & Ertzberger, 2013; Theodoropoulos & Lepouras, 2021). Today, with the developments in AI tools, their integration into education, customization for educational purposes, and adoption as a natural component of the teaching and learning process have become important topics of discussion. Although studies on AI have a long history, they have become increasingly integrated into daily life with the rapid development of AI chatbots over the last few years (Akpan et al., 2025; Hwang & Chang, 2023; Labadze et al., 2023; Owan et al., 2023; Ruiz-Rojas et al., 2023). Although the history of AI studies dates back to the 1950s (Turing, 1950; von Neumann, 1958; McCarthy, 2007), AI chatbots and other tools have increasingly integrated into our daily lives, particularly since 2022, starting with Weizenbaum's straightforward Eliza (Weizenbaum, 1966; Aker, 2025).

In the healthcare field, AI tools are used for medical imaging analyses, while in industry, autonomous vehicles are used more commonly. Similarly, in agriculture, it analyzes soil fertility. In the field of art, it can create unique combinations of works that will inspire artists. Many similar examples can be seen in the documentary "How Far Is Too Far?" The Age of AI, featuring fascinating and up-to-date AI projects from various fields (YouTube Originals, 2019).

Along with these, in education, students utilize AI tools for problem-solving, personalized learning, understanding complex topics, writing and content creation, exam preparation, test solutions, and literature reviews (Sun et al., 2025; Wang et al., 2025a, b). Google (2024) states in the Project Astra demo video: "Imagine a student solving a chemistry problem with the help of an AI that can see, hear, and reason in real time." Teachers use AI tools to create personalized lesson plans, develop, and enrich teaching materials, conduct assessments and grading, and increase classroom engagement (Al Nabhani et al., 2025). According to Laird et al. (2023), 58% of K–12 students reported using AI tools in the 2023 academic year, while nearly half of teachers reported using AI tools. Educational approaches are likely to be enriched and enhanced by the proper use of AI tools, which are actively utilized across various sectors, including by teachers and students.

Redefining the STEM education approach to make it suitable for the AI-driven age is crucial for its up-to-date and practical use (Zhai & Krajcik, 2024a). In the AI-driven age, STEM education is a process-oriented learning ecosystem that integrates multiple disciplines, encourages interdisciplinary knowledge driven by individual curiosity, questions, interprets information, asks the right questions, and creates descriptive systems to define and organize problems (Xu & Ouyang, 2022; Leon et al., 2025). It not only resolves existing issues but also transforms them into creative designs, promotes critical and analytical thinking in the context of data, enables predictions using algorithms, and allows individuals to interact effectively not just with each other but also with chatbots, viewing them as part of a team (El Fathi et al., 2025).

The Intersection of STEM Education and AI

The changes and developments affecting these disciplines can have a direct impact on the concept of STEM and the STEM education approach. AI has a profound impact on numerous fields and makes significant contributions to these specific disciplines. Recent progress in AI is fueled by more computing power, abundant data, and the development of advanced algorithms (Brown et al., 2024; Stanford Institute for Human-Centered Artificial Intelligence [Stanford HAI], 2025). Additionally, it has been noted that dedicating significant resources and encouraging collaboration across various groups are key factors in advancing AI. These collaborations are increasingly organized within each country's specific context, allowing nations to leverage their unique resources, institutions, and expertise for AI development (Marshall, 2025). For example, in the United States, OpenAI created ChatGPT; in China, High-Flyer AI Lab developed DeepSeek; and in France, Mistral AI SAS launched its own advanced chatbot models. These examples demonstrate that strong collaborations often remain within national borders but can still yield high-quality results, particularly when supported by public funding (Sayegh, 2025; Piquard, 2025; European Commission, 2025).

The data analysis capabilities and problem-solving abilities of AI tools in mathematics have advanced to the point where they can surpass human limitations. In the International Mathematical Olympiad, one of the world's most challenging mathematics competitions, AI models (OpenAI & Google) have achieved scores at the gold medal level, approaching the highest ranks attained by human contestants (Luong & Lockhart, 2025). They have not only excelled in numerical calculations but also demonstrated the ability to develop correct approaches to problems that require logical reasoning and formal proofs (Béchard & Visser, 2025). Moreover, they managed to break down complex problems into manageable parts, generate new solution strategies, and exhibit strategic thinking skills similar to those of human mathematicians. This suggests that when evaluated under the same standards as humans, AI can achieve results comparable to those of the highest levels of human performance. This achievement, along with others, is dedicated to the development of machine learning and deep learning algorithms, as well as the emergence of new fields in these areas of research (Gerrish, 2018; Mitchell, 2019).

In the past, classical chatbots were limited to simple text-based exchanges; however, the field has now taken a major leap forward with the emergence of multimodality. Today's AI systems can integrate text, images, audio, and even video, allowing for richer, more human-like interactions. Multi-modality is important because it enables AI to capture meaning across different forms of data, increasing accuracy, robustness, and applicability. At the same time, the shift from large language models (LLMs) to large action models (LAMs) marks another turning point: while LLMs primarily generate language, LAMs are designed to reason and take actions, bridging the gap between understanding and execution in real-world contexts.

However, the future of AI is not solely about transitioning to Large Action Models (LAMs). In certain contexts, Smaller Language Models (SLMs) are increasingly proving to be more valuable than traditional LLMs. Belcak et al. (2025) defended that small language models are more suitable and economical for many agentic AI applications. Their advantages lie in efficiency, lower computational cost, and the ability to be fine-tuned for highly specific tasks or domains. This makes SLMs particularly important for applications where resources are limited, data privacy is a concern, or domain expertise is crucial. Thus, the evolving landscape of AI suggests not a simple replacement of LLMs with LAMs, but rather a diversified ecosystem in which different model sizes and capabilities complement one another. These developments are especially significant for devices such as smartphones and laptops, as well as other battery-powered systems, where energy efficiency and compact deployment make SLMs even more valuable.

Implications of AI in science are far-reaching (World Economic Forum, 2024a; YouTube Originals, 2019; Xiang et al., 2025; Ying et al., 2025). Like mathematics, data analysis and pattern recognition can easily reveal details difficult for humans to recognize. It can predict many issues affecting people, such as climate change and global warming. It can utilize genetic data to treat diseases and even play a crucial role in developing medicines and vaccines (Gómez-González et al., 2020). Additionally, AI-integrated robotic technologies enable us to access scientific data in various fields, ranging from space science to underwater exploration. It also significantly increases the accessibility of scientific knowledge.

AI tools offer more personalized experiences also in the industry. The transformation of dummy robots into cobots (collaborative robots) has drastically enhanced the integration of robots into factories. A cobot is a collaborative robot designed to safely work alongside humans, assisting them in tasks rather than replacing them. With the development of the Robot Operating System (ROS), robots can now be reprogrammed multiple times, enabling them to evolve first into collaborative robots (cobots) and, with the integration of AI, further into autonomous agents. ROS provides a flexible and modular framework that standardizes software components, making it easier to adapt robots for different tasks and environments. When combined with AI, this adaptability allows robots not only to follow predefined instructions but also to learn, perceive, and act autonomously in dynamic contexts. Thus, the progression from robots to cobots and ultimately to intelligent agents (somehow for now) illustrates how ROS and AI together are reshaping automation into a human-centered, adaptive, and highly efficient ecosystem. The use of AI in technological tools enhances their capabilities, making them more adaptable and context-aware, and enabling AI to act as a catalyst for further development. The engineering discipline has inherently been given a problem-solving mission. AI solves many problems by producing effective, fast, and user-specific solutions (Chi et al., 2025). Automation of production and usage processes significantly reduces costs through less error-prone production. High-performance and sustainable solutions contribute to time savings. In the future, the engineering profession and AI are expected to reveal more in-depth problem-solving processes.

In the future, the engineering profession and AI are expected to reveal more in-depth problem-solving processes. This trend is closely linked to the rise of so-called "dark factories," which are fully automated production environments where little to no human presence is required. By integrating AI-driven decision-making with autonomous robotic systems, these factories demonstrate how advanced problem-solving capabilities can translate directly into higher efficiency, precision, and uninterrupted 24/7 operations.

Even AI has led to some changes in the production, consumption, and sharing processes of art. It helps to produce original works and create data-driven ones by bringing together data from many different platforms. It can provide an interactive art experience that transforms according to the viewer's wishes through interactive participation in the art-sharing process. It can add creative dimensions to the existing potential of artists.

In the future, humans and AI will likely collaborate to produce creative works, creating art in innovative and new ways (Nurton, 2023). Artists like Refik Anadol envision architecture as a canvas, where data become pigments and AI acts as a brush of imagination, turning buildings into living artworks (Anadol, 2023). One of his master pieces, inspired by Jorge Luis Borges' The Library of Babel and installations such as those at Frank Gehry's Walt Disney Concert Hall, explores the idea of structures that can dream and even carry memories. If the brain's ability to see is called imagination (Anadol, 2020), then AI-driven art invites us to ask what a machine might do with the memories of others.

While AI enriches disciplines, it also extends the boundaries; not only universities but also private tech companies are creating new science. This means that transformation and change will be inevitable across multiple stakeholders. In the age of AI, companies (Nvidia, OpenAI, Google, Microsoft, Meta, etc.) responsible for producing and disseminating knowledge are expected to replace scientific institutions and universities as primary sources of knowledge. However, this looks more feasible due to the high cost of advancing AI technologies. At universities, both experienced and early-career researchers often work with great passion despite relatively low salaries. However, many of them are being drawn to highly profitable private companies, resulting in a gradual shift of AI-related scientific production from universities to the corporate sector. This shift of fundamental STEM-related knowledge production toward the private sector also makes it more difficult to manage risks associated with accountability, as corporations often operate under different incentives and less transparent frameworks than academic institutions.

Although there are many ubiquitous advancements, the most appealing products of AI companies' chatbots are not perfect. Regardless of the companies that produce them, chatbots often come with numerous defects that fall short of perfection, such as exhibiting bias, providing superficial information, and lacking depth in their responses. Studies have shown that AI chatbots, the most prominent AI tools, often provide incorrect and incomplete answers to scientific questions (Elmas et al., 2024). For example, in the study by Elmas et al. (2024), scientific questions related to cell energy metabolism (a field of biochemistry) were posed to an AI chatbot. It was determined that most of the questions provided by the AI chatbot presented incorrect and superficial answers. Similarly, Dalkemper (2023) emphasized that chatbots generate incorrect responses. However, studies also compare which AI chatbots create more accurate answers (Cheong et al., 2023; Kaftan et al., 2024; Xie et al., 2024).

Such studies are crucial in highlighting the risks posed by AI, particularly in the fields of science and education. Zanetti et al. (2020) emphasized that biases found in AI tools pose a risk to using these tools for educational purposes. Because biased algorithms have the potential to significantly compromise the principle of neutrality in education, leading to unequal access and unfair outcomes. Pedro et al. (2019) noted that integrating AI into schooling can lead to economic and social divisions, particularly in societies with low socioeconomic status. Ensuring inclusion and equity in education will be a challenging task. Li and Gu (2023) propose a comprehensive risk framework for human-centered AI in education (HCAI), identifying significant risks including pedagogical mismatch, misuse of AI resources, privacy concerns, accountability issues, and bias. They emphasize that managing these risks through collaboration among stakeholders is essential to ensure the responsible, sustainable, and trustworthy integration of AI in education. While the positive impact of AI on STEM disciplines is widely recognized, the potential risks associated with AI integration into any pedagogy underscore the importance of incorporating clear guardrails, ensuring that similar safeguards are embedded within the AI-integrated STEM education approach.

Integrating AI into STEM Classrooms

First and foremost, there is an age limit in adapting and using AI tools—such as chatbots—within the STEM education approach. For instance, as stated in ChatGPT's Terms of Use, students must be at least 13 years old to access the system, and those under 18 are only allowed to use it with the consent and supervision of a legal guardian (OpenAI, 2024). This highlights the need to consider not only pedagogical but also legal and ethical boundaries when integrating AI into STEM education.

The widespread use of AI tools will transform the skills and expertise needed in the labor market of the 2020s. For example, AI technologies can automate some tasks that previously required human effort (Acemoglu & Restrepo, 2020). So, it will change the skill expectations expected from employees in the business sector (Babashahi et al., 2024). Alekseeva et al. (2021) emphasized that the demand for an AI-literate workforce has increased in recent years. In this context, STEM curricula must be adapted according to changing skills and areas of expertise (Hazarika, 2020).

According to Sidhu et al. (2024), STEM curricula that incorporate various aspects of AI are necessary to prepare the workforce for the AI technologies that have been developed (Kim & Kwon, 2025).

Studies on integrating AI into schooling are becoming more common (Guan et al., 2020; Zhai et al., 2021), and effective AI technologies are being implemented in classroom instruction. For example, the Khanmingo app offers students a personalized learning environment (Khan Academy, 2025). It is a totally different experience (Adaptive learning) from a computer-assisted instruction or video-based training (Khan, 2023); hence, they both have some capacity for learning (Khan, 2011). Adaptive learning enables students to learn at their own pace and at their own level, utilizing machine learning and algorithms. This allows students to transition from a fragmented, gap-filled form of learning toward a more comprehensive and mastery-based learning experience. It also empowers teachers to prepare and adapt lesson plans more conveniently, saving time while enhancing instructional quality.

In the STEM education approach, learning often begins with a scenario framed around a compelling question. To support this process, there are AI tools—such as chatbots, such as ChatGPT, DeepSeek, Claude, and Perplexity—that are designed to answer students' questions. After students upload their questions to these platforms, they can not only find solutions but also ask follow-up questions to clarify what they do not understand. These tools create opportunities for discussion that help students refine their questions and develop a deeper understanding, making them powerful companions in inquiry-based STEM learning. Moreover, when students encounter difficulties, they can present their questions visually and interact through voice or video. With the emergence of multimodal chatbots—such as Google's Project Astra—these interactions make it possible to address not only procedural tasks but also more conceptual questions (Google, 2025). This creates a richer and more engaging learning experience, offering students powerful support on their journey of inquiry.

Students are divided into groups in classes where the STEM education approach is applied. Students are expected to communicate effectively with their groupmates. Along with AI technologies such as chatbots (e.g., the Astra project by Google), being incorporated into classrooms, these chatbots might also be considered valuable group members with senior students who have previous experiences with them.

In the STEM education approach, real-life problems are addressed. These problems typically involve real-life scenarios that might attract student attention. New and engaging multiple scenarios can be created with chatbots more conveniently. Teachers have more time to focus on the context and details of the scenarios, especially if they have numerous options to choose the best-fit scenario for their students (Hooper, 2025). In a study conducted by Elmas (2020a, b), moderate students are more willing to learn about local problems that surround them, whereas more able students have a greater interest in extraordinary events, unusual incidents, or exceptional occurrences, such as the Northern Lights (also known as the Aurora Borealis).

In a standard STEM education classroom, student groups are given time to define the problem presented in this scenario and are expected to identify variables that help solve the inquiry. Students can discuss the problem among themselves while overthinking it. Moreover, students can enter some clues from the given scenario as a prompt into the chatbot and ask it to identify the problem in detail. Students can review the answers provided by the chatbot and discuss them within the group. Finally, the group can be thoroughly informed about the problem in some way.

Students must also collect information about the problem. However, in all processes of the STEM education approach, information collection continues, not in a single step. Previously, students could gather information by looking at the internet, books, and previous studies. Nowadays, students use chatbots to access knowledge right away. While implementing STEM education modules, students can utilize chatbots to actively collect information. Using context engineering will help students engage in an efficient conversation with the chatbot through high-quality prompts (Mei et al., 2025). However, the critical point here is that chatbots can sometimes provide incorrect information, regardless of how contextual or high-quality the prompt is; such inaccurate outputs are referred to as hallucinations or misconceptions. As mentioned in the first section, studies have shown that chatbots often provide incorrect or incomplete information. In this context, students must evaluate the accuracy of the information they obtain from Chatbots in every case. They must also critically evaluate the information obtained from chatbots and remain attentive to ethical considerations.

During the planning phase of the STEM education approach, students are expected to produce multiple solutions to the problem and choose the most convenient option, considering constraints such as cost and time.

Students can benefit from chatbots when creating solutions to problems. They can discuss their solution with the chatbots as if it were one of their group members. They can use tools such as Canva, Prezi, or Gama for creating AI-generated presentations. For example, students can engage in data visualization through tools like Microsoft Power BI, which incorporates AI integrations and functionalities (Tirupati et al., 2024). They may question the risks and disadvantages of the solution proposal they have developed, as well as the risks they foresee. These tools can offer students various perspectives. They can provide suggestions on areas that need improvement. Indeed, it may be asked to calculate factors such as cost and time.

Students must test the solution they have decided on to solve the problem in context (Elmas, 2020a, b). Before testing, they can ask AI tools whether the solution they have determined will be a good fit for the situation. This way, they can administer a virtual test before the actual test. After testing their prototype, students can decide whether it is suitable for solving the problem. They can repeat the process in case of possible errors. The teacher is in a guiding position for the students as before. In this context, AI does not replace the teacher (World Economic Forum, 2024b) but rather empowers students to work in teams as group members to solve problems. The guiding role of teachers is prominent. With the integration of AI tools, some changes occur in the STEM Education approach, which were mentioned above and summarized here:

- While students' analytical skills were more noticeable in the previous period, AI tools are now handling more analytical tasks and emphasizing data-driven decision-making skills in students.
- Technology no longer serves merely as a tool but acts as a team member, highlighting its role as an agent. This change encourages the active use of technology in all processes. It also helps develop students' digital literacy and tech skills, enabling them to use technology effectively.
- In the past, universities were the primary source of information; however, with the emergence of new stakeholders, databases utilized by AI are now also becoming sources of information. This means students can gather information from multiple resources, such as companies (Google, Meta, etc.), that are sharing academic research results in their research and development departments. At this point, it is up to the student to evaluate the bias, accuracy, and impartiality

of the information provided. In this context, students' critical thinking, research, and questioning skills are essential.

- AI tools can quickly address superficial problems. However, human creativity remains essential for addressing complex issues that are influenced by multiple factors. In this context, AI tools support human creativity.
- Simulations and virtual reality tools allow for prototype testing before development. Additionally, AI tools can assist in generating ideas for the prototype based on assumptions without requiring a virtual prototype.
- Since employees who can work with AI tools are essential, integrating AI technologies into the STEM curricula is valuable in preparing individuals to work effectively with AI technologies in various conditions.
- Prompting skills (context engineering) have become increasingly important not only for individuals but also in interactions with AI tools.

As AI increasingly transforms the ways humans learn, think, and produce knowledge, the traditional view of STEM education—combining science, technology, engineering, and math to foster interdisciplinary problem-solving, design thinking, and innovation—has become inadequate for addressing the realities of modern learning environments (Cai et al., 2025). In today's educational landscape, learners must not only grasp disciplinary content but also interact with intelligent systems that analyze data, create representations, and engage in complex reasoning. Therefore, AI should no longer be viewed as merely an external tool or added technology; instead, it serves as a cognitive and cultural partner that alters the experience of inquiry, creativity, planning, and collaboration (Jiang et al., 2024a, b; Celik et al., 2022). This shift requires redefining STEM education to recognize human–AI interaction as a new central area of learning, highlighting metacognition, ethical judgment, and responsible innovation as essential parts of modern scientific literacy (Yang et al., 2025; Lim, 2025).

AI-integrated STEM Education Approach describes a comprehensive framework that combines the foundational knowledge of STEM fields with the skills, literacies, and ethical considerations necessary for meaningful engagement with AI systems (Zhai & Krajcik, 2024b). In this approach, AI is viewed as an active partner that enhances human thinking in data analysis, modeling, problem-solving, and design tasks (Zhai & Krajcik, 2024a). Students are encouraged to develop algorithmic thinking, data literacy, and systems understanding while critically considering the social and moral effects of AI-involved inquiry (Infocomm Media Development Authority [IMDA] & AI Verify Foundation, 2024). More than just improving technological skills, the AI-integrated STEM Education Approach aims to foster adaptable, reflective, and ethically conscious individuals who can think with, through, and about AI, thereby transforming knowledge into responsible innovation in a rapidly changing world (European Commission, 2023). These three key aspects create a comprehensive model of AI-integrated STEM Education, based on the idea that learning in the age of AI is not just enhanced by technology but fundamentally reshaped through human–AI collaboration. This three-part framework views education as a dynamic system where thinking, ethics, and creativity converge to develop learners capable of reasoning, designing, and acting responsibly in AI-rich environments.

The new definition can be shaped around the discussions above and presented here as a final word: The *AI-integrated STEM Education Approach is an educational framework that combines the disciplines of science, technology, engineering, and mathematics with artificial intelligence literacy to address real-life problems through data-informed, creative, and ethically responsible solutions. It enables learners to engage with complex real-world challenges through motivating experiences where AI acts as an active collaborator in inquiry, design, and problem-solving processes. Emphasizing process- and skill-oriented learning rather than mere product creation, this approach cultivates cognitive, ethical, and creative competencies—such as data literacy, algorithmic reasoning, metacognitive reflection, and responsible innovation—aiming to develop individuals who can think critically, create meaningfully, and act responsibly in AI-augmented learning ecosystems.*

Disclaimer The authors gratefully acknowledge the support of chatbots such as Perplexity and ChatGPT in the preparation of this chapter. These chatbots were used as a scholarly assistant to facilitate preliminary literature exploration, conceptual organization, and linguistic refinement. Its contributions were limited to enhancing clarity, coherence, and translation accuracy, while all intellectual interpretations, theoretical insights, and final judgments presented herein remain solely those of the authors.

References

Acemoglu, D. (2025). The simple macroeconomics of AI. *Economic Policy, 40*(121), 13–58.

Acemoglu, D., & Restrepo, P. (2020). The wrong kind of AI? Artificial intelligence and the future of labour demand. *Cambridge Journal of Regions, Economy and Society, 13*(1), 25–35.

Adiguzel-Ulutas, M., Elmas, R., Karakaya, F., & Yilmaz, M. (2023). Examination of the studies on STEM education approach in the context of engineering design process in Türkiye. *Journal of Turkish Educational Sciences, 21*(2), 1111–1130.

Akarsu, M., Akçay, N. O., & Elmas, R. (2020). STEM eğitimi yaklaşımının özellikleri ve değerlendirilmesi. *Bogazici University Journal of Education, 37*, 155–175.

Aker, O. (2025). AI-assisted music mastering: An exploration of human and AI practices in contemporary music production. In *Understanding generative AI in a cultural context: Artificial myths and human realities* (pp. 17–50). IGI Global Scientific Publishing.

Akpan, I. J., Kobara, Y. M., Owolabi, J., Akpan, A. A., & Offodile, O. F. (2025). Conversational and generative artificial intelligence and human–chatbot interaction in education and research. *International Transactions in Operational Research, 32*(3), 1251–1281.

Al Nabhani, F., Hamzah, M. B., & Abuhassna, H. (2025). The role of artificial intelligence in personalizing educational content: Enhancing the learning experience and developing the teacher's role in an integrated educational environment. *Contemporary Educational Technology, 17*(2), ep573. https://doi.org/10.30935/cedtech/16089

Alekseeva, L., Azar, J., Giné, M., Samila, S., & Taska, B. (2021). The demand for AI skills in the labor market. *Labour Economics, 71*, 102002.

Anadol, R. (2020, July). *Art in the age of machine intelligence* [Video]. TED Conferences. https://www.ted.com/talks/refik_anadol_art_in_the_age_of_machine_intelligence?subtitle=tr

Anadol, R. (2023, April). *How AI art could enhance humanity's collective memory* [Video]. TED Conferences. https://www.ted.com/talks/refik_anadol_how_ai_art_could_enhance_humanity_s_collective_memory?subtitle=tr

Arlinwibowo, J., Retnawati, H., & Kartowagiran, B. (2021). How to integrate STEM education in the Indonesian curriculum? A systematic review. *Challenges of Science, IV*, 18–25. https://doi.org/10.31643/2021.03

Asghar, A., Ellington, R., Rice, E., Johnson, F., & Prime, G. M. (2012). Supporting STEM education in secondary science contexts. *Interdisciplinary Journal of Problem-Based Learning, 6*(2), 85–125. https://doi.org/10.7771/1541-5015.1349

Awinzeligo, M. H., Churcher, W. E., Oppong, C., Duut, N. G., Wemegah, R., & Kuupole, E. (2024). Effects of smartphone photography on commercial photographers in Bolgatanga, Ghana. *African Journal of Applied Research, 10*(2), 627–643. https://doi.org/10.26437/ajar.01.12.31

Babashahi, L., Barbosa, C. E., Lima, Y., Lyra, A., Salazar, H., Argôlo, M., de Almeida, M. A., & de Souza, J. M. (2024). AI in the workplace: A systematic review of skill transformation in the industry. *Administrative Sciences, 14*(6), 127. (28 pages).

Bagiati, A., & Evangelou, D. (2015). Engineering curriculum in the preschool classroom: The teacher's experience. *European Early Childhood Education Research Journal, 23*(1), 112–128. https://doi.org/10.1080/1350293X.2014.991099

Béchard, D. E., & Visser, D. (Eds.). (2025, July 22). *Can writing math proofs teach AI to reason like humans? OpenAI model earns gold medal score at International Math Olympiad.* Scientific American. https://www.scientificamerican.com

Belcak, P., Heinrich, G., Diao, S., Fu, Y., Dong, X., Muralidharan, S., Lin, Y. C., & Molchanov, P. (2025). Small language models are the future of agentic AI. NVIDIA Research. *arXiv.* https://arxiv.org/abs/2506.02153v2

Benitti, F. B. V. (2012). Exploring the educational potential of robotics in schools: A systematic review. *Computers & Education, 58*(3), 978–988.

Breiner, J., Harkness, M., Johnson, C. C., & Koehler, C. (2012). What is STEM? A discussion about conceptions of STEM in education and partnerships. *School Science and Mathematics, 112*(1), 3–11.

Brown, Z., Slattery, P., & Lyu, H. (2024, May 20). *What drives progress in AI? Trends in algorithms.* MIT FutureTech. https://futuretech.mit.edu/news/what-drives-progress-in-ai-trends-in-algorithms

Bryan, L. A., Moore, T. J., Johnson, C. C., & Roehrig, G. H. (2015). Integrated STEM education. In *STEM road map* (pp. 23–38). Routledge. https://doi.org/10.4324/9781315753157-5

Burrows, A. C., Breiner, J. M., Keiner, J., & Behm, C. (2014). Biodiesel and integrated STEM: Vertical alignment of high school biology/biochemistry and chemistry. *Journal of Chemical Education, 91*(9), 1379–1389.

Cai, C., Zhu, G., & Ma, M. (2025). A systemic review of AI for interdisciplinary learning: Application contexts, roles, and influences. *Education and Information Technologies, 30*, 9641–9687. https://doi.org/10.1007/s10639-024-13193-x

Cambria, E., Livingstone, A., & Hussain, A. (2012). The hourglass of emotions. In *Cognitive behavioural systems: COST 2102 international training school, dresden, Germany, February 21-26, 2011, revised selected papers* (pp. 144–157). Springer Berlin Heidelberg.

Celik, I., Dindar, M., Muukkonen, H., & Järvelä, S. (2022). The promises and challenges of artificial intelligence for teachers: A systematic review of research. *TechTrends, 66*(4), 616–630. https://doi.org/10.1007/s11528-022-00715-y

Cheong, R. C., Unadkat, S., Mcneillis, V., Williamson, A., Joseph, J., Randhawa, P., Andrews, P., & Paleri, V. (2023). Artificial intelligence chatbots as sources of patient education material for obstructive sleep apnea: ChatGPT versus Google bard. *European Archives of Oto-Rhino-Laryngology, 281*(2), 985–993. https://doi.org/10.1007/s00405-023-08319-9

Chi, N., Xu, F., Liu, Q., Zeng, X., Yang, F., & Gao, Y. (2025). AI for engineering. In Z. Xu, Y. Cheng, Y. Yang, Y. Xu, X. Qiu, Y. Fu, S. Wang, M. Yang, H. Zou, S. Lu, L. Shi, K. Wei, X. Zhu, W. Gao, Y. Li, W. Lin, W. Yang, H. Xiang, M. Ji, et al. (Eds.), *AI for science 2025* (pp. 53–56). Shanghai Academy of AI for Science & Fudan University.

Dahlkemper, M. N., Lahme, S. Z., & Klein, P. (2023). How do physics students evaluate artificial intelligence responses on comprehension questions? A study on the perceived scientific accuracy and linguistic quality of ChatGPT. *Physical Review Physics Education Research, 19*(1), 010142(1)–010142(25). https://doi.org/10.1103/physrevphyseducres.19.010142

Dare, E. A., Ellis, J. A., & Roehrig, G. H. (2018). Understanding science teachers' implementations of integrated STEM curricular units through a phenomenological multiple case study. *International Journal of STEM Education, 5*, 1–19.

El Fathi, T., Saad, A., Larhzil, H., Lamri, D., & Al Ibrahmi, E. M. (2025). Integrating generative AI into STEM education: Enhancing conceptual understanding, addressing misconceptions, and assessing student acceptance. *Disciplinary and Interdisciplinary Science Education Research, 7*(6), 1–21. https://doi.org/10.1186/s43031-025-00125-z

Elmas, R. (2020a). Bağlamın anlamı ve nitelikleri ve öğrencilerin fen eğitiminde bağlam tercihleri [The meaning and characteristics of context and students' context preferences in science education]. *Türkiye Kimya Derneği Dergisi, Kısım C: Kimya Eğitimi, 5*(1), 83–102.

Elmas, R. (2020b). The meaning and characteristics of context and students' context preferences in science education. *Journal of Turkish Chemical Society Section C: Chemistry Education (JOTCSC), 5*(1), 53–70.

Elmas, R., & Gul, M. (2020). Examination of the applicability of STEM education approach in the context of the 2018 science curriculum. *Journal of Turkish Chemical Society Section C: Chemistry Education (JOTCSC), 5*(2), 223–246. https://doi.org/10.37995/jotcsc.794547

Elmas, R., Adiguzel-Ulutas, M., & Yilmaz, M. (2024). Examining ChatGPT's validity as a source for scientific inquiry and its misconceptions regarding cell energy metabolism. *Education and Information Technologies, 29*, 25427–25456.

English, L. D. (2016). STEM education K-12: Perspectives on integration. *International Journal of STEM Education, 3*, 1–8.

European Commission. (2023). *AI and education: A policy perspective from the European Union*. Directorate-General for Education, Youth, Sport and Culture. Publications Office of the European Union. https://op.europa.eu/en/publication-detail/-/publication/9bb60fb1-b42a-11ee-b164-01aa75ed71a1/language-en

European Commission. (2025, February 11). *EU launches InvestAI initiative to mobilise €200 billion of investment in artificial intelligence* [Press release]. https://ec.europa.eu/commission/presscorner/detail/en/ip_25_467

Gerrish, S. (2018). *How smart machines think*. MIT Press.

Gómez-González, E., Márquez-Rivas, J., Guerrero-Claro, M., Fernández-Lizaranzu, I., Relimpio-López, M. I., Dorado, M. E., Mayorga-Buiza, M. J., & Izquierdo-Ayuso, G. (2020). Artificial intelligence in medicine and healthcare: A review and classification of current and near-future applications and their ethical and social impact. *arXiv*. https://arxiv.org/abs/2001.09778

Google. (2025, May). *Project Astra | Exploring the future of learning with an AI tutor research prototype* [Video]. YouTube. https://www.youtube.com/watch?v=MQ4JfafE5Wo

Google. (2024, May). *Project Astra | Exploring the future of learning with an AI* [Video]. YouTube. https://www.youtube.com/watch?v=MQ4JfafE5Wo

Guan, C., Mou, J., & Jiang, Z. (2020). Artificial intelligence innovation in education: A twenty-year data-driven historical analysis. *International Journal of Innovation Studies, 4*(4), 134–147.

Guzey, S. S., Moore, T. J., & Harwell, M. (2016). Building up STEM: An analysis of teacher-developed engineering design-based STEM integration curricular materials. *Journal of Pre-College Engineering Education Research (J-PEER), 6*(1), 2. https://doi.org/10.7771/2157-9288.1129

Hackman, S. T., Zhang, D., & He, J. (2021). Secondary school science teachers' attitudes towards STEM education in Liberia. *International Journal of Science Education, 43*(2), 223–246.

Hasanah, U. (2020). Key definitions of STEM education: Literature review. *Interdisciplinary Journal of Environmental and Science Education, 16*(3), e2217. https://doi.org/10.29333/ijese/8336

Hazarika, I. (2020). Artificial intelligence: Opportunities and implications for the health workforce. *International Health, 12*(4), 241–245.

Heba, E. D., Mansour, N., Alzaghibi, M., & Alhammad, K. (2017). Context of STEM integration in schools: Views from in-service science teachers. *Eurasia Journal of Mathematics, Science and Technology Education, 13*(6), 2459–2484.

Hooper, A. (2025, August). *Prompt engineering and in-context learning: A comprehensive survey of techniques, applications, and future directions* [Preprint]. University of Wisconsin–Madison. https://www.researchgate.net/publication/394775684_Prompt_Engineering_and_In-Context_Learning_A_Comprehensive_Survey_of_Techniques_Applications_and_Future_Directions

Hwang, G. J., & Chang, C. Y. (2023). A review of opportunities and challenges of chatbots in education. *Interactive Learning Environments, 31*(7), 4099–4112.

Ibrahim, M. S. N., Miard, P., Abdullah, N. A., & Julius, A. (2024). Embracing biodiversity: A perspective on transforming engineering education for sustainable innovation. *ASEAN Journal of Engineering Education, 8*(1), 30–40.

Infocomm Media Development Authority (IMDA) & AI Verify Foundation. (2024). *Model AI governance framework for generative AI.* Infocomm Media Development Authority (IMDA) and AI Verify Foundation. https://aiverify-foundation.sg/wp-content/uploads/2024/05/Model-AI-Governance-Framework-for-Generative-AI-May-2024-1-1.pdf

Jiang, T., Sun, Z., Fu, S., & Lv, Y. (2024a). Human-AI interaction research agenda: A user-centered perspective. *Data and Information Management, 8*, 100078. https://doi.org/10.1016/j.dim.2024.100078

Jiang, Y.-H., Li, R., Zhou, Y., Qi, C., Hu, H., Wei, Y., Jiang, B., & Wu, Y. (2024b). AI agent for education: Von Neumann multi-agent system framework. In *Proceedings of the 28th Global Chinese Conference on Computers in Education (GCCCE 2024)* (pp. 77–84). Global Chinese Conference on Computers in Education. https://arxiv.org/abs/2501.00083

Johnson, L., Adams Becker, S., Cummins, M., Estrada, V., Freeman, A., & Hall, C. (2016). *NMC horizon report: 2016 higher education edition.* The New Media Consortium. Retrieved January 6, 2025, from https://www.learntechlib.org/p/171478/

Kaftan, A. N., Hussain, M. K., & Naser, F. H. (2024). Response accuracy of ChatGPT 3.5 Copilot and Gemini in interpreting biochemical laboratory data a pilot study. *Scientific Reports, 14*(1), 8233. https://doi.org/10.1038/s41598-024-58964-1

Kamalov, F., Calonge, D. S., Smail, L., Azizov, D., Thadani, D. R., Kwong, T., & Atif, A. (2025). Evolution of AI in education: Agentic workflows. *arXiv.* https://arxiv.org/abs/2504.20082

Kayan-Fadlelmula, F., Sellami, A., Abdelkader, N., & Umer, S. (2022). A systematic review of STEM education research in the GCC countries: Trends, gaps and barriers. *International Journal of STEM Education, 9*, 1–24.

Kelley, T. R., & Knowles, J. G. (2016). A conceptual framework for integrated STEM education. *International Journal of STEM Education, 3*, 1–11.

Kennedy, T., & Odell, M. (2014). Engaging students in STEM education. *Science Education International, 25*(3), 246–258.

Khan Academy. (2025). *Khanmigo* [AI-teaching assistant application]. https://www.khanmigo.ai/

Khan, S. (2011, March). *Let's use video to reinvent education* [Video]. TED-Ed. https://ed.ted.com/lessons/let-s-use-video-to-reinvent-education-salman-khan

Khan, S. (2023, May 1). *How AI could save (not destroy) education* [Video]. TED. https://www.youtube.com/watch?v=hJP5GqnTrNo

Kim, K., & Kwon, K. (2025). From co-design to co-teaching: A comprehensive approach to an integrated AI curriculum in middle school STEM education. *Smart Learning Environments, 12*(57). https://doi.org/10.1186/s40561-025-00413-1

Labadze, L., Grigolia, M., & Machaidze, L. (2023). Role of AI chatbots in education: Systematic literature review. *International Journal of Educational Technology in Higher Education, 20*(1), 56.

Laird, E., Dwyer, M., & Grant-Chapman, H. (2023). *Off task: EdTech threats to student privacy and equity in the age of AI* [Report]. Center for Democracy and Technology.https://cdt.org/insights/report-off-task-edtech-threats-to-student-privacy-and-equity-in-the-age-of-ai/

Lea-Smith, D. J., Hassard, F., Coulon, F., Partridge, N., Horsfall, L., Parker, K. D. J., Smith, R. D. J., McCarthy, R. R., McKew, B., Gutierrez, T., Kumar, V., Dotro, G., Yang, Z., EBIC partners, & Krasnogor, N. (2025). Engineering biology applications for environmental solutions: Potential and challenges. *Nature Communications, 16*, 3538. https://doi.org/10.1038/s41467-025-58492-0

Leon, C., Lipuma, J., & Oviedo-Torres, X. (2025). Artificial intelligence in STEM education: A transdisciplinary framework for engagement and innovation. *Frontiers in Education, 10*, 1619888. https://doi.org/10.3389/feduc.2025.1619888

Li, S., & Gu, X. (2023). A risk framework for human-centered artificial intelligence in education. *Educational Technology & Society, 26*(1), 187–202.

Li, Y., Wang, K., Xiao, Y., Froyd, J. E., & Nite, S. B. (2020). Research and trends in STEM education: A systematic analysis of publicly funded projects. *International Journal of STEM Education, 7*(1), 1–16. https://doi.org/10.1186/s40594-020-00213-8

Lim, C. (2025, April 26–May 1). DeBiasMe: De-biasing human-AI interactions with metacognitive AIED (AI in education) interventions. In *Proceedings of the 2025 ACM Workshop on Human-AI integration for Augmented Reasoning* (Yokohama, Japan). arXiv. https://arxiv.org/html/2504.16770v1

Luong, T., & Lockhart, E. (2025, July 21). Advanced version of Gemini with deep think officially achieves gold-medal standard at the International Mathematical Olympiad. *DeepMind*. https://deepmind.google/blog/advanced-version-of-gemini-with-deep-think-officially-achieves-gold-medal-standard-at-the-international-mathematical-olympiad/

Madani, R. A. (2020). Teaching challenges and perceptions on STEM implementation for schools in Saudi Arabia. *European Journal of STEM Education, 5*(1), 3.

Margot, K. C., & Kettler, T. (2019). Teachers' perception of STEM integration and education: A systematic literature review. *International Journal of STEM Education, 6*(1), 1–16.

Mark, G., & Semaan, B. (2008). Resilience in collaboration: Technology as a resource for new patterns of action. In B. Begole & D. W. McDonald (Eds.), *Proceedings of the 2008 ACM Conference on Computer Supported Cooperative Work* (pp. 127–136). Association for Computing Machinery. https://doi.org/10.1145/1460563.1460585

Marshall, C. (2025, January 18). Here's what's in "Stargate," the $500-billion Trump-endorsed plan to power U.S. AI. *Scientific American*. https://www.scientificamerican.com/article/heres-whats-in-stargate-the-usd500-billion-trump-endorsed-plan-to-power-u-s/

Martín-Páez, T., Aguilera, D., Perales-Palacios, F. J., & Vílchez-González, J. M. (2019). What are we talking about when we talk about STEM education? A review of literature. *Science Education, 103*(4), 799–822. https://doi.org/10.1002/sce.21522

Martin, F., & Ertzberger, J. (2013). Here and now mobile learning: An experimental study on the use of mobile technology. *Computers & Education, 68*, 76–85.

Mathis, C. A., Siverling, E. A., Glancy, A. W., & Moore, T. J. (2017). Teachers' incorporation of argumentation to support engineering learning in STEM integration curricula. *Journal of Pre-College Engineering Education Research (J-PEER), 7*(1), 6.

McBrien, J. L., Cheng, R., & Jones, P. (2009). Virtual spaces: Employing a synchronous online classroom to facilitate student engagement in online learning. *The International Review of Research in Open and Distributed Learning, 10*(3), 1–17.

McCarthy, J. (2007). From here to human-level AI. *Artificial Intelligence, 171*(18), 1174–1182. https://doi.org/10.1016/j.artint.2007.10.009

McDonnell, C. H., Adams, I. N., Hynes, M. M., Guzey, S. S., Pilotte, M. K., Strimel, G. J., Tank, K. M., & Moore, T. J. (2025). From context to connection: Client letters in STEM integration curricula. *Education Sciences, 15*(6), 696. https://doi.org/10.3390/educsci15060696

Mei, L., Yao, J., Ge, Y., Wang, Y., Bi, B., Cai, Y., Liu, J., Li, M., Li, Z.-Z., Zhang, D., Zhou, C., Mao, J., Xia, T., Guo, J., & Liu, S. (2025, July). A survey of context engineering for large language models. *arXiv*. https://arxiv.org/abs/2507.13334

Minsky, M. (2007). *The emotion machine: Commonsense thinking, artificial intelligence, and the future of the human mind*. Simon and Schuster.

Mitchell, M. (2019). *Artificial intelligence: A guide for thinking humans*. Farrar, Straus and Giroux.

Mohd Shahali, E. H., Halim, L., Rasul, M. S., Osman, K., & Mohamad Arsad, N. (2019). Students' interest towards STEM: A longitudinal study. *Research in Science & Technological Education, 37*(1), 71–89.

Moore, T. J., & Smith, K. A. (2014). Advancing the state of the art of STEM integration. *Journal of STEM Education: Innovations and Research, 15*(1), 5–10.

Moore, T., Stohlmann, M., Wang, H., Tank, K., Glancy, A., & Roehrig, G. (2014). Implementation and integration of engineering in K-12 STEM education. In S. Purzer, J. Strobel, & M. Cardella (Eds.), *Engineering in pre-college settings: Synthesizing research, policy, and practices* (pp. 35–60). Purdue University Press.
Nadelson, L. S., & Seifert, A. L. (2017). Integrated STEM defined: Contexts, challenges, and the future. *The Journal of Educational Research, 110*(3), 221–223. https://doi.org/10.1080/00220671.2017.1289775
Nadelson, L. S., Seifert, A., Moll, A. J., & Coats, B. (2012). i-STEM Summer Institute: An integrated approach to teacher professional development in STEM. *Journal of STEM Education: Innovation and Outreach, 13*(2), 69–83.
National Academies of Sciences, Engineering, and Medicine. (2025). *Transforming undergraduate STEM education: Supporting equitable and effective teaching*. The National Academies Press. https://doi.org/10.17226/28268
National Science Board. (2018). Science and engineering indicators 2018 (NSB-2018-1). National Science Foundation.
Nugraha, M. G., Kidman, G., & Tan, H. (2024). Interdisciplinary STEM education foundational concepts: Implementation for knowledge creation. *EURASIA Journal of Mathematics, Science and Technology Education, 20*(10), em2523. https://doi.org/10.29333/ejmste/15471
Nurton, J. (2023, October 30). 'Painting' with data: How media artist Refik Anadol creates art using generative AI. *WIPO Magazine*. World Intellectual Property Organization. https://www.wipo.int/web/wipo-magazine/articles/painting-with-data-how-media-artist-refik-anadol-creates-art-using-generative-ai-67301
Ogegbo, A. A., & Aina, A. Y. (2024). Exploring young students' attitude towards coding and its relationship with STEM career interest. *Education and Information Technologies, 29*(8), 9041–9059.
OpenAI. (2024, December 11). *Kullanım Şartları* [Terms of Use]. OpenAI. https://openai.com/tr-TR/policies/row-terms-of-use/
Organization for Economic Cooperation and Development [OECD]. (1999). *The economic and social impact of electronic commerce*. OECD.
Owan, V. J., Abang, K. B., Idika, D. O., Etta, E. O., & Bassey, B. A. (2023). Exploring the potential of artificial intelligence tools in educational measurement and assessment. *Eurasia Journal of Mathematics, Science and Technology Education, 19*(8), em2307.
Park, H., Byun, S. Y., Sim, J., Han, H. S., & Baek, Y. S. (2016). Teachers' perceptions and practices of STEAM education in South Korea. *Eurasia Journal of Mathematics, Science and Technology Education, 12*(7), 1739–1753.
Pedro, F., Subosa, M., Rivas, A., & Valverde, P. (2019). *Artificial intelligence in education: Challenges and opportunities for sustainable development*. UNESCO.
Piquard, A. (2025, February 10). AI: With announcement of investments worth €109 billion, Macron intends to take on US. *Le Monde*. https://www.lemonde.

fr/en/economy/article/2025/02/10/ai-with-the-announcement-of-a-109-billion-investment-macron-intends-to-take-on-the-us_6737985_19.html

Roehrig, G. H., Moore, T. J., Wang, H. H., & Park, M. S. (2012). Is adding the E enough? Investigating the impact of K-12 engineering standards on the implementation of STEM integration. *School Science and Mathematics, 112*(1), 31–44.

Ruiz-Rojas, L. I., Acosta-Vargas, P., De-Moreta-Llovet, J., & Gonzalez-Rodriguez, M. (2023). Empowering education with generative artificial intelligence tools: Approach with an instructional design matrix. *Sustainability, 15*(15), 11524.

Sanders, M. (2009). STEM, STEM education, STEM mania. *Technology Teacher, 68*(4), 20–26.

Sayegh, E. (2025, January 22). Stargate AI project: The $500 billion gamble to dominate the future. *Forbes*. https://www.forbes.com/sites/emilsayegh/2025/01/22/stargate-ai-project-the-500-billion-gamble-to-dominate-the-future/

Schipper, E. L. F., Dubash, N. K., & Mulugetta, Y. (2021). Climate change research and the search for solutions: Rethinking interdisciplinarity. *Climatic Change, 168*(18), 1–18. https://doi.org/10.1007/s10584-021-03237-3

Sellami, A., Santhosh, M., Bhadra, J., & Ahmad, Z. (2024, April). Teachers' perceptions of the barriers to STEM teaching in Qatar's secondary schools: A structural equation modeling analysis. *Frontiers in Education, 9*, 1333669. Frontiers Media SA.

Shepherd, T. G., & Sobel, A. H. (2020). Localness in climate change. *Comparative Studies of South Asia, Africa and the Middle East, 40*(1), 7–34. Duke University Press. https://doi.org/10.1215/1089201X-8185983

Sidhu, G. S., Sayem, M. A., Taslima, N., Anwar, A. S., Chowdhury, F., & Rowshon, M. (2024). AI and workforce development: A comparative analysis of skill gaps and training needs in emerging economies. *International Journal of Business and Management Sciences, 4*(08), 12–28.

Stanford Institute for Human-Centered Artificial Intelligence. (2025). *AI Index report 2025*. Stanford University. https://hai.stanford.edu/ai-index/2025-ai-index-report

Stohlmann, M., Moore, T., & Roehrig, G. (2012). Considerations for teaching integrated STEM education. *Journal of Pre-College Engineering Education Research, 2*(1), 28–34. https://doi.org/10.5703/1288284314653

Struyf, A., De Loof, H., Boeve-de Pauw, J., & Van Petegem, P. (2019). Students' engagement in different STEM learning environments: Integrated STEM education as promising practice? *International Journal of Science Education, 41*(10), 1387–1407. https://doi.org/10.1080/09500693.2019.1607983

Stump, S. L., Bryan, J. A., & McConnell, T. J. (2016). Making STEM connections. *Mathematics Teacher, 109*(8), 576–583.

Sun, D., Cheng, G., Yu, P. L. H., Jia, J., Zheng, Z., & Chen, A. (2025). Personalized STEM education empowered by artificial intelligence: A comprehensive review and content analysis. *Interactive Learning Environments, 33*(7), 4419–4441. https://doi.org/10.1080/10494820.2025.2462156

Taylor, J. A., Kowalski, S. M., Askinas, K. M., & Watson, K. (2023). Analyzing funding patterns in science education research: Establishing a baseline for improved monitoring of research to advance science education. *Social Sciences & Humanities Open, 7*(1), 100445. https://doi.org/10.1016/j.ssaho.2023.100445

Theodoropoulos, A., & Lepouras, G. (2021). Augmented reality and programming education: A systematic review. *International Journal of Child-Computer Interaction, 30*, 100335.

Tirupati, K. K., Joshi, A., Singh, S. P., Chhapola, A., Jain, S., & Gupta, A. (2024). Leveraging power BI for enhanced data visualization and business intelligence. *Universal Research Reports, 10*(2), 676–711.

Tsai, L. T., Chang, C. C., & Cheng, H. T. (2021). Effect of a STEM-oriented course on students' marine science motivation, interest, and achievements. *Journal of Baltic Science Education, 20*(1), 134–145.

Turing, A. M. (1950). Computing machinery and intelligence. *Mind, 59*(236), 433–460. https://doi.org/10.1093/mind/LIX.236.433

Twomey, A. J., Hyman, J., Indraswari, K., Kotz, M., Morgans, C. L., & Bairos-Novak, K. R. (2025). From silos to solutions: Navigating transdisciplinary conservation research for early career researchers. *Perspectives in Ecology and Conservation, 23*(1), 26–31. https://doi.org/10.1016/j.pecon.2024.12.002

Tytler, R. (2020). STEM education for the twenty-first century. In J. Anderson & Y. Li (Eds.), *Integrated approaches to STEM education* (Advances in STEM Education). Springer. https://doi.org/10.1007/978-3-030-52229-2_3

von Neumann, J. (1958). *The computer and the brain*. Yale University Press.

Wallace, R. M. (2003). Online learning in higher education: A review of research on interactions among teachers and students. *Education, Communication & Information, 3*(2), 241–280.

Wang, H., Moore, T. J., Roehrig, G. H., & Park, M. S. (2011). STEM integration: Teacher perceptions and practice. *Journal of Pre-College Engineering Education Research, 1*(2), 1–13. https://doi.org/10.5703/1288284314636

Wang, J. C., Hsieh, C. Y., Tsai, T. Y., & Chung, S. C. (2025a). Utilization of government grants for funding: Insights into STEM education teachers in Taiwan. *Cogent Education, 12*(1). https://doi.org/10.1080/2331186X.2025.2530904

Wang, K. D., Wu, Z., Tufts, L. N., Wieman, C., Salehi, S., & Haber, N. (2025b, April). Scaffold or crutch? Examining college students' use and views of generative AI tools for STEM education. In *2025 IEEE Global Engineering Education Conference (EDUCON)* (pp. 1–10). IEEE. https://doi.org/10.1109/EDUCON62633.2025.11016406

Weizenbaum, J. (1966). ELIZA — A computer program for the study of natural language communication between man and machine. *Communications of the ACM, 9*(1), 36–45. https://doi.org/10.1145/365153.365168

Widya, W., Rifandi, R., & Rahmi, Y. L. (2019). STEM education to fulfil the 21st century demand: A literature review. *Journal of Physics: Conf. Series, 1317*, 012208. https://doi.org/10.1088/1742-6596/1317/1/012208

World Economic Forum. (2024a, April). *Shaping the future of learning: The role of AI in Education 4.0*. World Economic Forum. https://www.weforum.org/publications/shaping-the-future-of-learning-the-role-of-ai-in-education-4-0

World Economic Forum. (2024b, June). *Top 10 emerging technologies of 2024*. In collaboration with Frontiers. World Economic Forum. https://www.weforum.org/publications/top-10-emerging-technologies-2024/

Xiang, H., Ji, M., Liu, Z., Cao, F., & Gao, Y. (2025). AI for physical sciences. In Z. Xu, Y. Cheng, Y. Yang, Y. Xu, X. Qiu, Y. Fu, S. Wang, M. Yang, H. Zou, S. Lu, L. Shi, K. Wei, X. Zhu, W. Gao, Y. Li, W. Lin, W. Yang, H. Xiang, M. Ji, et al. (Eds.), *AI for science 2025* (pp. 28–33). Shanghai Academy of AI for Science & Fudan University.

Xie, Y., Seth, I., Hunter-Smith, D. J., Rozen, W. M., & Seifman, M. A. (2024). Investigating the impact of innovative AI chatbot on post-pandemic medical education and clinical assistance: A comprehensive analysis. *ANZ Journal of Surgery, 94*(1–2), 68–77. https://doi.org/10.1111/ans.18666

Xu, W., & Ouyang, F. (2022). The application of AI technologies in STEM education: A systematic review from 2011 to 2021. *International Journal of STEM Education, 9*(59). https://doi.org/10.1186/s40594-022-00377-5

Yan, L. (2025). From passive tool to socio-cognitive teammate: A conceptual framework for agentic AI in human-AI collaborative learning. *arXiv preprint arXiv:2508.14825.*

Yang, Y., Sun, W., Sun, D., & Salas-Pilco, S. Z. (2025). Navigating the AI-enhanced STEM education landscape: A decade of insights, trends, and opportunities. *Research in Science & Technological Education, 43*(3), 693–717. https://doi.org/10.1080/02635143.2024.2370764

Ying, T., Yu, J., Liu, L., Cheng, Y., Zhu, S., Peng, H., & Xu, S. (2025). AI for life sciences. In Z. Xu, Y. Cheng, Y. Yang, Y. Xu, X. Qiu, Y. Fu, S. Wang, M. Yang, H. Zou, S. Lu, L. Shi, K. Wei, X. Zhu, W. Gao, Y. Li, W. Lin, W. Yang, H. Xiang, M. Ji, et al. (Eds.), *AI for science 2025* (pp. 37–42). Shanghai Academy of AI for Science & Fudan University.

YouTube Originals. (2019, December 18). *How far is too far? The age of AI* [Video]. YouTube. https://www.youtube.com/watch?v=UwsrzCVZAb8

Zanetti, M., Rendina, S., Piceci, L., & Peluso Cassese, F. (2020). Potential risks of artificial intelligence in education. *Open Journal Per La Formazione in Rete, 20*(1), 368–378. https://doi.org/10.13128/form-8113

Zhai, X., & Krajcik, J. (2024b). Artificial intelligence-based STEM education. In X. Zhai & J. Krajcik (Eds.), *AI and STEM education: Emerging research and innovations* (pp. 3–14). Springer. https://doi.org/10.1007/978-3-031-51452-5_1

Zhai, X., & Krajcik, J. (Eds.). (2024a). *Uses of artificial intelligence in STEM education.* Oxford University Press. https://doi.org/10.1093/oso/9780198882077

Zhai, X., Chu, X., Chai, C. S., Jong, M. S. Y., Istenic, A., Spector, M., et al. (2021). A review of artificial intelligence (AI) in education from 2010 to 2020. *Complexity, 2021*(1), 8812542.

Zhu, J., & Wang, M. (2022). Analyzing the effect of people utilizing mobile technology to make banking services more accessible. *Frontiers in Public Health, 10*, 879342. https://doi.org/10.3389/fpubh.2022.879342

CHAPTER 4

The Janus-Faced Nature of AI in STEM Education

Peter Wulff

Introduction

From the early stages of human civilization, people sought to design and implement automata and machines that imitate the capabilities of living creatures on Earth. Progress resulted from imitating the complex functioning of mammalian brains, particularly in the context of studying the perceptron by Rosenblatt, McCulloch, Pitts, Hebb, and others, where it was shown that input-output mappings can be learned by artificial neurons (Domingos, 2015). The term artificial intelligence (AI) was coined for the study "to describe and build agents that receive percepts from the environment and perform actions" (Russell & Norvig, 1995, p. 7).[1] Due

[1] Of course, these actions should be rational/intelligent, even though the term is contested ever since, and the critique of anthropomorphic language "machines reason/think/feel/understand…" was uttered also from early on; see: McDermott (1976).

P. Wulff (✉)
Ludwigsburg University of Education, Ludwigsburg, Germany
e-mail: peter.wulff@ph-ludwigsburg.de

R. Elmas, M. Akarsu (eds.), *STEM Education*, Maarif Global Education Series,
https://doi.org/10.1007/978-981-95-9023-0_4

to the intricate relationship between education and intelligence, the development of AI will likely have a profound impact on the study of education across all disciplines. Moreover, science, technology, engineering, and mathematics (STEM) play pivotal roles in the development of AI as recently (2024) attested by the Nobel Prize in Physics, which awarded physics research from the 1980s that laid the foundations for later AI developments. In this chapter, we will explore some of the fundamental principles that enable AI methods, such as machine learning (ML) and natural language processing (NLP), to acquire capabilities that make them useful for STEM education. Against this background, we review various studies in STEM education that demonstrate potentials and particular challenges associated with using AI in research and instruction. This will enable us to address some more fundamental questions related to ML, which are also relevant to the application of AI in STEM education and research.

Foundations of AI, ML, and NLP

ML and NLP can be viewed as methodological subsets of AI (Goodfellow et al., 2016). ML and NLP encompass various techniques, concepts, algorithms, models, and so on that allow machines to acquire capabilities that make them seem to behave intelligently. In fact, "[m]ost of what is now classified as AI is really a case of machine learning" (Krauss, 2023, p. 223). Mitchell (1997, p. 2) provided a widespread definition of ML: "A computer program is said to learn from experience E with respect to some classes of task T and performance measure P if its performance can improve with E on T measured by P." Hence, ML is an inherently inductive learning approach (grounded in examples and training data), and a key goal of ML is to detect patterns in data. Moreover, the breadth of tasks that can be learned are intricately constrained by the chosen algorithms and learning procedures. For example, a linear model can only learn linear mappings in the data, which, for many real-world, complex problems, is likely an oversimplification (Brunton & Kutz, 2019). Especially so for learning-related data (Koopmans & Stamovlasis, 2016). A crucial decision in using ML includes the transformation of the complex data at hand to make it suitable for computer-processing. For example, language data should be transformed into numeric data in order to process it with ML algorithms. The transformation and representation of language data (and other modalities such as images) makes the problem at hand amenable to the machine and plays a pivotal role in the machine's actual learning. Incomplete or suboptimal representation of this data might constrain the learning. Particularly for language-related tasks, NLP tools enable researchers to represent language in meaningful ways, allowing ML algorithms to process them effectively and efficiently. In recent years, deep learning (e.g., the multi-layer perceptron as an example of a deep artificial neural network architecture) broadened the possibility of language processing, and input-output mappings, given that deep artificial neural networks can automatically learn the feature representations (representation learning) and exempt human researchers from this task (Bishop & Bishop, 2024). Moreover, different ML-learning types enable different capabilities of ML models. For example, supervised ML requires labeled data and enables the machine to learn the labeling of similar data. Unsupervised ML does not require labeled data but rather enables the machine to detect patterns in complex, unstructured data. Other learning types include self-supervised learning, which enables the machine to learn patterns in sequence data. This, among other things, gave rise to the predictive capabilities of recently developed large language models (LLMs) that are frequently grounded in the transformer architecture (Manning, 2022). With LLMs, researchers built so-called foundation models (Bommasani et al., 2022) that can form the backbone of specific applications and advance development processes for LLMs. For example, foundation LLMs can be further trained (e.g., through reinforcement learning with human feedback) to accommodate for dialog-based interactions with humans, or mathematical calculations (Lewkowycz et al., 2022), making them suitable for these purposes. Moreover, targeted instruction (so-called prompting) enables LLMs to adjust and constrain their outputs according to user-defined requirements, and the addition of information documents can enhance the specificity of outputs through retrieval-augmented generation (Wei et al., 2022). Ever more specific architectures also allow for domain-specific, expert processing, such as the mixture-of-experts architecture (Dai et al., 2024).

Applications of ML and NLP in STEM Education

Before 2022, STEM researchers and instructors primarily used ML and NLP methods for classification, regression, and clustering of complex data. Many studies within STEM disciplines engaged in the classification of textual responses from student data according to one or another underlying framework (supervised ML). For example, Yan (2014) utilized ML and NLP for misconception identification after an ontological schema training, finding that a supervised ML classifier yielded substantial human–machine agreement (F1 of 0.75), and was superior to a query-based approach, for a certain coding of conceptual change and investigating students' conceptual change after the ontological schema training. Similar approaches have been conducted with other underlying frameworks, such as modeling (Krüger & Krell, 2020; Linn et al., 2014; Zhai et al., 2022b), argumentation (Liu et al., 2014; Mao et al., 2018; Zhai et al., 2022a; Zhu et al., 2017), conceptual understanding and explaining (Donnelly et al., 2015; Nehm & Härtig, 2012), reflection (Carpenter et al., 2020; Wulff et al., 2021; Wulff et al., 2022b), professional knowledge (Zhai et al., 2020a), or generality statements (Rosenberg & Krist, 2021). Training ML models with supervised learning allowed implementing automated assessment systems that can be used to scale educational interventions and streamline resources in educational institutions (Moharreri et al., 2014).

Other studies utilized ML and NLP to explore patterns in complex data in STEM fields (unsupervised ML). For example, Odden et al. (2021) apply a clustering algorithm on 100 years of articles from the journal *Science Education* to track changes in themes and cross-pollinations among them throughout time. Another application of unsupervised ML was done by Sherin (2013), who analyzed students' interview transcripts with NLP and a clustering approach to identify conceptions of seasonal change explanations. Unsupervised ML has further been applied to many different areas in STEM, such as detecting competency profiles of teachers' related to professional knowledge (Zeller & Riese, 2025), reasoning in chemistry (Martin et al., 2024), distinguish phases in problem-solving processes (Tschisgale et al., 2023), curriculum assessment (Anderson et al., 2020), or themes in written reflections (Wulff et al., 2022a).

The abovementioned studies, and many others (W. Xu & Ouyang, 2022; Zhai et al., 2020b), document that ML and NLP have become methods in the toolbox for researchers in STEM education. Supervised and unsupervised ML, alongside NLP, provide valuable means of automating classification, regression, and clustering tasks to some extent. However, supervised and unsupervised ML as well as NLP also provide specific challenges. For example, studies that utilize supervised ML need to assure that valid and reliable coding of the data at hand is possible. Often, only one-dimensional coding problems (e.g., multi-class problems) are considered; however, language data and other complex data are multi-dimensional, and ambiguous. Language often leaves information implicit, and machines trained with supervised or unsupervised ML only respond to explicitly provided information. This results in the shortcoming that ML and NLP coding problems might be constrained in the analytical depth, and further research needs to take this onto a next level. For example, ML algorithms for hierarchical labeled data can be used to advance existing methods (Wang et al., 2017).

For unsupervised ML, researchers often face the difficulty to make sense of the algorithm's outputs, for example, the resulting clusters in high-dimensional space. While this might be feasible for simple algorithms (Wulff et al., 2025), it certainly is more involved for complex algorithms, or LLM-enhanced clustering algorithms (Odden et al., 2024). It was shown that the representations learned by unsupervised ML often suffer from spurious correlations, that is, using non-informative features to form clusters that do not generalize well (Kauffmann et al., 2025). Consequently, unsupervised ML research in STEM education, as well as supervised ML research, should also utilize methods of explainable AI to interpret complex patterns incorporated in the classification and clustering models. Explainable AI (xAI) research seeks to understand ML model decisions in various ways and will be integral to trustworthy AI.

Finally, utilizing NLP methods in STEM education often constrains the input features that are crucial for downstream processing. For example, the widely used sentence segmentation of student response data might not be suitable for certain research problems, given that arguments span across sentences. Yet, segmenting language into meaningful chunks remains a difficult challenge. Complex procedures such as elementary discourse unit identification exist (Stede & Neumann, 2014). However, they barely found entrance into STEM education research. Progress in the meaningful processing of language data should come from STEM research as well as AI research, given the difficult, domain-related problem of transforming language data into meaningful units.

Applications of LLMs in STEM Education

Since the advent of LLMs (and especially OpenAI's ChatGPT in 2022), also STEM researchers extended their use of AI with generative LLMs, and vision models, oftentimes based on the transformer architecture.[2] LLMs such as generative pretrained transformer (GPT) were found to be capable of flexibly answering on conceptual questions in STEM fields, successfully solving concept inventory tests (Kieser et al., 2023; Kortemeyer, 2023a; West, 2023), meaningfully answering questions related to STEM teaching and learning (Cooper, 2023), or writing essays on par with student performance (Yeadon et al., 2023; Tschisgale et al., 2025). Moreover, vision models were found to be capable—to some extent—to solve tasks that involve diagrams (Polverini & Gregorcic, 2024), and parse and assess handwritten exams by students (Kortemeyer, 2023b; Kortemeyer et al., 2024). At the same time, STEM education researchers noted intricate problems related to generative LLMs. For example, hallucinations and verbose outputs hamper their expert performance (Gregorcic & Pendrill, 2023; Kieser & Wulff, 2024), and testset leakage (i.e., the issue that these generative LLMs already saw the questions in the training data) renders many of the acquired capabilities suspicious (C. Xu et al., 2024). This did not stop researchers from pondering whether (some) LLMs already showed "sparks of artificial general intelligence" (Bubeck et al., 2023).

Of particular value to partly tackle the hallucination (more accurately: confabulation) problem was the provision of prompts that ground the LLM's outputs. For example, Kieser et al. (2023) showed that instructing an LLM to answer a concept inventory as-if it were a participant with a certain misconception could reasonably simulate responses. Moreover, Wan and Chen (2024) showed that simple feedback for a physics problem could be outsourced to a generative LLM that provided useful and correct guidance for students. The authors designed an elaborate feedback-prompt that guided the LLM's responses. However, automating complex coding to generative LLMs does not seem to function off-the-shelf (zero-shot). Kieser et al. (2024) showed that for a specific problem of coding concepts in students' responses, more traditional supervised ML algorithms outperformed a generative LLM. This might be enhanced with specific prompting (Hu et al., 2023). Designing domain-specific prompts

[2] STEM researchers also used large language models and deep learning before 2022; see: Carpenter et al. (2020) or Wulff et al. (2022a, b).

and systematically testing them with LLMs therefore becomes part of the responsibilities for STEM education researchers that implement generative LLMs.

Concerning are also findings related to the actual learning of students that use generative LLMs as tutors. For example, Bastani et al. (2024) provided (in a randomized controlled trial) students access or no access to LLMs while solving math problems, and found the following: "[c]onsistent with prior work, [] results show that access to GPT-4 significantly improves performance [...]. However, we additionally find that when access is subsequently taken away, students actually perform worse than those who never had access [...]" (Bastani et al., 2024, p. 1). The authors conclude "that while access to generative AI can improve performance, it can substantially inhibit learning" (Bastani et al., 2024, p. 4). Moreover, carefully designed learning environments are crucial to adhere to established principles of effective learning, such as constructivist learning principles. Another study compared two cohorts (one with ChatGPT, and one with traditional teaching methods) of engineering physics students, and similarly found that "the use of ChatGPT led to a significant decrease in student performance, as evidenced by lower grades and negative Hake factors compared to the control group" (Forero & Herrera-Suárez, 2023, p. 1). Students were concerned that the LLM could "reduce critical thinking and independent learning" (Forero & Herrera-Suárez, 2023, p. 1). Further studies in computer science and physics education research yield also rather concerning results: "The ChatGPT group [in a comparison design] 'remembered nothing, and they all failed.'" Reminding us that "[w]orking hard and struggling is actually an important way of learning. When you're given an answer, you're not struggling and you're not learning. And when you get more of a complex problem, it's tedious to go back to the beginning of a large language model and troubleshoot it and integrate it." (as cited in: Shein, 2024) Krupp et al. (2023) found that students often copy-paste LLM outputs (referred to as "unreflected acceptance" of responses), which can equally hinder genuine learning in STEM disciplines.

These findings present first use-cases of LLMs as tutors in STEM education learning and teaching processes. We do not think that they suggest that LLMs are no help in STEM-related teaching and learning, but rather that the meaningful integration of LLMs requires substantive expertise in STEM education. Take as a counter example the study by Wan and Chen

(2024) who showed that LLM-based tutors can have value in providing individualized feedback for students.

Challenges of ML, NLP, and LLMs

We already outlined some challenges above, namely that ML critically depends on the provided training data. This also may lead to biases in ML responses, especially also in generative LLMs (Caliskan et al., 2017). LLMs were reported to output gendered choice options when certain names are provided. Due et al. (2024) investigated an LLM for STEM-related gendered choices and report: "We find that there are significant and large differences in the ratio of STEM to non-STEM suggested education paths provided by ChatGPT when using typical girl vs boy names to prompt lists of suggested things to become" (p. 1). Such inequalities also manifest (less overtly, and more covertly) with regard to race and ethnicity. Hofmann et al. (2024) showed that African-American-English dialect as compared to other dialects results in far more negative suggestions, and cannot easily be mended by explicit prompting. Given that LLMs are increasingly used in high-stakes environments and recommender systems in STEM education, regulatory frameworks such as the EU AI Act may function as important guideposts for educators to not uncritically utilize flawed systems in practice.

Moreover, intricate challenges related to flawed epistemologies and ontologies of LLMs, as well as the advancement of a general-purpose AI: "AI has always had as its goal the creation of machines with general powers of intelligence"—also referred to as artificial general intelligence (AGI) (AAAI, 2025). One prerequisite for genuine intelligence is the reasoning and acting upon so-called world models. Researchers showed that ML algorithms can develop what they call a "world model" and reason upon it in the board game Othello (Li et al., 2023). Actually, from an early stage, researchers have demonstrated that simple world models can be derived from training ML algorithms (Ha & Schmidhuber, 2018), and it is an interesting question to what extend different LLMs eventually settle on universal representations as stated in the Platonic Representation Hypothesis (Huh et al., 2024). However, LLMs learn (what might be called) their "world models" based solely on text data, and possibly image data for vision models or multimodal models. Restricting the input "percepts" to language and vision (and possibly some other modalities) is certainly a restricted way to learn about the world, and many researchers doubt that it can leverage far-reaching capabilities such as a reliable understanding of how the world works (Browning & LeCun, 2022). At a minimum, AI likely needs a grounding in the world (embodiment) to develop more general capabilities and accurate world models. Surveys show that the majority of a group of leading AI researchers (76%) does not believe that current AI systems develop AGI merely by upscaling (AAAI, 2025). In this line, Wolfram (2024) estimates AI capabilities as follows: "[W]ith more training, there is improvement, but by the end it seems like it probably won't get much better. (Though its loss curve does show some sudden downward jumps during the course of training, presumably as 'discoveries' are made—and we can't be sure there won't be more of these.) It's extremely typical of ML that it manages to do a good job of getting things 'roughly right'. But nailing the details is not what ML tends to be good at." In fact, adversarial examples and minute changes in LLM prompts can easily take them off track, showcasing the brittleness and flawed domain understanding that LLMs learn. Perhaps, the training of LLMs should be more aligned with STEM educational knowledge on how domain understanding can be acquired, which would call for more communication across the involved disciplines.

For STEM educators and researchers it is important to be aware of the inherent limitations of current AI tools with regard to true domain expertise. While many AI tools can augment learning and teaching processes even without genuine domain expertise, researchers and instructors need to maintain a critical stance toward outputs of AI tools.

Conclusions, Implications, and Future Directions

AI has a Janus-faced nature in STEM education: AI systems can imitate many human capabilities that are essential for advancing STEM teaching and learning. However, when it comes to more knowledge-intensive tasks in STEM education, the capabilities of AI systems are brittle. True domain understanding is still lacking, and it is uncertain whether upscaling current architectures, such as LLMs based on transformers, can overcome more fundamental issues, including developing a consistent and operational epistemology and ontology of the real world.

We showed above that ML, NLP, and LLMs are widely used in STEM education, and for good reason. After all, true understanding and expertise are relatively high standards to meet, considering the current state of STEM education in many countries worldwide. Workforce shortages and rising student populations, coupled with stagnant or declining investment in STEM education, urgently require supportive tools for STEM educators and researchers to ensure a quality education. We consider AI tools to be part of the solution, given that researchers and instructors develop relevant skills in critically using and advancing such tools. Trustworthy implementation of AI-based tools hinges on a robust understanding of the potential and challenges of ML, NLP, and LLMs in STEM education. We outlined above that STEM education researchers have already established a plethora of findings that texture and refine our understanding of how AI can be utilized in STEM education. Given this somewhat disparate picture of findings and approaches, supplemented by the reproducibility crisis in ML research (Kapoor & Narayanan, 2023), STEM education researchers are well advised to develop standards for conducting and reporting AI research in STEM. The REFORMs (Reporting Standards For Machine Learning Based Science) framework can provide initial insights into areas to focus on when conducting and reporting AI research (Kapoor et al., 2023).

We also outlined fundamental challenges that are related to utilizing AI in STEM education. These range from rather technical challenges, such as ensuring proper validation of the AI model's capabilities, to epistemological and ontological challenges related to the domain understanding of LLMs. While the former can be mitigated by adhering to standards such as the one's outlined in the REFORMs framework, the latter requires domain expertise in STEM. Among other things, guardrails in the form of prompting techniques and templates are needed to ensure reliable and valid outputs for the LLMs. As such, it will become crucial for STEM education researchers to engage in the development of methodologies for integrating LLMs into inquiry processes, as well as for instructing LLMs to ensure efficient and effective STEM education.

References

AAAI. (2025). *AAAI 2025 Presidential panel on the future of AI research*. https://aaai.org/wp-content/uploads/2025/03/AAAI-2025-PresPanel-Report-FINAL.pdf

Anderson, D., Rowley, B., Stegenga, S., Irvin, P. S., & Rosenberg, J. M. (2020). Evaluating content-related validity evidence using a text-based machine learning procedure. *Educational Measurement: Issues and Practice, 39*(4), 53–64. https://doi.org/10.1111/emip.12314

Bastani et al. (2024). *Generative AI can harm learning.* SSRN.

Bishop, C. M., & Bishop, H. (2024). *Deep learning: Foundations and concepts.* Springer.

Bommasani, R., Hudson, D. A., Adeli, E., Altman, R., Arora, S., Arx, S. V., Bernstein, M. S., Bohg, J., Bosselut, A., Brunskill, E., Brynjolfsson, E., Buch, S., Card, D., Castellon, R., Chatterji, N., Chen, A., Creel, K., Davis, J. Q., Demszky, D., … Liang, P. (2022). On the opportunities and risks of foundation models. *ArXiv.*

Browning, J., & LeCun, Y. (2022). AI and the limits of language. https://www.noemamag.com/ai-and-the-limits-of-language/

Brunton, S. L., & Kutz, J. N. (2019). *Data-driven science and engineering.* Cambridge University Press. https://doi.org/10.1017/9781108380690

Bubeck, S., Chandrasekaran, V., Eldan, R., Gehrke, J., Horvitz, E., Kamar, E., Lee, P., Lee, Y. T., Li, Y., Lundberg, S., Nori, H., Palangi, H., Ribeiro, M. T., & Zhang, Y. (2023). Sparks of artificial general intelligence: Early experiments with GPT-4. *ArXiv.*

Caliskan, A., Bryson, J. J., & Narayanan, A. (2017). Semantics derived automatically from language corpora contain human-like biases. *Science* (New York, N.Y.), 356(6334), 183–186. https://doi.org/10.1126/science.aal4230.

Carpenter, D., Geden, M., Rowe, J., Azevedo, R., & Lester, J. (2020). Automated analysis of middle school students' written reflections during game-based learning. In I. I. Bittencourt, M. Cukurova, K. Muldner, R. Luckin, & E. Millán (Eds.), *Artificial intelligence in education* (pp. 67–78). Springer International Publishing.

Cooper, G. (2023). Examining science education in ChatGPT: An exploratory study of generative artificial intelligence. *Journal of Science Education and Technology, 32*(3), 444–452. https://doi.org/10.1007/s10956-023-10039-y

Dai, D., Deng, C., Zhao, C., Xu, R. X., Gao, H., Chen, D., Li, J., Zeng, W., Yu, X., Wu, Y., Xie, Z., Li, Y. K., Huang, P., Luo, F., Ruan, C., Sui, Z., & Liang, W. (2024). DeepSeekMoE: Towards ultimate expert specialization in mixture-of-experts language models. *ArXiv.*

Domingos, P. (2015). *The master algorithm: How the quest for the ultimate learning machine will remake our world.* Basic Books.

Donnelly, D. F., Vitale, J. M., & Linn, M. C. (2015). Automated guidance for thermodynamics essays: Critiquing versus revisiting. *Journal of Science Education and Technology, 24*(6), 861–874. https://doi.org/10.1007/s10956-015-9569-1

Due, S., Das, S., Andersen, M., López, B. P., Nexø, S. A., & Clemmensen, L. (2024). Evaluation of large language models: STEM education and gender stereotypes. *ArXiv.*

Forero, M. G., & Herrera-Suárez, H. J. (2023). ChatGPT in the classroom: Boon or bane for physics students' Academic performance? *ArXiv.*

Goodfellow, I., Bengio, Y., & Courville, A. (2016). *Deep learning*. MIT Press. http://www.deeplearningbook.org/

Gregorcic, B., & Pendrill, A.-M. (2023). ChatGPT and the frustrated Socrates. *Physics Education, 58*(3), 35021. https://doi.org/10.1088/1361-6552/acc299

Ha, D., & Schmidhuber, J. (2018). World models. *ArXiv*. Advance online publication. https://doi.org/10.5281/zenodo.1207631

Hofmann, V., Kalluri, P. R., Jurafsky, D., & King, S. (2024). Ai generates covertly racist decisions about people based on their dialect. *Nature, 633*(8028), 147–154. https://doi.org/10.1038/s41586-024-07856-5

Hu, Y., Parolin, E. S., Khan, L., Brandt, P. T., Osorio, J., & D'Orazio, V. J. (2023). Leveraging codebook knowledge with NLI and ChatGPT for zero-shot political relation classification. *ArXiv*.

Huh, M., Cheung, B., Wang, T., & Isola, P. (2024). *The Platonic representation hypothesis*. arXiv. https://doi.org/10.48550/arXiv.2405.07987

Kapoor, S., & Narayanan, A. (2023). Leakage and the reproducibility crisis in machine-learning-based science. *Patterns* (New York, N.Y.), 4(9), 100804. https://doi.org/10.1016/j.patter.2023.100804

Kapoor, S., Cantrell, E., Peng, K., Pham, T. H., Bail, C. A., Gundersen, O. E., Hofman, J. M., Hullman, J., Lones, M. A., Malik, M. M., Nanayakkara, P., Poldrack, R. A., Raji, I. D., Roberts, M., Salganik, M. J., Serra-Garcia, M., Stewart, B. M., Vandewiele, G., & Narayanan, A. (2023). REFORMS: Reporting standards for machine learning based science. *ArXiv*.

Kauffmann, J., Dippel, J., Ruff, L., Samek, W., Müller, K.-R., & Montavon, G. (2025). Explainable AI reveals Clever Hans effects in unsupervised learning models. *Nature Machine Intelligence*. Advance online publication. https://doi.org/10.1038/s42256-025-01000-2

Kieser, F., & Wulff, P. (2024). Using large language models to probe cognitive constructs, augment data, and design instructional materials. In M. S. Khine (Ed.), *Machine learning in educational sciences: Approaches, Applications and Advances*. Springer Nature. https://doi.org/10.1007/978-981-99-9379-6_14

Kieser, F., Wulff, P., Kuhn, J., & Küchemann, S. (2023). Educational data augmentation in physics education research using ChatGPT. *Physical Review Physics Education Research, 19*(2), 1–13. https://doi.org/10.1103/PhysRevPhysEducRes.19.020150

Kieser, F., Tschisgale, P., Rauh, S., Bai, X., Maus, H., Petersen, S., Stede, M., Neumann, K., & Wulff, P. (2024). David vs. Goliath: Comparing conventional machine learning and a large language model for assessing students' concept use in a physics problem. *Frontiers in Artificial Intelligence, 7*, 1408817. https://doi.org/10.3389/frai.2024.1408817

Koopmans, M., & Stamovlasis, D. (2016). Introducation to education as a complex dynamical system. In M. Koopmans & D. Stamovlasis (Eds.), *Complex dynamical systems in education*. Springer.

Kortemeyer, G. (2023a). Could an Artificial-Intelligence agent pass an introductory physics course? *Physical Review Physics Education Research, 19*(1), 010132, 15. https://doi.org/10.1103/PhysRevPhysEducRes.19.010132

Kortemeyer, G. (2023b). Toward AI grading of student problem solutions in introductory physics: A feasibility study. *Physical Review Physics Education Research, 19*(2). https://doi.org/10.1103/PhysRevPhysEducRes.19.020163

Kortemeyer, G., Nöhl, J., & Onishchuk, D. (2024). Grading assistance for a handwritten thermodynamics exam using artificial intelligence: An exploratory study. *ArXiv*.

Krauss, L. M. (2023). *The known unknowns: The unsolved mysteries of the cosmos*. Head of Zeus.

Krüger, D., & Krell, M. (2020). Maschinelles Lernen mit Aussagen zur Modellkompetenz. *Zeitschrift Für Didaktik Der Naturwissenschaften, 26*(1), 157–172. https://doi.org/10.1007/s40573-020-00118-7

Krupp, L., Steinert, S., Kiefer-Emmanouilidis, M., Avila, K. E., Lukowicz, P., Kuhn, J., Küchemann, S., & Karolus, J. (2023). Unreflected acceptance — Investigating the negative consequences of ChatGPT-assisted problem solving in physics education. *ArXiv*.

Lewkowycz, A., Andreassen, A., Dohan, D., Dyer, E., Michalewski, H., Ramasesh, V., Slone, A., Anil, C., Schlag, I., Gutman-Solo, T., Wu, Y [Yuhuai], Neyshabur, B., Gur-Ari, G., & Misra, V. (2022). Solving quantitative reasoning problems with language models. *ArXiv*.

Li, K., Hopkins, A. K., Bau, D., Viégas, F., Pfister, H., & Wattenberg, M. (2023). Emergent world representations: Exploring a sequence model trained on a synthetic task. *ArXiv*.

Linn, M. C., Gerard, L., Ryoo, K., McElhaney, K., Liu, O. L., & Rafferty, A. N. (2014). Education technology. Computer-guided inquiry to improve science learning. *Science* (New York, N.Y.), 344(6180), 155–156. https://doi.org/10.1126/science.1245980

Liu, O. L., Brew, C., Blackmore, J., Gerard, L., Madhok, J., & Linn, M. C. (2014). Automated scoring of constructed-response science items: Prospects and obstacles. *Educational Measurement: Issues and Practice, 33*(2), 19–28. https://doi.org/10.1111/emip.12028

Manning, C. D. (2022). Human language understanding & reasoning. *Daedalus, 151*(2), 127–138. https://doi.org/10.1162/daed_a_01905

Mao, L., Liu, O. L., Roohr, K., Belur, V., Mulholland, M., Lee, H.-S., & Pallant, A. (2018). Validation of automated scoring for a formative assessment that employs scientific argumentation. *Educational Assessment, 23*(2), 121–138. https://doi.org/10.1080/10627197.2018.1427570

Martin, P. P., Kranz, D., & Graulich, N. (2024). Revealing rubric relations: Investigating the interdependence of a research-informed and a machine learning-based rubric in assessing student reasoning in chemistry. *International*

Journal of Artificial Intelligence in Education. Advance online publication. https://doi.org/10.1007/s40593-024-00440-y

McDermott, D. (1976). Artificial intelligence meets natural stupidity. *SIGART Newsletter, 57*, 4–9.

Mitchell, T. (1997). *Machine learning*. McGraw-Hill Education.

Moharreri, K., Ha, M., & Nehm, R. H. (2014). EvoGrader: An online formative assessment tool for automatically evaluating written evolutionary explanations. *Evolution: Education and Outreach,* 7(15). http://www.evolution-outreach.com/content/7/1/15

Nehm, R. H., & Härtig, H. (2012). Human vs. computer diagnosis of students' natural selection knowledge: Testing the efficacy of text analytic software. *Journal of Science Education and Technology, 21*(1), 56–73. https://doi.org/10.1007/s10956-011-9282-7

Odden, T. O. B., Marin, A., & Rudolph, J. L. (2021). How has Science Education changed over the last 100 years? An analysis using natural language processing. *Science Education, 105*(4), 653–680. https://doi.org/10.1002/sce.21623

Odden, T. O. B., Tyseng, H., Mjaaland, J. T., Kreutzer, M. F., & Malthe-Sørenssen, A. (2024). Using text embeddings for deductive qualitative research at scale in physics education. *Physical Review Physics Education Research, 20*(2). https://doi.org/10.1103/PhysRevPhysEducRes.20.020151

Polverini, G., & Gregorcic, B. (2024). Performance of ChatGPT on the test of understanding graphs in kinematics. *ArXiv*.

Rosenberg, J. M., & Krist, C. (2021). Combining machine learning and qualitative methods to elaborate students' ideas about the generality of their model-based explanations. *Journal of Science Education and Technology, 30*, 255–267.

Russell, S., & Norvig, P. (1995). *Artificial intelligence: A modern approach*. Prentice Hall.

Shein, E. (2024). The impact of AI on computer science education. *Communications of the ACM,* 67(9), 13–15. https://doi.org/10.1145/3673428

Sherin, B. (2013). A computational study of commonsense science: An exploration in the automated analysis of clinical interview data. *Journal of the Learning Sciences, 22*(4), 600–638. https://doi.org/10.1080/10508406.2013.836654

Stede, M., & Neumann, A. (2014). Potsdam commentary corpus 2.0: Annotation for discourse research. In *Proceedings of the Language Resources and Evaluation Conference (LREC)*.

Tschisgale, P., Wulff, P., & Kubsch, M. (2023). Integrating artificial intelligence-based methods into qualitative research in physics education research: A case for computational grounded theory. *Physical Review Physics Education Research, 19*(020123), 1–24. https://doi.org/10.1103/PhysRevPhysEducRes.19.020123

Tschisgale, P., Maus, H., Kieser, F., Kroehs, B., Petersen, S., & Wulff, P. (2025). Evaluating GPT- and reasoning-based large language models on physics Olympiad problems: Surpassing human performance and implications for educational assessment. *Physical Review Physics Education Research*. https://doi.org/10.1103/6fmx-bsnl

Wan, T., & Chen, Z. (2024). Exploring generative AI assisted feedback writing for students' written responses to a physics conceptual question with prompt engineering and few-shot learning. *Physical Review Physics Education Research, 20*(1). https://doi.org/10.1103/PhysRevPhysEducRes.20.010152

Wang, D., Huang, H., Lu, C., Feng, B.-S., Nie, L., Wen, G., & Mao, X.-L. (2017). Supervised deep hashing for hierarchical labeled data. *ArXiv*.

Wei, J., Bosma, M., Zhao, V. Y., Guu, K., Yu, A. W., Lester, B., Du Nan, Dai, A. M., & Le V, Q. (2022). Finetuned language models are zero-shot learners. *ArXiv*. https://arxiv.org/pdf/2109.01652v5

West, C. G. (2023). AI and the FCI: Can ChatGPT project an understanding of introductory physics? *ArXiv*.

Wolfram, S. (2024). Can AI solve science? https://writings.stephenwolfram.com/2024/03/can-ai-solve-science/

Wulff, P., Buschhüter, D., Nowak, A., Westphal, A., Becker, L., Robalino, H., Stede, M., & Borowski, A. (2021). Computer-based classification of preservice physics teachers' written reflections. *Journal of Science Education and Technology, 30*, 1–15. https://doi.org/10.1007/s10956-020-09865-1

Wulff, P., Buschhüter, D., Westphal, A., Mientus, L., Nowak, A., & Borowski, A. (2022a). Bridging the gap between qualitative and quantitative assessment in science education research with machine learning — A case for Pretrained language models-based clustering. *Journal of Science Education and Technology, 31*, 490–513. https://doi.org/10.1007/s10956-022-09969-w

Wulff, P., Mientus, L., Nowak, A., & Borowski, A. (2022b). Utilizing a Pretrained language model (BERT) to classify preservice physics teachers' written reflections. *International Journal of Artificial Intelligence in Education*. Advance online publication. https://doi.org/10.1007/s40593-022-00290-6.

Wulff, P., Kubsch, M., & Krist, C. (Eds.). (2025). *Springer Texts in Education. Applying Machine Learning in Science Education Research: When, How, and Why?* (1st ed. 2025). Springer Nature Switzerland, Imprint: Springer. https://doi.org/10.1007/978-3-031-74227-9

Xu, W., & Ouyang, F. (2022). The application of AI technologies in STEM education: A systematic review from 2011 to 2021. *International Journal of STEM Education, 9*(1). https://doi.org/10.1186/s40594-022-00377-5

Xu, C., Guan, S., Greene, D., & Kechadi, M.-T. (2024). Benchmark data contamination of large language models: A survey. *ArXiv*.

Yan, J. (2014). A computer-based approach for identifying student conceptual change. *Open Access Theses., 289*.

Yeadon, W., Inyang, O.-O., Mizouri, A., Peach, A., & Testrow, C. P. (2023). The death of the short-form physics essay in the coming AI revolution. *Physics Education, 58*(3), 35027. https://doi.org/10.1088/1361-6552/acc5cf

Zeller, J., & Riese, J. (2025). Competency profiles of PCK using unsupervised learning: What implications for the structures of pPCK emerge from non-

hierarchical analyses? *Journal of Research in Science Teaching*, Article tea.70001. Advance online publication. https://doi.org/10.1002/tea.70001
Zhai, X., Haudek, K. C., Stuhlsatz, M. A., & Wilson, C. (2020a). Evaluation of construct-irrelevant variance yielded by machine and human scoring of a science teacher PCK constructed response assessment. *Studies in Educational Evaluation, 67*, 100916. https://doi.org/10.1016/j.stueduc.2020.100916
Zhai, X., Yin, Y., Pellegrino, J. W., Haudek, K. C., & Shi, L. (2020b). Applying machine learning in science assessment: A systematic review. *Studies in Science Education, 56*(1), 111–151. https://doi.org/10.1080/03057267.2020.1735757
Zhai, X., Haudek, K. C., & Ma, W. (2022a). Assessing argumentation using machine learning and cognitive diagnostic modeling. *Research in Science Education*. Advance online publication. https://doi.org/10.1007/s11165-022-10062-w.
Zhai, X., He, P., & Krajcik, J. S. (2022b). Applying machine learning to automatically assess scientific models. *Journal of Research in Science Teaching., 59*, 1765.
Zhu, M., Lee, H.-S., Wang, T., Liu, O. L., Belur, V., & Pallant, A. (2017). Investigating the impact of automated feedback on students' scientific argumentation. *International Journal of Science Education, 39*(12), 1648–1668. https://doi.org/10.1080/09500693.2017.1347303

CHAPTER 5

Integrating Artificial Intelligence into the Engineering Design Process: Transforming STEM Education

Murat Akarsu and Brooke Max

Introduction

Real-life problems often require the integration of multiple disciplines within a unified framework to be effectively addressed. Recognizing this, researchers such as Brophy et al. (2008), and Honey et al. (2014) have emphasized the importance of holistic educational approaches that combine science, technology, engineering, and mathematics. One well-established approach is STEM education, which aims to cultivate students' integrative thinking skills by providing interdisciplinary learning environments that connect these domains meaningfully. Within STEM education, engineering holds a particularly pivotal role, especially through the Engineering Design Process (EDP), a process that fosters students'

M. Akarsu (✉)
Faculty of Education, Agri İbrahim Cecen University, Ağrı, Türkiye
e-mail: makarsu@agri.edu.tr

B. Max
Purdue University, West Lafayette, IN, USA
e-mail: bmax@purdue.edu

R. Elmas, M. Akarsu (eds.), *STEM Education*, Maarif Global Education Series,
https://doi.org/10.1007/978-981-95-9023-0_5

problem-solving abilities, creativity, and collaboration skills, as highlighted by Johnston et al. (2019). EDP is an iterative process central to STEM education, empowering learners to devise thoughtful solutions to real-world challenges rather than simply applying isolated concepts.

In recent years, the integration of artificial intelligence (AI) technologies into the EDP has further enriched the engineering design experience, transforming how students approach complex problems (Chandrasekera et al., 2025). AI equips students with powerful tools to analyze large datasets, construct predictive models, optimize design alternatives, and iterate solutions more efficiently (Liao et al., 2023; Olewnik & Svihla, 2025). These capabilities encourage more informed, creative, and data-driven decision-making. Importantly, the use of AI not only enhances students' technical competencies but also strengthens their capacity to address multifaceted interdisciplinary problems in a systematic way (Osasona et al., 2024). In line with these perspectives, we will outline the relationship of STEM with AI, the EDP, and how AI can enhance each stage of the EDP.

The Interrelationship of Engineering, Design, and Technology: A Perspective with Artificial Intelligence

Engineering and design play a pivotal role in the advancement of technology, which in turn drives societal development. Historically, these processes have consistently acted as catalysts for technological innovation, shaping both the products we use and the tools we rely on to create them. In this context, technology can be understood not only as the outcome of engineering and design but also as the collection of methods and tools, such as testing, modeling, and simulation systems, that make these processes possible and more effective. For example, the International Technology and Engineering Educators Association (ITEEA, 2020) emphasizes that technology includes the development of products, systems, and processes that modify the natural world to meet human needs, while engineering seeks to optimize these technologies to ensure they remain efficient and responsive to real-world demands.

This integrated perspective is further reinforced in the Engineering in K–12 Education report, which defines technology as the knowledge, techniques, and products people create to address everyday challenges (National Academy of Engineering & National Research Council [NAE &

NRC], 2009). Such definitions underscore the close, dynamic relationship among engineering, design, and technology, showing how each continuously influences the other. Today, this relationship is undergoing a significant transformation with the rise of AI. No longer limited to final technological products, AI capabilities are becoming deeply embedded within the design and engineering processes themselves. Techniques like machine learning, for instance, help analyze vast datasets to uncover patterns that would otherwise remain hidden (Joseph & Uzondu, 2024), while predictive analytics enables designers to anticipate performance outcomes before building physical prototypes (Figoli et al., 2022). Similarly, advances in image recognition and computer vision tools now allow engineers to perform real-time monitoring and quality control throughout iterative design cycles (Liao et al., 2023). When thoughtfully integrated into the engineering design process, these AI tools enrich adaptive feedback loops and enable more informed, agile responses to design challenges. This means that both students and professionals are better equipped to engage in creative, innovative, and efficient problem-solving, skills that are increasingly critical in a world where technological advancements continuously reshape how we live and work.

The systematic and meaningful integration of technology, particularly artificial intelligence (AI), into education significantly enhances students' digital literacy. This literacy extends beyond the ability to use digital tools; it encompasses a critical awareness of the broader cultural, economic, and societal impacts of emerging technologies (ITEEA, 2020). Well-designed educational experiences that include structured, hands-on engagement with technology can foster both technical competence and ethical reflection. Through these experiences, learners are encouraged to explore real-world problems, engage in interdisciplinary thinking, and consider the societal consequences of technological design and implementation. By actively involving students in the responsible use and understanding of AI and other technologies, such approaches empower them to become not just consumers of technology but informed and thoughtful contributors to a rapidly evolving digital world.

Research on these programs has shown that they effectively support learning in science, mathematics, and technology, while also promoting interdisciplinary and critical thinking (Nawaz, 2024). By incorporating AI into such curricula, a new dimension of learning is introduced. Students not only design and build technological artifacts but also evaluate how AI systems function, analyzing the impact of algorithms, the importance of data quality, and the consequences of machine-based decisions (George, 2023).

Through these AI-enhanced learning experiences, students are encouraged to reflect on ethical questions, consider the societal impact of technological choices, and develop critical thinking and problem-solving skills (Holmes et al., 2023). This convergence of engineering, design, and AI opens powerful opportunities in STEM education. It allows learners to build technical proficiency while simultaneously cultivating creativity, collaboration, and ethical awareness, essential skills for addressing the complex challenges of the twenty-first century.

Introduction to the Engineering Design Process in STEM Education

Engineering is often defined as the process of producing practical and functional solutions to societal needs by utilizing knowledge from the sciences and mathematics (Purzer et al., 2018). While this definition captures the technical core of engineering, it does not fully reflect the discipline's broader, multidimensional character. In practice, engineering encompasses an interdisciplinary way of thinking that extends beyond technical knowledge to engage with social, cultural, and ethical dimensions (Figueiredo, 2008; Johri & Olds, 2011). This means that engineering thinking emerges as a cognitive process that enables the systematic, creative, and iterative handling of complex, real-world problems (Litzinger et al., 2011). It involves not only acquiring and applying knowledge but also continuously testing and reshaping it through critical inquiry and reflection (Olewnik & Svihla, 2025). Accordingly, engineering plays a vital role in generating holistic, sustainable, and innovative solutions to contemporary challenges (National Academy of Engineering, 2021).

One of the most prominent manifestations of engineering thinking in practice is the EDP. This process acts as a bridge between theoretical knowledge and real-world application, making it a central element in engineering education (Lachapelle & Cunningham, 2014). Although design is not unique to engineering (Lawson, 2005) and is widely used in fields such as architecture, fashion, and information technology (Adams et al., 2016), it remains a cornerstone for developing problem-solving and creative capacities in engineering contexts. To understand its importance, it is essential to first clarify what is meant by "design" in this context. Design is defined as a structured, conceptual, and iterative process aimed at improving existing systems or developing novel solutions to open-ended problems (Dym et al., 2005). It supports the generation of multiple possible outcomes and is particularly effective in addressing ill-structured, real-world challenges.

In educational settings, engineering design is framed as a hands-on, cyclical learning experience that integrates science, technology, engineering, and mathematics in meaningful ways (Akarsu & Guzey, 2022). It promotes systems thinking, analytical reasoning, and modeling skills, while also providing opportunities for interdisciplinary collaboration (Moore et al., 2014). Given these characteristics, engineering design serves as a powerful pedagogical tool within integrated STEM education, especially in developing students' ability to tackle complex and authentic problems (English, 2016).

Recognizing this potential, many countries have revised their educational standards to incorporate engineering practices into STEM curricula (NRC, 2012). To meet the need for a clear and structured instructional approach, Moore et al. (2013) proposed a detailed model of the Engineering Design Process. Their framework positions the EDP as a robust foundation for addressing real-life problems through STEM education and outlines six key stages: Define, Learn, Plan, Try, Test, and Decide. Unlike traditional linear models, this framework emphasizes iteration, encouraging students to revisit earlier stages to refine their understanding and improve their designs. Moreover, the process is strengthened through collaborative teamwork, communication strategies, and the use of design notebooks, all of which foster creativity, critical thinking, and problem-solving skills essential for innovation in the twenty-first century (Pleasants & Olson, 2019) (Fig. 5.1).

The EDP begins with a thorough problem definition, followed by the acquisition, synthesis, and application of relevant knowledge. However, the process extends beyond mere knowledge application, emphasizing continuous cycles of learning and refinement to ensure high-quality outcomes. Each stage of the EDP is interconnected and requires meticulous attention. The iterative nature of the EDP ensures that solutions are not static but evolve through repeated cycles of evaluation and improvement.

Ultimately, Moore's model offers more than just a sequence of steps; it presents a comprehensive instructional approach that supports student engagement in authentic, hands-on learning experiences. By guiding learners through iterative cycles of design and reflection, the EDP nurtures an engineering mindset and prepares students to tackle real-world

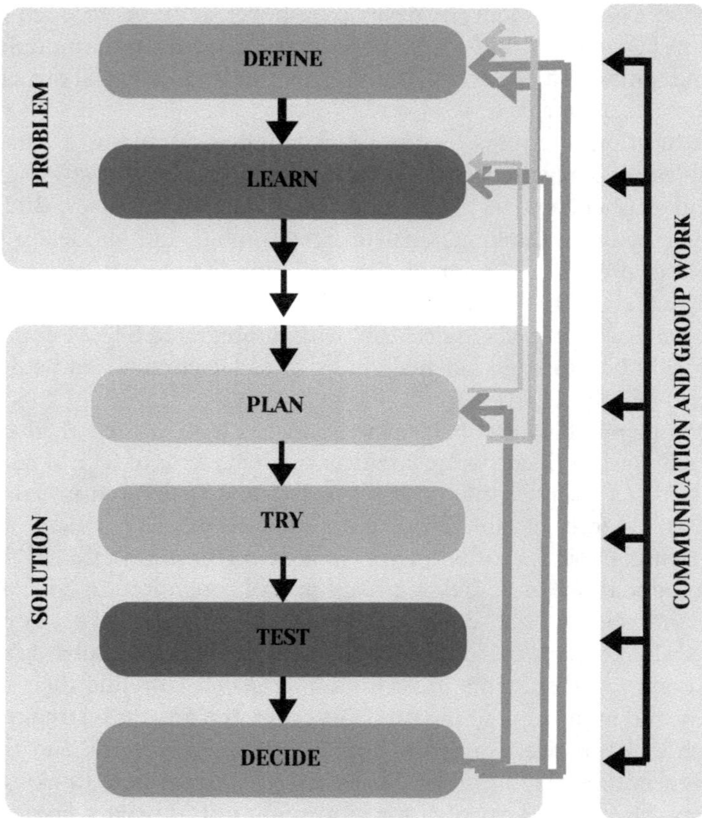

Fig. 5.1 Engineering design process (Moore et al., 2013)

challenges with confidence and creativity. We will use the EDP as our framework and outline the individual stages of the EDP, along with how AI can be integrated into each of them, below.

Integration of AI into Each Stage of the Engineering Design Process

The integration of AI into the six stages of the EDP offers transformative potential (Chandrasekera et al., 2025; Liao et al., 2023). By leveraging AI, the EDP becomes a more dynamic and impactful framework, equipping individuals with the tools and methodologies needed to tackle complex, real-world problems effectively. In the following section, the role of AI in each stage of the EDP will be explored, highlighting how AI can enhance and transform learning and problem-solving within this framework.

Enhancing the Problem Defining Stage of the Engineering Design Process with AI

The EDP typically begins with the definition of a real-world problem that needs to be resolved. Students are presented with a context-rich scenario (Elmas, 2020) that includes implicit constraints and criteria, prompting them to ask questions, explore the scenario, and collaboratively refine the problem within their groups (Atman et al., 2007). Allowing students to uncover these elements through inquiry not only fosters critical thinking but also lays a solid foundation for the subsequent steps in the EDP. When problems are well-structured, they necessitate interdisciplinary thinking and open the door to multiple possible solutions. This approach helps students draw meaningful connections across STEM domains while simultaneously enhancing their motivation and engagement.

However, a common misconception in STEM education is that its primary focus is merely the creation of a final product (Akarsu et al., 2020). In reality, great value lies in cultivating essential twenty-first-century skills such as problem-solving, collaboration, and design thinking skills that are highly relevant to real-world industry demands. Unfortunately, this initial phase of the learning process is often overlooked, as students may rush through it in their eagerness to build something tangible. This can result in a shallow understanding of the problem and difficulties in later phases, such as planning or prototyping. Teachers, therefore, play a vital role in this stage by posing open-ended, guiding questions that prompt deeper thinking and ongoing reflection, helping students maintain focus on understanding the problem before attempting to solve it.

Using AI-supported exploration transforms a vague, everyday frustration into a clearly defined, complex design challenge that demands both technical analysis and ethical consideration. To further support this process, AI-driven virtual assistants can pose guiding questions that push students to consider diverse stakeholder perspectives: for example, "How might claw machine owners balance profitability with fair play?" "What design changes could make outcomes more transparent for players?" or "How could new policies prevent manipulative settings?" (Chandrasekera et al., 2025). This approach encourages learners to expand their understanding of the issue beyond its mechanical aspects, engaging them in critical reflection about consumer trust and ethical business practices (Floridi et al., 2018).

To further support this process, AI-driven virtual assistants can pose guiding, Socratic questions that push students to consider diverse stakeholder perspectives: for example, "How might claw machine owners balance profitability with fair play?" "What design changes could make outcomes more transparent for players?" or "How could new policies prevent manipulative settings?" This approach encourages learners to expand their understanding of the issue beyond its mechanical aspects, engaging them in critical reflection about consumer trust and ethical business practices. However, it is crucial to ensure that AI does not replace students' authentic inquiry; overreliance could hinder the very problem-solving mindset that STEM education aims to foster. Used thoughtfully, AI serves as a scaffold that helps students examine real-world constraints and stakeholder needs as they develop their own solutions. When implemented with contextual relevance and cognitive challenge, AI can transform the problem defining stage into a rigorous, meaningful launch point for designing a fairer, more transparent claw machine anchoring deep STEM learning throughout the entire design process (Liao et al., 2023; Olewnik & Svihla, 2025).

Enhancing the Learning Stage of the Engineering Design Process with AI

The learning stage of the EDP involves acquiring the knowledge and conceptual foundations necessary to solve the real-world problem identified during the define stage. Here, students are expected to uncover and organize relevant scientific and mathematical concepts that will enable them to address the problem effectively. However, if students spend excessive time on irrelevant or tangential information, they can easily lose focus, resulting in shallow engagement and wasted effort. Therefore, the learning process must be purposeful and focused on information that directly supports the design challenge, approached holistically to ensure depth and relevance.

AI can play a transformative role in this phase by making the learning process more targeted, adaptive, and engaging. AI-supported learning environments can provide students with personalized pathways to essential content, helping them focus on the variables and concepts that matter most for their design goals. For example, consider a unit where students explore how to design and optimize an electromagnet for a specific purpose, such as lifting the most hex nuts. In these lessons, students first build foundational knowledge by rotating through four stations that introduce key concepts in magnetism: they test which materials are magnetic, verify that all magnets have north and south poles, examine how size and shape affect strength, and determine whether magnetic fields can pass through various materials. This station-based approach provides rich, hands-on experiences that AI can complement by recommending interactive simulations or short, targeted videos to reinforce or clarify tricky concepts students encounter at each station.

As students delve into testing-specific variables, such as how the number of coils affects the strength of an electromagnet, AI can further enhance the learning process. For example, when students collect data on how many hex nuts their electromagnet can lift with different coil numbers, an AI-powered tool could instantly generate scatterplots, calculate trendlines, and help them interpret the relationship between variables. Natural Language Processing (NLP) features could guide them through this analysis by asking probing questions, such as, "What does the shape of this scatterplot suggest about the relationship between coils and strength?" or "How strong is the correlation here?" This level of support helps students develop a deeper understanding of data interpretation without taking away their agency as young researchers.

AI can also support interdisciplinary integration and collaboration, which are essential for meaningful STEM learning. In the electromagnet example, AI might suggest relevant mathematical models that help students quantify their findings or highlight connections to real-world engineering contexts, such as how industrial cranes use electromagnets. By connecting science concepts with applied mathematics, students begin to see how different disciplines work together to solve authentic problems. Furthermore, the collaborative nature of the learning stage benefits from AI-enhanced digital platforms that support teamwork. When students work in groups to generate hypotheses, like which variables could further improve their electromagnet design, AI tools can help them document ideas, organize evidence, and synthesize diverse perspectives. For example, an AI chatbot could prompt students to consider overlooked variables, such as the gauge of the wire or the power source, keeping the inquiry process dynamic and student-driven.

However, it is crucial that AI does not replace authentic, hands-on exploration. Students still need to physically wrap coils, test electromagnets, and grapple with real data to develop a deep, embodied understanding of scientific concepts. While AI can enhance these experiences by providing instant feedback and adaptive support, it also carries the risk of diverting students into unproductive or superficial pathways if not carefully guided. Without intentional scaffolding, students may become overly reliant on AI tools, diminishing opportunities for critical thinking and problem-solving. Teachers, therefore, play a vital role in ensuring that AI is used to support rather than shortcut student learning. By balancing experiential learning with thoughtfully integrated AI, educators can help students build strong inquiry skills while responsibly navigating the digital tools that shape their understanding.

In summary, when AI is used strategically, it transforms the learning stage from a simple information-gathering process into an interactive, inquiry-rich experience. Students engage more deeply with scientific concepts, develop stronger analytical skills, and learn to apply interdisciplinary knowledge creatively to real-world problems, like designing a more effective electromagnet. This not only improves the quality of the learning experience itself but also reinforces the essential function of learning as a bridge between problem definition and solution generation within the EDP.

Enhancing the Planning Stage of the Engineering Design Process with AI

The planning stage is where students develop practical solutions to real-world problems. To begin, students sketch their initial design ideas in their engineering notebooks, applying the scientific and mathematical knowledge acquired during the learning phase. For example, in a unit where students design and build an electromagnetic claw arm prototype for a company, each student first uses what they learned about magnets and electromagnets from their previous experiments to create at least two different design ideas. They justify their design choices using data and evidence from lessons on variables such as the number of coils and material selection. By using evidence-based reasoning, they critically analyze the pros and cons of each idea, ensuring that their plan is grounded in evidence rather than assumptions.

Once students have planned individually, they come together as teams to share and discuss their proposals. In these small groups, they evaluate each design option by considering the strengths and weaknesses identified in their protocols. They must draw on data collected from lessons, like testing how the number of coils affects electromagnet strength or how different magnetic materials perform. Before finalizing their team design for the claw arm, they engage in evidence-based reasoning discussions to ensure that the chosen solution addresses all identified constraints and criteria. Teachers facilitate using observation protocols to guide students' collaborative thinking and keep them focused on purposeful decision making.

The integration of AI at this stage can enhance both the depth and efficiency of the planning process. AI tools can help students digitize their sketches of the electromagnetic claw arm, analyze them, and explore alternative solutions using AI-supported design and modeling platforms. For instance, the AI system might simulate the claw arm's performance, highlighting how adjusting the number of coils, changing wire gauge, or modifying the claw's shape could improve lifting capacity while minimizing energy use. This allows students to compare multiple versions of their prototype efficiently, deepening their understanding of the tradeoffs involved.

Furthermore, AI can assess the strengths and weaknesses of different plans using metrics such as material cost, durability, or power efficiency, generating clear visualizations that support students' evidence-based reasoning. For example, an AI dashboard could display a side-by-side comparison of each design's predicted lifting force and energy consumption, helping teams justify why they might opt for a slightly more complex design if it significantly enhances performance. Natural Language Processing (NLP) tools can also act as virtual design coaches by answering students' "what if" questions, such as, "What happens if we add more coils but reduce current?" or "How does claw shape affect grip stability on irregular objects?" Such AI prompts encourage students to refine their thinking rather than settle for superficial plans.

AI-assisted tools can even support students in preparing clear, professional presentations of their final electromagnetic claw arm design. For example, they can help generate annotated diagrams, simulation videos, or infographics for posters that teams use to present their solutions to peers. This strengthens both technical and communication skills and reinforces the practice of justifying design decisions with data.

However, the use of AI also presents certain pedagogical risks. The convenience of AI-generated suggestions may tempt students to bypass their own creative exploration or rely too heavily on automated advice. To prevent this, teachers must stress that AI is an aid for generating and testing ideas, not a shortcut for original problem-solving. Students should be encouraged to propose initial claw arm designs independently before comparing them with AI insights, critically discussing any differences in performance or feasibility. Teachers can guide conversations around whether an AI-generated idea aligns with the ethical, environmental, and contextual considerations of their project. For instance, ensuring that the final design is safe, energy-efficient, and practical for real-life amusement arcade use.

When thoughtfully integrated, AI enhances students' ability to work systematically, make data-driven decisions, and approach the planning stage with both creativity and analytical rigor. Through a rich cycle of hands-on experimentation, collaborative reasoning, and AI-supported modeling, students designing the electromagnetic claw arm prototype not only strengthen their technical design skills but also learn to justify their choices using clear, evidence-based arguments. This transforms the planning phase from a simple sketching exercise into a robust, meaningful step that bridges learning and real-world application within the EDP.

Enhancing the Trying Stage of the Engineering Design Process with AI

In the EDP, the trying stage plays a crucial role in transforming the ideas formulated during the planning phase into tangible prototypes. This stage allows students to model their designs, refine ideas, identify flaws early on, and test potential solutions in practice (Yang, 2005). Key guiding questions for students include: "Does the product function as intended?" and "Have alternative solutions been tested before proceeding to full scale production?" Each prototype created during the trying phase reflects not only the students' problem-solving strategies but also serves as a visible manifestation of their creativity, one that can be evaluated, tested, and improved with peer and teacher feedback.

For example, imagine students working on redesigning a fairer claw machine for an arcade company. Building on what they learned in earlier lessons, such as how grip strength and electromagnets work, students might use what they planned to start constructing a working claw arm prototype that can lift objects more consistently and transparently. Teams could begin building immediately, guided by their evidence-based reasoning and the pros and cons discussed in their design ideas.

During the trying stage, STEM oriented design processes emphasize adhering to defined criteria and constraints. For the claw machine project, this might include grip strength consistency, player fairness, energy efficiency, cost, and safety. AI can play a transformative role here. For example, students could use AI-powered simulation tools to model the electromagnetic claw arm's performance before physical testing. An AI system could simulate how the claw grips different objects, showing potential failure points such as a claw that accidentally drops plush toys too soon or uses too much current, overheating the system.

In addition, AI can help students compare different design alternatives. Suppose one group tests two versions of their claw arm, one using more coils on the electromagnet for stronger grip and another using fewer coils for energy efficiency. AI tools could process test data instantly, generating graphs and highlighting performance tradeoffs to support evidence-based decisions. This gives students a deeper, data-driven understanding of how their design works under real-world conditions. AI can also give students access to databases of prior claw machine designs and studies, helping them learn from industry examples without simply copying them. For instance, an AI tool might demonstrate how specific settings are utilized to strike a balance between profitability and fairness in existing claw games, prompting students to explore methods to enhance transparency and player trust.

However, it is essential that AI does not do the designing for students. Teachers should emphasize that AI is a support, not a substitute for students' hands on work and creative thinking. Students must still build, test, and refine their claw arms physically, grappling with the realities of electromagnet strength, claw alignment, and mechanical motion. This ensures they develop authentic problem-solving skills. Teachers can further support this by circulating during the trying phase, using observation protocols to check whether teams are iterating thoughtfully and using AI insights responsibly.

Finally, AI-powered feedback tools can guide students to strengthen their prototypes' functionality, role, and aesthetic appeal (Houde & Hill, 1997). For the claw game, functionality focuses on whether the claw consistently picks up prizes as intended. The role relates to how the redesigned claw promotes fair play and customer trust. The aesthetic appeal could include how the claw looks inside the machine, appearing sturdy and reliable to players.

In summary, the trying stage is much more than simply building a model. It is a rich opportunity for students to apply interdisciplinary knowledge, test real constraints, and build essential twenty-first-century skills like iteration, evidence based reasoning, and ethical design thinking. When used wisely, AI provides students with powerful tools to simulate, analyze, and improve their ideas, but it must always be balanced with authentic, hands-on experimentation and thoughtful teacher guidance. In this way, students designing a better claw machine or any prototype can experience the full depth and rigor of the EDP.

Enhancing the Testing Stage of the Engineering Design Process with AI

After the trying stage, students move to the testing stage, where they systematically evaluate how well their developed prototypes meet the predefined criteria and constraints outlined during the problem definition and planning stages. As in our fairer claw machine example, the testing stage involves moving beyond the initial claw arm build and rigorously evaluating whether the claw can reliably pick up and transport a small toy with an attached washer from one container to another—mirroring the actual function of a real arcade claw machine. To support this, the teacher prepares clear, accessible testing stations in the classroom. Each testing station might consist of two shoe boxes (or other small containers) and a duck counter toy or similar medium-sized toy with a washer securely glued to it. This setup provides students with a consistent, replicable environment to test their electromagnetic claw arms. Students discuss and agree on what success looks like: for example, does the claw consistently lift the toy without dropping it? Does it align accurately and release the toy into the second box? They might also use a simple class-developed rubric or testing plan to ensure each group applies the same criteria during evaluation.

Integrating AI into this testing stage can significantly enhance its effectiveness and depth. For instance, while students repeatedly attempt to pick up the toy at their testing station, AI-powered vision recognition tools or simple sensors could record the number of successful lifts versus drops and analyze grip consistency. Students can compare how different variables, such as the number of coils on the electromagnet, affect performance. AI tools could instantly generate graphs and heat maps that show where the claw is gripping inconsistently or where energy usage spikes, highlighting points of failure that may not be visible through observation alone.

Beyond data collection, AI can also simulate the prototype's performance under different conditions that would be difficult to replicate in class. For the claw machine, AI-powered simulations could model how the claw performs if the toy's weight changes or if the machine operates for extended periods, helping students think about durability and fairness in real-world scenarios. This virtual extension pushes students to test not just "Does it work once?" but "Will it work reliably for players over time?" Additionally, AI can give students access to past design data or common performance benchmarks for similar arcade claw mechanisms. A group could, for example, compare their claw's success rate to data from actual commercial machines. This exposure helps them understand how industry settings balance profitability and fairness, prompting them to consider whether their own designs meet ethical standards.

Analyzing the data is just as vital as collecting it. AI-supported visualization tools make patterns easier to see and interpret. For example, students can compare the success rates of different prototype versions, see grip strength trends across multiple trials, and use this evidence to engage in collaborative group discussions. These discussions might ask: "Where does our claw fail most often?" "How can we redesign the claw arm or electromagnet to increase consistency?" "What trade-offs exist between energy efficiency and grip reliability?"

Throughout this stage, the teacher plays a crucial role in guiding students to utilize AI as a tool for deeper thinking, rather than just a shortcut for answers. Students must remain actively involved in physically testing their claw machine arms at the station, troubleshooting mechanical issues, and iterating on their designs. Formative checks like a "Fist to 5" help students reflect on progress and give teachers a snapshot of where additional support or questioning may be needed that AI might not be able to provide. These human-centered moments allow educators to pick up on subtle misconceptions, emotional cues, or group dynamics that AI may overlook, ensuring that learning remains responsive and personalized.

The testing stage is much more than simply confirming whether a prototype works; it is a rich opportunity for students to apply scientific reasoning, data literacy, and critical thinking. When intentionally integrated, AI strengthens this experience by providing instant, data-driven insights and visualizations that help students connect evidence to design decisions.

Combined with authentic hands-on trials, like repeatedly moving toys between containers using their electromagnetic claw, AI ensures that students gain not only technical proficiency but also a deeper appreciation for iteration, ethical design, and real-world problem-solving.

Enhancing the Decide Stage of the Engineering Design Process with AI

The decision stage of the EDP is when students evaluate how well their tested prototypes performed, identify weaknesses, and determine how they will improve or redesign their solutions. At this stage, students deepen their understanding of engineering as an iterative process, learning to view failure or unexpected results not as setbacks but as opportunities for growth.

For example, imagine students working on the arcade claw machine redesign project. After building and testing their electromagnetic claw arms using a simple testing station with two containers and a toy with a washer attached, students gather data on how reliably the claw picks up and moves the toy. Some teams may discover that their claw drops the toy too soon, overheats, or struggles to align correctly. Others may feel satisfied with their initial results and question whether redesign is needed, while a few might feel discouraged if their first design fell short of the criteria.

In these moments, the teacher plays a crucial role by reinforcing that learning from failure is a natural part of real-world engineering. Teachers can guide students to use structured reflection prompts to help them think through their next steps. For instance, students might record in their notebooks what worked well, what did not, and what they would change, giving them a chance to process the test results and compare insights during class discussion before deciding on changes.

AI can add depth to this process by helping students analyze test data more effectively. An AI-powered dashboard could compile success rates and failure points for each team's claw arm, then display this information through simple graphs or heat maps. AI tools might highlight trends that students could easily overlook, such as the relationship between the number of coil wraps and overheating issues. Seeing these patterns helps students base redesign choices on clear evidence rather than guesswork.

In addition, AI-based support tools can help students weigh different redesign options. For example, one group may consider adding more coils to strengthen the claw's grip, but an AI analysis might show that this would increase energy use and the risk of overheating. The system could suggest alternatives, like using thicker wire or adjusting the power source, encouraging students to evaluate trade-offs thoughtfully instead of making changes blindly.

Sharing AI-supported insights can also make the decision-making process more collaborative. Through shared dashboards or class discussions, teams can learn from each other's data. For instance, if one group finds an effective balance between grip strength and energy efficiency, they can share how they did it, helping other teams improve their own designs. This openness supports a culture of peer learning and continuous improvement. AI can also encourage students who feel stuck by suggesting pathways they might not have considered. For example, an AI tool might indicate that adjusting the claw's alignment could be more effective than changing the electromagnet's strength alone. This helps students visualize new directions and stay motivated to test improvements.

As stated previously, it is essential that students do not rely solely on AI tools. Teachers should help students interpret AI-generated insights critically and ensure they justify their choices based on real data. Clear evaluation rubrics can help students explain not just what they decided to change, but why. Guiding questions such as, "What trade-offs does this redesign introduce?" or "Does this version solve the original problem without creating new ones?" keep students engaged in careful reasoning rather than simply accepting AI suggestions at face value.

By the end of the decision stage, each team documents its revised plan with clear, evidence-based reasoning. They might update design sketches, plan notes, or slides to explain how they will improve their claw machine design in the next cycle. For instance, they might decide to modify the claw to better hold the toy, adjust the electromagnet to use less energy, or make the claw's grip strength more visible to players to increase fairness and trust. These decisions would be a direct result of AI and teacher support.

In summary, the decide stage is more than just choosing what to build next; it is where students connect test results to meaningful redesign decisions. It builds their capacity to use data to drive decisions, reflect on what went wrong or right, and approach problem-solving with a mindset of continuous improvement. When combined with thoughtful teacher guidance and authentic reflection, AI can make this process clearer and more informed, helping students develop real-world engineering skills like critical thinking, evidence-based reasoning, and collaboration.

Summary

A significant pedagogical and practical enhancement can be achieved by integrating AI into the EDP, from the initial steps of problem definition to the final stage of decision-making. This chapter presents a comprehensive overview of how AI tools can be incorporated at each stage of the EDP, thereby facilitating more effective, data-driven, and student-centered learning experiences.

At each stage of the EDP, there is an opportunity for AI to enhance the students' experiences. At the problem definition stage, students can use AI-powered tools such as NLP models to interpret complex problem statements, identify key variables, and reframe ill-structured problems into more manageable forms. These technologies help learners clarify constraints and objectives, fostering deeper engagement with the problem context and laying the groundwork for meaningful inquiry. During the research and learning stage, AI enhances students' access to relevant scientific content, past design cases, and technical data through intelligent search systems and recommendation engines. Personalized AI-based platforms further support this process by assessing students' prior knowledge and delivering tailored content that strengthens their conceptual foundations for design. In the planning stage, AI facilitates idea generation and organization through tools that support brainstorming, idea clustering, and feasibility analysis. Generative AI models enable students to create preliminary sketches, propose alternative solutions, and anticipate potential challenges. AI-driven decision-support systems help evaluate various strategies within specific design constraints, empowering students to make informed, strategic choices.

Prototyping is significantly accelerated through AI-supported design automation platforms. Generative design algorithms offer a range of prototype options for students to assess in terms of functionality, efficiency, and sustainability. AI assists in selecting optimal shapes, materials, and configurations that align with performance goals, streamlining the design iteration process. AI also transforms the testing stage by providing access to powerful simulation environments, which reduce material costs and time consumption. Machine learning algorithms can analyze large datasets in real time, detect performance issues, and deliver immediate, data-informed feedback. This allows students to refine their prototypes based on concrete evidence, rather than trial and error. At the deciding stage, AI supports robust, evidence-based conclusions through visualization tools, multi-criteria evaluation frameworks, and predictive analytics. Students can justify their design choices with empirical data, compare alternatives, and articulate the rationale behind their decisions. Additionally, AI-powered collaborative platforms promote peer feedback and encourage reflective thinking, enhancing both individual and group learning outcomes.

In conclusion, integrating AI throughout the EDP creates a dynamic, iterative, and reflective learning environment. By leveraging intelligent, data-driven design practices, students not only develop their problem-solving abilities but also strengthen critical thinking, creativity, and collaboration—key competencies essential for addressing real-world engineering challenges in the twenty-first century.

Disclaimer Similar to how we expect teachers to use AI to support their classrooms, we used ChatGPT to assist with writing this paper. Specifically, we consulted ChatGPT throughout the research and drafting process, and the AI helped clarify concepts during editing. These tools provided helpful suggestions for alternative terminology and assisted in refining the language. Ultimately, as writers and researchers, we created the final version of the paper.

References

Adams, R. S., Cardella, M., & Purzer, Ş. (2016). Analyzing design review conversations: Connecting design knowing, being and coaching. *Design Studies, 45*(Part A), 1–8.

Akarsu, M., & Guzey, S. S. (2022). Mühendislik Tasarım Süreci Temelli STEM Eğitimi Yaklaşımı, (Engineering design process-based STEM education approach). In M. Akarsu, N. Okur Akçay, & R. Elmas (Eds.), *STEM Eğitimi Yaklaşımı (STEM education approach)* (pp. 121–136). Pegem Akademi.

Akarsu, M., Okur Akçay, N., & Elmas, R. (2020). STEM eğitimi yaklaşımının özellikleri ve değerlendirilmesi. *Bogazici Journal of Education, 37*(Special Issue), 156–175.

Atman, C. J., Adams, R. S., Cardella, M. E., Turns, J., Mosborg, S., & Saleem, J. (2007). Engineering design processes: A comparison of students and expert practitioners. *Journal of Engineering Education, 96*(4), 359–379.

Brophy, S., Klein, S., Portsmore, M., & Rogers, C. (2008). Advancing engineering education in P-12 classrooms. *Journal of Engineering Education, 97*(3), 369–387.

Chandrasekera, T., Hosseini, Z., Perera, U., & Bazhaw Hyscher, A. (2025). Generative artificial intelligence tools for diverse learning styles in design education. *International Journal of Architectural Computing, 23*(2), 358–369.

Dym, C. L., Agogino, A. M., Eris, O., Frey, D. D., & Leifer, L. J. (2005). Engineering design thinking, teaching, and learning. *Journal of Engineering Education, 94*(1), 103–120.

Elmas, R. (2020). Bağlamın anlamı ve nitelikleri ve öğrencilerin fen eğitiminde bağlam tercihleri. *Turkiye Kimya Dernegi Dergisi Kısım C: Kimya Egitimi, 5*(1), 53–70.

English, L. D. (2016). STEM education K-12: Perspectives on integration. *International Journal of STEM Education, 3*(1), 3.

Figoli, F. A., Rampino, L., & Mattioli, F. (2022). AI in the design process: Training the human-AI collaboration. In *Proceedings of the 24th international conference on Engineering and Product Design Education (E&PDE 2022)* (pp. 1–6).

Figueiredo, A. D. D. (2008, November). Toward an epistemology of engineering. In *2008 workshop on philosophy and engineering*. The Royal Academy of Engineering.

Floridi, L., Cowls, J., Beltrametti, M., Chatila, R., Chazerand, P., Dignum, V., et al. (2018). AI4People—An ethical framework for a good AI society: Opportunities, risks, principles, and recommendations. *Minds and Machines, 28*(4), 689–707.

George, A. S. (2023). Preparing students for an AI-driven world: Rethinking curriculum and pedagogy in the age of artificial intelligence. *Partners Universal Innovative Research Publication, 1*(2), 112–136.

Holmes, W., Bialik, M., & Fadel, C. (2023). *Artificial intelligence in education: Promises and implications for teaching and learning*. Center for Curriculum Redesign.

Honey, M., Pearson, G., & Schweingruber, H. (2014). *STEM integration in K-12 education: Status, prospects, and an agenda for research*. National Academies Press.

Houde, S., & Hill, C. (1997). What do prototypes prototype? In *Handbook of human-computer interaction* (pp. 367–381). North-Holland.

International Technology and Engineering Educators Association. (2020). *Standards for technological and engineering literacy: The role of technology and engineering in STEM education*. Author.

Johnston, A. C., Akarsu, M., Moore, T. J., & Guzey, S. S. (2019). Engineering as the integrator: A case study of one middle school science teacher's talk. *Journal of Engineering Education, 108*(3), 418–440. https://doi.org/10.1002/jee.20286

Johri, A., & Olds, B. M. (2011). Situated engineering learning: Bridging engineering education research and the learning sciences. *Journal of Engineering Education, 100*(1), 151–185.

Joseph, O. B., & Uzondu, N. C. (2024). Integrating AI and machine learning in STEM education: Challenges and opportunities. *Computer Science & IT Research Journal, 5*(8), 1732–1750.

Lachapelle, C. P., & Cunningham, C. M. (2014). Engineering in elementary schools. In *Engineering in pre-college settings: Synthesizing research, policy, and practices* (pp. 61–88).

Lawson, B. (2005). *How designers think: The design process demystified.* Routledge.

Liao, H., He, Y., Wu, X., Wu, Z., & Bausys, R. (2023). Reimagining multi-criterion decision making by data-driven methods based on machine learning: A literature review. *Information Fusion, 100*, 101970.

Litzinger, T. A., Lattuca, L. R., Hadgraft, R. G., & Newstetter, W. C. (2011). Engineering education and the development of expertise. *Journal of Engineering Education, 100*(1), 123–150.

Moore, T. J., Glancy, A. W., Tank, K. M., Kersten, J. A., Stohlmann, M. S., Ntow, F. D., & Smith, K. A. (2013). A framework for implementing quality K–12 engineering education. In *Paper presented at the American Society for Engineering Education annual conference. Atlanta, GA.*

Moore, T. J., Stohlmann, M. S., Wang, H. H., Tank, K. M., Glancy, A. W., & Roehrig, G. H. (2014). Implementation and integration of engineering in K-12 STEM education. In S. Purzer, J. Strobel, & M. Cardella (Eds.), *Engineering in precollege settings: Research into practice* (pp. 35–60). Purdue Press.

National Academy of Engineering. (2021). *The engineering of 2020: Visions of engineering in the new century.* The National Academies Press.

National Academy of Engineering and National Research Council. (2009). *Engineering in K-12 education: Understanding the status and improving the prospects.* The National Academies Press.

Nawaz, M. (2024). AI in STEM education: Enhancing problem-solving and critical thinking skills. *Journal of AI Integration in Education, 1*(4), 38–48.

NRC. (2012). *A framework for K-12 science education: Practices, crosscutting concepts, and core ideas.* National Research Council.

Olewnik, A., & Svihla, V. (2025). Engineering requirements and their role in engineering undergraduates' design decision making. *International Journal of Technology and Design Education, 35*(2), 673–698.

Osasona, F., Amoo, O. O., Atadoga, A., Abrahams, T. O., Farayola, O. A., & Ayinla, B. S. (2024). Reviewing the ethical implications of AI in decision making processes. *International Journal of Management & Entrepreneurship Research, 6*(2), 322–335.

Pleasants, J., & Olson, J. K. (2019). Refining an instrument and studying elementary teachers' understanding of the scope of engineering. *Journal of Pre-College Engineering Education Research (J-PEER), 9*(2), 1.

Purzer, S., Moore, T. J., & Dringenberg, E. (2018). *Engineering cognition: A process of knowledge acquisition and application.* In Cognition, metacognition, and culture in STEM education: Learning, teaching and assessment (pp. 167–190). Cham: Springer International Publishing.

Yang, M. C. (2005). A study of prototypes, design activity, and design outcome. *Design Studies, 26*(6), 649–669.

CHAPTER 6

Design of STE(A)M Learning in the Context of Modern Technologies

Anssi Lindell and Anna-Leena Kähkönen

Introduction to Twenty-First-Century Skills

In the twenty-first century, education faces the profound challenge of preparing learners for a globalized world with sustainability demands, and the rapid technological advancements. The Fourth Industrial Revolution (4IR) exemplifies this transformation through the fusion of biological, physical, and digital systems driven by technologies like biotechnology, robotics, Internet of Things (IoT), artificial intelligence (AI), and quantum computing (Chaka, 2020). It isn't only the 4IR that challenges our ability to cope and thrive, as the change is not stepwise but a continuous process. The World Economic Forum (WEF) (Leaders, 2016) has categorized essential skills into three interconnected domains: literacies, competencies, and character qualities. These categories reflect the diverse skill sets required to navigate everyday life, thrive in working life, and adapt to a complex ever-evolving world.

Literacies encompass the foundational knowledge necessary for functioning in modern society, starting from traditional academic skills of

A. Lindell (✉) • A.-L. Kähkönen
University of Jyväskylä, Jyväskylä, Finland
e-mail: anssi.e.lindell@jyu.fi; anna-leena.m.kahkonen@jyu.fi

R. Elmas, M. Akarsu (eds.), *STEM Education*, Maarif Global Education Series,
https://doi.org/10.1007/978-981-95-9023-0_6

reading, writing, and numeracy, commonly referred to as the "3Rs" (Reading, wRiting, and aRithmetic) (National Research Council, 2012). These are complemented by scientific, digital, financial, and cultural literacies. Technological advancements have shifted the emphasis of "old school" literacies, and even routine tasks now demand the integration of multiple literacies. Can you imagine the literacies that are NOT needed in an apparently easy and trivial task of purchasing a public transportation ticket, for example? Everyday activities rely on a convergence of diverse literacies, reflecting the evolving demands of functioning effectively in technology-aided everyday environments. Today, this integration is increasingly mediated by the routines and automated processes of AI systems.

Transversal competencies, in contrast, represent higher-order cognitive abilities that AI and machines cannot yet easily replicate. The "Four Cs" for *critical thinking*, *communication*, *collaboration*, and *creativity* are widely recognized as essential for solving complex, real-world problems (Partnership for 21st Century Skills, 2009a, b). Critical thinking is vital for addressing complex problems where the results are interdependent on multiple factors. Creativity feeds innovation and novel solutions. Communication and collaboration are essential competencies while working in interdisciplinary teams tackling multidimensional challenges. In an information-saturated world, these skills are also key to evaluating information, avoiding misinformation, and fostering respectful dialogue.

The Future of Jobs Report by the WEF outlines a broad set of skills that are expected to grow in importance in working life (Masterson, 2023). Table 6.1 presents the results of a survey reflecting businesses' views on the key skills workers will need to master by 2027.

These competencies—particularly analytical thinking, systems thinking, and AI literacy—highlight the growing importance of thoughtfully adapting educational approaches, especially within STEM disciplines, to align with this evolving landscape. As science and technology evolve together, curricula should treat them as integrated, not separate domains. Learning tools and about tools like microscopes reflects this, blending scientific understanding with technological practice. Preparing students to be not just users but contributors to technology requires recognizing this interdependence.

Table 6.1 Skills growing in importance of working life

Skill	*Description*
Creative thinking	The ability to generate new and original ideas and approaches
Analytical thinking	The ability to critically analyze problems and data to find effective solutions
Technological literacy	Proficiency in using and understanding modern (often digital) technologies
Curiosity and lifelong learning	Commitment to ongoing education and adaptability to new information
Resilience, flexibility, and agility	Ability to adapt to change, recover from setbacks, and adjust quickly
Systems thinking	Understanding and managing complex interrelated systems and processes
AI and big data	Skill in using artificial intelligence tools and analyzing large data sets for insights
Motivation and self-awareness	Recognizing and managing one's motivations and emotions for effective self-leadership
Talent management	Ability to recruit, develop, and retain organizational talent effectively
Service orientation and customer service	Providing high-quality support and services that meet client needs
Leadership and social influence	The ability to lead, inspire, and influence others

Adapted from Masterson (2023)

The Role of Modern Technologies in STEM Learning

Push for STEM in Technology, Society, and Education

The evolution of STEM (Science, Technology, Engineering, and Mathematics) as an educational and societal concept is deeply connected to technological advancements and their influence on economies, industries, and global priorities. STEM education finds its roots in the mid-twentieth century, when technological competition among global superpowers began to shape national education agendas. A defining moment was the launch of the Soviet satellite *Sputnik 1* in 1957, prompting Western nations to invest heavily in science, mathematics, and engineering education, often geared toward the most academically able students (Jolly, 2009).

The development of microprocessors and microelectronics in the 1970s and 1980s accelerated the rise of consumer electronics, automation, and computing. These innovations laid the groundwork for innovation-driven economies, where continuous technological development became a central pillar of economic growth. STEM education evolved accordingly, emphasizing engineering design, problem-solving, and innovation, particularly in countries focused on technological leadership. The rise of digital technologies in the late twentieth century created new demands for digitally literate workers. In response, schools and universities integrated computing and information technology into STEM education. Simultaneously, educational technology (EdTech) transformed teaching practices through simulations, online learning, and coding.

This development of processor manufacturing technology and digital solutions paved the way for new scientific process. By the early twenty-first century, breakthroughs in nanotechnology, biotechnology, microsensors, and materials science opened new possibilities in medicine, energy, defense, and sustainability. These fields exemplify the interdisciplinary nature of modern science, blurring boundaries between disciplines.

Initially driven by economic and geopolitical imperatives, the STEM educational approach now addresses broader societal challenges such as climate change, digital equity, and sustainable development. This shift has influenced education policies worldwide, encouraging models that link STEM with ethical reasoning, global citizenship, and authentic problem-solving.

Sustainability Widening Scope Beyond STEM

Striving for Sustainable Development

Zheng (2021) points out that STEM is born at the same time as the story of the STEM crisis is born—the story of a world crisis that is solved through a technoscientific vision. It follows that the world crisis is rapidly worsening, as the lack of STEM professionals or workforce makes it impossible to work out the technical and scientific solution. It took a long time to challenge this narrow technoscientific vision. Ibrahim et al. (2022, p. 34) see that the interplay of science, technology, societies, and environments was an underemphasized, "little foci" area in education. Sustainability issues are interconnected and complex, including climate change, biodiversity loss, and resource depletion (UNESCO, 2010).

Addressing these wicked problems requires not only scientific and technological expertise but also insights from the humanities, social sciences, economics, ethics, and cultural studies. Wiek et al. (2011) identify five key competencies essential for sustainability professionals. These competencies are represented and aligned with key future workforce skills identified in Table 6.2. Skills such as systems thinking, creative and analytical thinking, and AI and technological literacy complement sustainability goals by preparing individuals to navigate complexity, anticipate change, and lead ethical, strategic initiatives. This integration ensures that future professionals are equipped not only to thrive in evolving job markets but also to contribute meaningfully to sustainable development.

STEAM as an Educational Movement

Ultimately, preparing learners for the future of work and sustainable living in everyday and global settings should not be viewed as separate aims. Both demand the ability of scientific thinking and technological literacy. They aim at the capacity to lead and influence others effectively and commitment to self-development and adaptability.

This reconceptualization reflects the growing understanding that today's societal, economic, and environmental challenges demand interdisciplinary collaboration that goes beyond technoscientific fields (Ibrahim et al., 2022). Expanding STEM into a STEAM approach allows for the inclusion of diverse knowledge systems and perspectives. The European Commission's Science Education for Responsible Citizenship (Hazelkorn et al., 2015) stresses the need for science education to foster responsible citizenship by engaging students in authentic, real-world

Table 6.2 Alignment of Wiek's (2011) sustainability competencies with predicted future workforce skills (Masterson, 2023)

Sustainability competency	*Matching selected workforce skills*
Systems thinking	Systems thinking, analytical thinking, AI and big data
Anticipatory (futures) thinking	Creative thinking, curiosity and lifelong learning, AI and big data
Normative competence	Motivation and self-awareness, leadership and social influence
Strategic competence	Leadership and social influence, AI and big data
Interpersonal competence	Service orientation and customer service, talent management, leadership and social influence

problem-solving. A transdisciplinary approach enables learners to examine values, cultural perspectives, and ethical considerations alongside scientific knowledge. While the letter "A" in STEAM was initially seen as the integration of the (visual) arts into STEM, it has evolved via A for all humanistic disciplines to position the "A" as "Any" relevant discipline that enriches learning in complex, real-world contexts.

The integration does not shift the focus away from science or STEM, but rather grounds it by bringing in perspectives that connect science to society: ethics, communication, design, cultural context, and social and environmental impact. Therefore, science education becomes more relevant, reflective, and responsible, equipping learners to engage with real-world challenges in meaningful ways. It is a way to equip learners with skills to navigate uncertainty, assess risk, and collaborate across disciplines and sectors (Laurie et al., 2016).

This integrated approach is echoed in UNESCO's Education for Sustainable Development (ESD) framework, which emphasizes systems thinking, critical thinking, and participatory skills as foundations for co-creating sustainable futures (UNESCO, 2017). In practice, this calls for moving beyond siloed instruction and fostering learning ecologies that value and mobilize diverse, and inclusive ways of knowing (C4s Communities for Sciences, 2018).

Entering the Age of AI

Shift in Technology and Society

Information technology (IT) is concerned with the technological means of developing information processing, communication, and management. To view IT as only an instrument to increase effectiveness of traditional projects, is selling short the range of connections to information. While there exists information *about* reality (science) and information *for* reality (engineering), information *as* reality is diminishing the importance of interacting with the fundamental realities (Mitcham, 2004). In STEM education, computer simulations can help students explore complex systems like climate change or organism behavior. These simulations present simplified models of reality and can make learning efficient and engaging. However, if students or scientists only interact with these simulations, they may begin to see the simulated model as the reality itself.

In contrast, computer-aided experiments, such as using sensors to collect real-time data from a plant growing under different light conditions, keep students more connected to the physical world. For example, the CarbonTree project[1] at Hyytiälä field measurement station provides real-time data from microsensors monitoring the tree's exhalation and inhalation of carbon dioxide. Its educational use is promoted by University of Helsinki science outreach. Such examples make advanced technology and data collection open through citizen science principles. This allows anybody to study effects of natural phenomena through technological interfaces. This example highlights the concern: while both simulations and experimental data are valuable, relying only on simulations risks distancing learners from the real world, reinforcing the idea that "information is reality."

Mitcham (2004) writes about the modern information technology doing with language what modern technology does with matter: making it a resource for human manipulation. The development of sharing and processing information and making it available has had a profound impact on the way people view knowledge and information, as well as skills and competencies regarding information processing. Whereas information storage and processing load can increasingly be shifted onto a machine, the demands for humans shift towards ways of retaining the desired information, critical appraisal of information, the abilities to synthesize or locate discrepancies, as well as to effectively communicate the information to varied human audiences.

The traditional IT tools, such as graphical user interfaces (GUIs), have been analyzed through their symbolic and cultural dimensions (e.g., Markus & Silver, 2008; Orlikowski & Iacono, 2001, Cheikh-Ammar, 2018). These frameworks have not been applied to the AI tools, in particular the LLMs (large language models). Unlike GUIs, which are rich in symbolic expressions that visually convey system functionality, LLMs lack such immediate symbolic scaffolds. This difference significantly affects how users interpret their purpose and functionality. The symbolic expressions guide the users' expectations and actions with the tool, working also as a scaffold for the uninitiated user (Markus & Silver, 2008). Therefore, while the affordances of LLMs are vast, because of the users' unfamiliarity with this novel presentation and mode of communication of IT artifacts, they are rarely realized to the fullest.

[1] The project is accessible at https://www.hiilipuu.fi/index.php/en/home/

Building on the concept of symbolic expressions and Orlikowski and Iacono's (2001) emphasis on the cultural and interpretive dimensions of IT, Cheikh-Ammar (2018) reformulates a unified notion of the 'IT Spirit'—a culturally grounded, symbolically mediated understanding of technological artifacts. The first group of AI users, the "early adopters," are defining the AI "spirit" with their experiences and actions, but also their values, expectations, and applications define the appropriate areas in which AI is to serve.

The authorities' stakehold to the approaches and affordances necessarily comes after. This is why, despite subscribing to Responsible Research and Innovation (RRI) in businesses and universities alike (European Commission, 2013), the ethical use guidelines of innovation trickle down to the teachers through the initially established values, cultures, and appropriations. The EU and UNESCO indeed have released recommendations on the ethical use of AI early on (UNESCO, 2021; European Commission, 2022), addressing the narrow set of features of LLMs that have been used the longest.

Casner-Lotto and Barrington (2006) observed that employers began to place a higher value on applied skills—such as teamwork, critical thinking, and communication—over basic academic skills like reading comprehension and mathematics. The twenty-first-century skills (e.g., Partnership for 21st Century Skills, 2009a, b; Binkley et al. 2012) are heavy in applied skills. Nevertheless, there is an emphasis on a robust STEM skillset at the core of the twenty-first-century skills, particularly highlighting the (information) technological ability (Valtonen et al., 2021).

Teaching in the Age of AI

The educational shift from the transmission model of learning, crucial in a world where information was scarce, has been painful. Laurillard (2011) describes academia as failing this task as late as the 2010s, calling for reflective teaching practices in the age of the internet and knowledge society. The ideal has shifted from the knowledge of each individual to collaborative projects and sharing expertise, and the ideal of school has become a place for knowledge production rather than transmission (Bereiter, 2002).

Alongside this shift towards information production and use, the new technology warrants a reaction from schools and officials, how to approach it. Seeing the role of new technology beyond being content to learn means that the teachers need parallel approaches to the pedagogical use of AI technologies—teaching about AI and with AI. Seeing detailed and carefully outlined examples of such teaching helps teachers incorporate it in their Technological, Pedagogical, and Content Knowledge (TPACK; Özgün-Koca et al., 2010). Mikeladze et al. (2024a, b) discuss the teacher competencies related to using AI (see Fig. 6.1). Their overview of teacher AI competency frameworks (AI CFTs) highlights the varied purposes for which they are developed. In this chapter, we mean any references to AI competencies from a descriptive standpoint and as opportunities to train in competencies (evaluative), rather than attempts to prescribe ways for a teacher to give evidence of their AI competencies (assessment).

UNESCO has outlined a pedagogical use of AI CFT (Cukurova & Miao, 2024) for teachers adopting AI in their work, aligned for new adopters of AI and not advanced teaching. It places high importance on ethical decision-making. With a clearer understanding of the relationship between science and society, there are many more policies in place to ensure technoscientific progress does not go against the moral and ethical consensus, such as the European Commission, EU, and UNESCO codes (European Commission, 2013, 2022; UNESCO, 2021). It includes teachers' capability to use AI to stay involved in lifelong professional

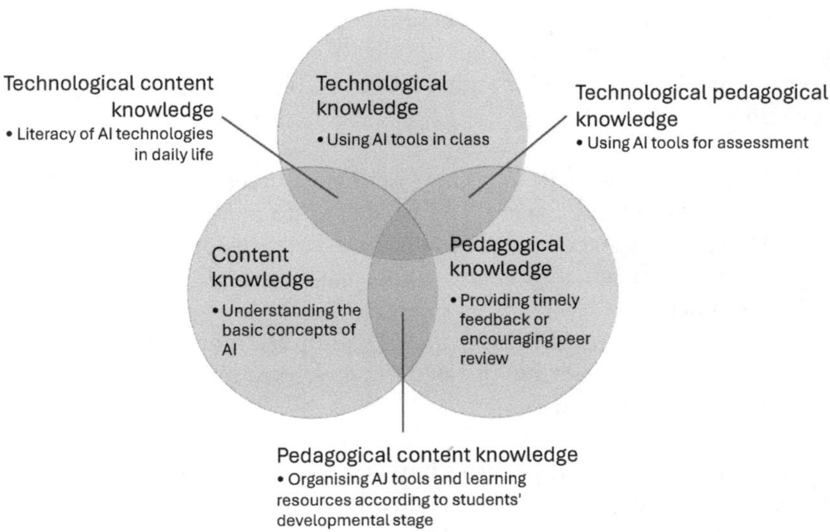

Fig. 6.1 Examples of AI competencies in a TPACK classification after Mikeladze et al. (2024a, b)

learning—such as design of the teacher's learning progressions, or support to lesson design and reflection. Lastly, it includes the teachers' technological and pedagogical knowledge of AI. The re-packaging of teachers' AI competences within a TPACK framework (Mikeladze et al., 2024a, b; Kim & Kwon, 2023a, b) gives detail to these two aspects.

From the exploration of both the recommendations to teachers and the current resources for teachers, it is clear that "basic concepts" or "using AI tools" are loaded with each author's or authorities' subjective window into the AI use cases.

Designing Modern STEAM Education

Starting with the Outcomes

Designing effective STEAM instruction requires a consistent, structured approach that enables learners to demonstrate their skills and understanding clearly. One such strategy is "backward" design of Understanding by Design framework, developed by Wiggins and McTighe (2005). Backward design turns the traditional instructional planning process around by instead of beginning with activities or content, educators first define clear, measurable learning objectives. These outcomes reflect the essential skills and understandings that students should show by the end of an instructional unit (Wiggins & McTighe, 2005). The process involves three key stages:

1. Identifying Desired Outcomes: Educators determine what students should know, understand, and be able to do by the end of the unit.
2. Determining Acceptable Evidence: Design assessments that provide valid evidence of student understanding and achievement of the outcomes.
3. Planning Instructional Activities: Develop learning experiences that will lead to achieving those outcomes, ensuring alignment with assessments and goals.

Similarly to the STEAM approach, this method is better suited for large instructional units rather than individual lessons, making sense in a more comprehensive integration of skills and knowledge across disciplines.

This methodology has been widely supported in educational research for the use of an effective, research-based instructional 5E model, which can help students learn fundamental concepts in science and other domains (Bybee et al., 2006). However, Koyunlu and Dökme (2022) found in their systematic review that the 5E model in science education focuses mainly on the lower Bloom's level thinking skills—remembering and understanding. Their conclusion suggests a skill-based STEM instructional model built on the 5E framework, emphasizing real-world applications, and specifically designed to enhance general skills. Situated learning is critical in this process, as students learn best when knowledge is embedded in real-world contexts (Lave & Wenger, 1991). Backward design supports this by enabling the creation of authentic learning experiences that mirror complex, real-life situations. Additionally, backward design fosters flexible understanding by encouraging students to apply their knowledge across various contexts. That promotes deeper comprehension and the ability to transfer learning to new situations (Bransford et al., 2000).

AI can support every step of the backward design process by helping teachers predict students' misconceptions, align the evidence of learning with the learning goals, and personalize the learning activities, for example.

Balancing AI Use and Individual Contribution

WEF's broader list of skills for the working life includes many related to thought: critical, analytical and systems thinking. Many others relate to self-reflection: curiosity, resilience, and self-awareness (Masterson, 2023). These human traits are not—as is—replaceable by computing or AI models. Still, many teachers worry about students' use of AI to skip learning rather than enhance it.

Molenaar (2022a) introduces the concept of Hybrid Human-AI Regulation (HHAIR) to explore the different routes towards self-regulated learning: they separate the adaptive learning technology environments. The adaptive environment (Molenaar, 2022b) itself may vary in the level of control that the teacher and AI share. Adaptive learning environments are those where the system of teacher and AI is pre-designed to regulate the students' learning. However, the possibility of gradually shifting the responsibilities from the AI to the user, allowing for independence and gain in self-regulation capacities, can be thought of as a goal for mature citizens' mode of using AI in their learning (2022a). Von Garrel and Mayer (2023) show that university students of science and mathematics

Table 6.3 Using AI supports to support students' 4Cs

Critical thinking	*Communication*	*Collaboration*	*Creativity*
Asking AI for comments or improvements on a plan	Asking AI to engage in a debate about a claim, to practice argumentation and evidence use	Asking AI for ideas to seed human group brainstorming	Expanding from an idea or prompt from AI
Comparing their own work with the AI execution of a task, locating the differences	Asking AI to create questions that they may use to gauge their level of understanding	Using AI to help sequence, structure or divide human project work	Prompting for creative outputs with generative image AI
Comparing AI responses with research text *or* comparing an AI summary of a research text with the original text		Practicing human–AI collaborative skills and responsibilities through prompting and interpretation	

Examples adapted from Farrokhnia et al. (2024), Nikolic et al. (2023), Lee et al. (2024)

do use AI tools for checking their understanding of a concept or for exam preparation, rather than "only" problem-solving for them.

The 4Cs of critical thinking, communication, collaboration, and creativity cannot become the responsibility of a pre-designed learning environment, but tools that AI and LLMs can offer may be used to help students in learning the 4Cs. Table 6.3 explores some procedures relevant to the 4Cs where AI tools aid students rather than solve problems for them.

While it is relatively easy to come up with strategies such as Table 6.3, the teacher needs to remain practical with their suggestions. Which strategies would the students actually find appealing and which might actually be put to use? This depends on the context and the learning goals these students are setting for themselves. For a student striving for mastery or a special interest, it may make sense to ask AI for practice questions. For a student who is seeking to pass, asking AI for improvements on their work may make more sense. Still, both of these ways of engaging with AI require self-regulated practices and metacognition (Molenaar, 2022a).

Choosing Contexts and Outcomes

National curricula in many countries are built to address concurrent needs and ensure skills for young people growing up in a time of sustainability crisis. In this chapter, we explore four examples of preparing learners for the future of work and sustainable living, which is the common aim of the design task. Figure 6.2 shows the examples and their relation to the skills needed for working life (Masterson, 2023).

Designing for Systems Thinking Through Smart Sensor Projects

Systems thinking can be learned in the context of modern technology, helping students understand cause-and-effect dynamics in environmental systems, and applying project-based learning in designing and testing

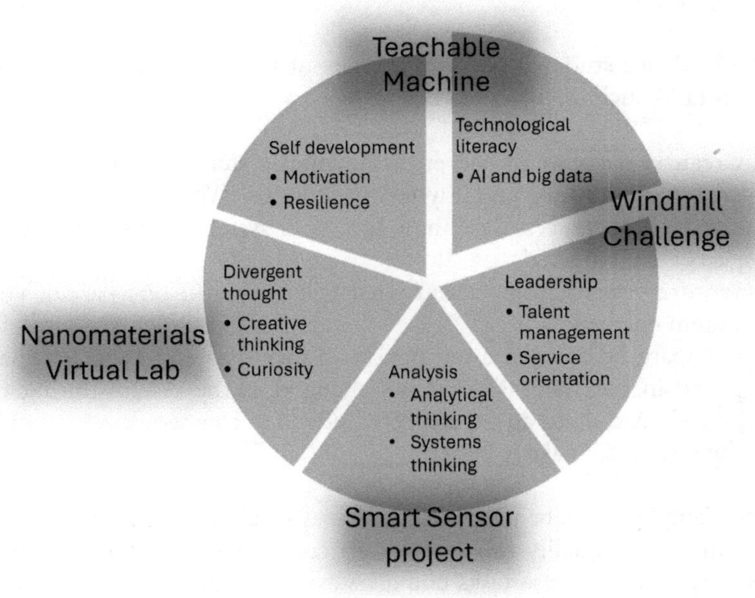

Fig. 6.2 Four examples and their target skills within Masterson's (2023) skills for working life: The Smart Sensor projects (5.1.), Nanomaterials Virtual Lab (5.2.), the Windmill Challenge (5.3.) and the Teachable Machine project (5.4)

environmental sensor systems. Weintrop et al. (2016) argue that the computational tools can deepen mathematics and science learning, and likewise the science and mathematics offer contexts for application of computational thinking. D'eon et al. (2021) reported a project-based learning experiment using portable air sensors to assess indoor and outdoor air quality.

The backward design begins with a clearly defined learning outcome: students will be able to recognize, analyze, and reflect on how different variables interact within an environmental system. This includes the abilities of:

- Identify key system components (inputs, processes, outputs, and feedback).
- Predict how changes in one part of the system affect the whole system.
- Evaluate the effects of different interventions or inputs on system behavior.

The evidence students need to give about their learning may include practical tasks such as:

1. Create a diagram of the environmental system. Label the inputs, processes, outputs, and any feedback loops. Write a short explanation (5–8 sentences) describing how your system works and how the parts are connected.
2. Write 2–3 predictions about what will happen if one part of your system changes.
3. Compare before and after data for a given system. What has happened in the system? What has changed in the system during the period? Are there any unexpected effects? How would you modify the system to make it better?

For example, the data in the third task may originate from a greenhouse thermometer, air quality monitor, or water circulation system. The students' responses to these tasks could be assessed through a simple three level rubric (Table 6.4), which is easy for teachers to use.

The learning activities in this case may be project-based learning (PBL), where students in Grades 9–12 collaboratively design a smart sensor intervention to monitor or manage a real-world or a model environmental system (e.g., temperature control in a greenhouse, air quality monitoring,

Table 6.4 A three-level rubric of criteria for systems thinking competences

Criteria	*Needs improvement*	*Meets expectations*	*Exceeds expectations*
Identifying system components (inputs, processes, outputs, feedback)	Identifies few or unrelated components; misses key system elements or feedback	Identifies key components and basic feedback loops; shows partial system understanding	Demonstrates clear understanding of all components and feedback; maps system structure holistically
Predicting system behavior based on changes in components	Provides unclear or unsupported predictions; lacks connection to system structure	Explains plausible system responses to changes; recognizes cause-effect links	Predicts direct and indirect outcomes with rationale; integrates feedback and time delays
Evaluating inputs effects on overall system performance	Describes outputs without evaluating or reflecting on systemic effects	Analyzes intervention outputs with some depth; considers impact on system behavior	Evaluates intervention outputs critically; reflects on system-wide implications and complexity

or water regulation in a school garden). The project groups use microcontrollers (e.g., Arduino or Raspberry Pi), sensors (e.g., CO_2, temperature, moisture), and basic programming to build a functioning prototype. They define the environmental variables they want to influence, explore their interrelationships, and simulate how different inputs, such as increased sunlight, rain, or temperature, alter system outputs.

Throughout the project, students are encouraged to ask questions connecting the experiment to real life: Who benefits from this system? Who might be left out? What are the social or environmental impacts of your solution? Students reflect on these questions by discussing equity, accessibility, or unintended consequences of automation or tech in farming and urban design. A teacher may also connect the intervention with the usability of the systems by asking: How would you design the user interface or physical system to be clear, intuitive, and user-friendly? Students then create more user-friendly interfaces, manuals, or physical casings for sensors. Language may be integrated by assigning the students to explain their system to people who don't know much about technology or science. In this case, students may design multiple representations, such as explainer videos, infographics, or podcasts, to communicate their findings to the community.

In this example of STEM intervention, the "A" in STEAM was not an explicit learning objective on its own, but it plays a vital role in making STEM learning more relevant and authentic. By drawing naturally from fields like design, communication, and ethics, the project situates STEM within real-life technological and societal contexts, helping students see the broader purpose and impact of their learning.

Artificial intelligence (AI) can be integrated into this intervention to enhance both the functionality and analysis of the environmental system. For example, students can use simple machine learning models to predict environmental conditions (e.g., temperature trends or moisture levels) based on previous sensor data. AI can also support automated decision-making, such as adjusting gardening schedules based on patterns detected in the data. Introducing AI tools not only enriches the technical depth of the project but also helps students understand the role of intelligent systems in solving real-world and future sustainability challenges.

Designing for Modeling and Interpretation Through Nanomaterials Virtual Lab

Simulation and modeling activities have a vital role in science education, particularly in domains such as molecular dynamics and nanotechnology (see, for example, Kilani et al., 2018; nanoHUB, n.d.). These "classical" interventions of visualizing molecular interactions, material properties, or environmental systems have proven especially valuable in helping students grasp phenomena invisible to our senses (Montalbano, 2014). When situated within the context of education for sustainable development, simulations are powerful tools for exploring urgent global challenges of climate change, pollution, and resource efficiency, for example.

In this intervention, students explore and optimize nanomaterials in a virtual lab. It aims to develop students' competence in scientific modeling and interpretation through simulation-based exploration of nanomaterials. The learning outcome is the ability to model, optimize, and interpret the behavior of nanomaterials in virtual environments, applying scientific reasoning and critical analysis to evaluate materials' potential applications.

Students will show their competence in a task where they are asked to design and justify an optimized material configuration for a selected real-world application.

To gain this competence, students work in teams to explore a virtual nanoscience lab (e.g., PhET simulations, NanoHub.org, or custom tools) where they manipulate particle size, surface area, and thermal and/or electrical conductivity of nanomaterials. They test their material's behavior under different simulated conditions and try to optimize it for a chosen application, such as improving water filtration or creating efficient solar coatings. Teachers guide and assess students through a design–test–reflect cycle:

- Define a goal (e.g., maximize/minimize heat/electric conductivity or minimize toxicity)
- Simulate multiple configurations and record results
- Select an optimized design
- Present findings with evidence and argumentation

In this intervention, STEAM integration is context specific. To communicate scientific results effectively using multiple representations, students may create clear visualizations, graphs, charts, and animations. They may be assigned to write blog posts or to create video pitches to explain their innovation to a public or policy audience, supporting science communication skills. Students also explore the societal implications of nanotechnology, environmental risks, accessibility, and bioethics. While not assessed as stand-alone objectives, these disciplines contextualize STEAM content, deepening relevance and engagement.

AI can be used in several ways to extend learning. AI-based simulation assistants (e.g., generative tools or pattern recognition models) can help students predict optimal configurations based on input parameters. Students can use other AI tools like ChatGPT or Wolfram Alpha engine to generate hypotheses, explain complex concepts, or refine their lab reports.

For advanced learners, teachers can introduce basic machine learning models (e.g., regression or classification) to predict simulation outcomes based on past data, building data literacy and analytical thinking.

Designing for Collaboration and Technological Literacy through the Windmill Challenge for Sustainable Development

This third example activity, structured according to the backward design framework, engages students in collaborative, problem-solving tasks linked to education for sustainable development and real-world challenges. It is anchored in Sustainable Development Goal 7: Affordable and Clean Energy (UNESCO, 2017). The activity can help students overcome the Eurocentric view of scientific innovation as they are finding examples about wind turbines or windmills also outside of their everyday knowledge (C4S, 2018). The activity invites students to explore the potential of wind energy through engineering design, mathematical modeling, and teamwork. The design is grounded in theoretical foundations for effective STEM learning environments (Glancy & Moore, 2013).

The learning outcomes are set so that after completing the activity, students will be able to:

- Understand and apply scientific and mathematical concepts.
- Construct experimental setups to study the efficiency of wind turbines.
- Use digital tools to design, simulate, and present windmill models and performance results.
- Connect windmill performance to energy efficiency and the broader context of clean, sustainable energy solutions.

Students will demonstrate their learning through the following artifacts:

- A presentation or video explaining the design and experimenting process, collaborative roles, and sustainability implications.
- A written reflection on the student's own performance and the contributions of other group members, focusing on how everyone worked together on both the final product and the learning during the project.

This learning activity follows the principles of the Inquiry based science education (IBSE) and Gallery Walk learning methods, which has been shown to be effective for supporting conceptual learning (Sunarti & Septiana, 2019). Students work in groups of four. In the first phase of the activity, they are asked to share their group's prior knowledge about wind turbines, considering different STEAM perspectives. This phase is based on the constructivist principle, which emphasizes that students' knowledge should build on their prior experiences and understandings (Dienes, 1960).

In the second phase, aligned with the dynamic principle, which states that the underlying structure of a system becomes visible only through change, and the principle of structural variability, students systematically explore variables such as blade area, length, shape, number, angle, and placement. They continue developing their posters and conducted experiments to determine the optimal setup and location for a wind turbine.

In the final phase, students annotate their posters with the concepts they connected in their investigation. These include surface area to length, pitch, angle, gear ratio, friction, flow, fluid dynamics, and material texture and stiffness. Students are asked to represent these concepts in multiple forms, appropriate to the task of optimizing turbine design and placement within the school area. This aligns with the principle of perceptual variability and Lesh's Translation Model (Lesh et al., 1987), which emphasizes that conceptual understanding is demonstrated through students' ability to represent ideas in various forms and to translate between these representations.

This intervention demonstrates how STEAM disciplines converge in real-world problem-solving within the context of sustainable development. In this case:

- Science supports understanding the possibilities and environmental impact of wind energy.
- Technology enables testing, simulation, and data interpretation.
- Engineering is addressed through iterative design and experimental construction.
- Mathematics, especially ratios, is central to analyzing and optimizing windmill performance.
- Other disciplines can be integrated by discussing the sustainability of materials, the aesthetic and acoustic impact of wind farms, or the economic trade-offs related to energy production and consumption (efficiency), for example.

Here, AI may be incorporated to support deeper learning and iteration, helping students explore variables that are difficult to test physically. This promotes experimentation, innovation, and systems thinking, equipping students to approach sustainability challenges both creatively and collaboratively.

Designing for AI Utilization in STEAM Through Teachable Machine and Machine Learning

Artificial intelligence is increasingly applied in real-world classification tasks, including high-impact domains such as medical diagnostics, where machine learning models assist in identifying cancerous cells and tumors from medical imaging (Hosny et al., 2018). Similar classification techniques are used in environmental monitoring (e.g., species recognition from camera traps (Tabak et al., 2019), automated waste sorting (Hannan et al., 2020), and even in authenticating works of art (Ostmeyer et al., 2024). These systems rely on large and varied datasets to recognize patterns and make predictions, but their effectiveness depends on the quality of the data, the clarity of the defined categories, and human decisions made during the design and training process.

This interdisciplinary learning activity uses a teachable machine, developed in Generation AI (2025) project, as an accessible entry point to explore how machine learning can be used to classify images across a range of scientific and sustainability-related themes. By designing and training a custom classifier, students gain practical insight into how machine learning models are created, refined, and interpreted (Generation AI, 2025). The activity integrates core concepts from the STEAM disciplines, digital literacy, and ethical reasoning.

The learning objectives are designed to help students understand how classification systems operate across scientific disciplines and how these can be recontextualized through machine learning. Students learn how to collect, label, and organize visual data into meaningful categories. Beyond the technical process, students learn to analyze and communicate the performance of their trained model, recognize its limitations, and reflect critically on the ethical and practical consequences of using AI or human experts for the final decision-making.

As evidence of learning, students are expected to design a classification task that is both educationally and ethically relevant. For example, students might explore the classification of edible versus inedible mushrooms and consider the potential real-world consequences of a misclassification. After selecting a focus, they gather and label an appropriate image dataset to train their model, paying attention to variation, quality, and category clarity. They then test the trained model using new examples and evaluate its performance using both qualitative reflection and basic tools such as confusion matrices. This phase of the activity encourages students to examine the risks of model bias, data limitations, and unintended consequences, particularly in high-stakes contexts. Reflections on these issues are framed in terms of both human and machine responsibility, fostering ethical awareness alongside technical skills.

The activity is guided but open-ended, allowing students to select a topic of personal or curricular relevance, such as classifying cloud types for weather prediction, sorting waste materials for sustainability, or analyzing visual styles in art history. Students are encouraged to iterate on their models and to reflect on the educational potential of AI-supported classification tasks in their future classrooms. Throughout the process, peer discussion and instructor feedback support collaborative learning and critical dialogue.

This activity provides foundational knowledge about how machine learning works, including how data trains algorithms and what classification errors can reveal about model limitations. It also invites students to consider how humans interact with, guide, and interpret AI systems. By engaging in the full cycle of designing, training, testing, and evaluating a machine learning model, students develop STEAM transferable skills and ethical understanding that prepare them to both use and critique AI technologies in educational and societal contexts.

Conclusions

Lessons for Everyone

The four examples from the previous chapters tie up loose ends in terms of how to design teaching that is relevant and rigorous for today's student. Hopefully, they have convinced both the bold explorer teacher and the slow-to-adapt teacher to consider strategically enriching their curricula with new technologies and learning activities that go beyond the STEM skillset.

The science teacher and teacher educator alike may find that starting the learning activity design from the expected learning outcomes makes it easier to keep focus on the important foundational and work–life skills, as it can be easier to keep adjusting around the science goals and science skills that we are more familiar with teaching (rather than the other way around). Starting with the expected learning outcomes also ensures the exploration of novel technologies such as nanoscience, programmable sensors, or machine learning, does not deduct from the "regular" curriculum and become exploration just for explorations' sake.

The teacher students' technical-pedagogical competencies vary. Modeling the use of technologies in educational settings, in particular in K–12 classroom experiences, is a major predictor of applications also in their teaching and confidence to try new technologies (Wang et al., 2018).

The teaching practicum and other teaching opportunities during student teachers' training offer a safe training ground for teaching with and for technology. The teacher students, who have learned the latest content knowledge, are in a position to both learn from and train the teachers in the field in classroom use of novel technologies. Both teachers and teacher students may need support and encouragement from the teacher educators in these fresh attempts, meaning the teacher educators have to have the T-PACK to be helpful.

The rapidly evolving technological and professional landscape call for schools to stay connected in the now. The school and society can move forward in unison, teaching the skills that are in demand as part of the classroom activities—as creativity, collaboration, or critical thought simply are meaningless to train at without a context. The "digital natives" narrative is based on anecdotal evidence at best—the generational technology gap is more related to one's media environment than chronological age, and the perceived gap is perceived to be the largest in use of novel technologies (Vaterlaus et al., 2015). The presence of technologies affects everyone whether it is part of a particular person's life growing up or not—the exosystem of societal values and generalities, as well as the mesosystem of the school or friend group socialize and normalize the existence of technology as a large part of everyday life in today's children and youth (McHale et al., 2009). Without immersion in any technologies for a significant time, it is a learning curve for us all regardless of age. So, while the students and young teachers may enjoy their limited time having an advantage in getting acquainted with AI technologies, their media environment is unlikely to be the same as the one the next generation grows up in.

Discussion

The pressure to try new strategies in education stems from three influences: the changes in our everyday life, our working life, and the complex global challenges and their effects on individuals. The challenge to STEAM education is not falling for the aforementioned STEM crisis story, but in including everyone in solving global challenges. It requires tailoring science, engineering, and mathematics education of today to address the global challenges and also teaching children and youth to learn the connections of these disciplines to the lived realities of people near and far.

The gap between our current education goals and the ideal or necessary goals is tackleable. As the four examples we shared tell us, it is not outrageous to design education that is both rigorous in its science or technology content and that teaches general work–life skills. And the shift in how we approach AI tools will slowly go from content of our studies (what is AI) to a medium of our studies (using AI to study and produce new knowledge)—the same process that the abacus, calculator, electron tubes, magnetic memory, Internet protocol, qubits, or cloud services underwent.

As a society, we are able to deal with more complex issues and problems as technology aids us in processing information. This obviously requires new skills that are unrelated to fact retention. Perhaps they are the twenty-first-century skills, or perhaps we need to look even further than that. The science educator who stays in the now of technology is well equipped to deal with change.

Disclaimer In the development of this chapter, ChatGPT4 was used in the idea generation phase to run a completeness check and gap analysis, as well as for spell-checking the initial version of the manuscript.

References

Bereiter, C. (2002). Education in a knowledge society. In *Liberal education in a knowledge society* (pp. 11–34).

Binkley, M., Erstad, O., Herman, J., Raizen, S., Ripley, M., Miller-Ricci, M., & Rumble, M. (2012). Defining twenty-first century skills. In *Assessment and teaching of 21st century skills* (pp. 17–66). Springer.

Bransford, J. D., Brown, A. L., & Cocking, R. R. (2000). *How people learn: Brain, mind, experience, and school* (Expanded ed.). National Academy Press.

Bybee, R. W., Taylor, J. A., Gardner, A., Van Scotter, P., Powell, J. C., Westbrook, A., & Landes, N. (2006). *The BSCS 5E instructional model: Origins, effectiveness, and applications.* BSCS.

C4s Communities for Sciences (2018). *The white book on inclusive science education.* Retrieved on 4 April, 2025, from http://www.communities-for-sciences.eu/wp-content/uploads/2024/07/White-Book-on-Inclusive-Science-Education_C4S_ONLINE.pdf

Casner-Lotto, J., & Barrington, L. (2006). Are they really ready to work? Employers' perspectives on the basic knowledge and applied skills of new entrants to the 21st century U.S. In *The conference board, corporate voices for working families, partnership for 21st century skills, & society for human resource management.*

Chaka, C. (2020). Skills, competencies and literacies attributed to 4IR/Industry 4.0: Scoping review. *IFLA Journal, 46*(4), 369–399.

Cheikh-Ammar, M. (2018). The IT artifact and its spirit: A nexus of human values, affordances, symbolic expressions, and IT features. *European Journal of Information Systems, 27*(3), 278–294.

Cukurova, M., & Miao, F. (2024). *AI competency framework for teachers.* UNESCO Publishing.

D'eon, J. C., Stirchak, L. T., Brown, A.-S., & Saifuddin, Y. (2021). Project-based learning experience that uses portable air sensors to characterize indoor and outdoor air quality. *Journal of Chemical Education, 98*(2), 445–453. https://doi.org/10.1021/acs.jchemed.0c00268

Dienes, Z. P. (1960). *Building up mathematics.* Hutchinson Educational.

European Commission. (2013). *Responsible Research and Innovation (RRI), science and technology: Report.* Publications Office. Retrieved February, 2025, from https://data.europa.eu/doi/10.2777/45726.

European Commission. (2022). *Ethical guidelines on the use of artificial intelligence (AI) and data in teaching and learning for educators.* Publications Office of the European Union. Retrieved February, 2025, from https://data.europa.eu/doi/10.2766/153756

Farrokhnia, M., Banihashem, S. K., Noroozi, O., & Wals, A. (2024). A SWOT analysis of ChatGPT: Implications for educational practice and research. *Innovations in Education and Teaching International, 61*(3), 460–474.

Generation AI. (2025). *Generation AI—Education and artificial intelligence.* Retrieved 13 June, 2025, from https://www.generation-ai-stn.fi/en/

Glancy, A. W., & Moore, T. J. (2013). *Theoretical foundations for effective STEM learning environments.* School of Engineering Education Working Papers, Paper 1.

Hannan, M. A., Al-Sakkaf, A., Ker, P. J., Begum, R. A., Basri, H., Hussain, A., & Mahlia, T. M. I. (2020). A review of intelligent sensor-based waste materials recognition and sorting systems. *Resources, Conservation and Recycling, 152*, 104506. https://doi.org/10.1016/j.resconrec.2019.104506

Hazelkorn, E., Ryan, C., & Shiel, G. (2015). *Science education for responsible citizenship.* European Commission. Retrieved 13 June, 2025, from https://www.researchgate.net/publication/280831573_Science_Education_for_Responsible_Citizenship

Hosny, A., Parmar, C., Quackenbush, J., Schwartz, L. H., & Aerts, H. J. W. L. (2018). Artificial intelligence in radiology. *Nature Reviews Cancer, 18*(8), 500–510. https://doi.org/10.1038/s41568-018-0016-5

Ibrahim, S., Del Gobbo, D., El Halwany, S., Zouda, M., Milanovic, M., Hassan, N., et al. (2022). STEAM not STEaM: Revisioned pedagogies prioritizing social justice and ecological sustainability in STEAM education. *Journal for Activist Science and Technology Education, 12*(1).

Jolly, J. L. (2009). Historical perspectives: The national defense education act, current STEM initiative, and the gifted. *Gifted Child Today, 32*(2), 50–53.

Kilani, M., Torabi, K., & Mao, G. (2018). Application of virtual laboratories and molecular simulations in teaching nanoengineering to undergraduate students. *Computer Applications in Engineering Education, 26*(6), 1527–1538.

Kim, K., & Kwon, K. (2023a). Exploring the AI competencies of elementary school teachers in South Korea. *Computers and Education: Artificial Intelligence, 4*, 100137.

Kim, K., & Kwon, K. (2023b). Exploring the AI competencies of elementary school teachers in South Korea. *Computers and Education: Artificial Intelligence, 4*, Article 100137. https://doi.org/10.1016/j.caeai.2023.100137

Koyunlu Ünlü, Z., & Dökme, İ. (2022). A systematic review of 5E model in science education: Proposing a skill-based STEM instructional model within the 21st-century skills. *International Journal of Science Education, 44*(13), 2110–2130. https://doi.org/10.1080/09500693.2022.2114031

Laurie, R., Nonoyama-Tarumi, Y., McKeown, R., & Hopkins, C. (2016). Contributions of Education for Sustainable Development (ESD) to quality education: A synthesis of research. *Journal of Education for Sustainable Development, 10*(2), 226–242.

Laurillard, D. (2011). Rethinking teaching for knowledge society. In *eDemocracy & eGovernment: Stages of a democratic knowledge society* (pp. 191–204). Springer Berlin Heidelberg.

Lave, J., & Wenger, E. (1991). *Situated learning: Legitimate peripheral participation*. Cambridge University Press.

Leaders, Y. G. (2016). World economic forum annual meeting 2016 mastering the fourth industrial revolution. In *World economic forum: Cologny, Switzerland*.

Lee, U., Han, A., Lee, J., Lee, E., Kim, J., Kim, H., & Lim, C. (2024). Prompt Aloud!: Incorporating image-generative AI into STEAM class with learning analytics using prompt data. *Education and Information Technologies, 29*(8), 9575–9605.

Lesh, R., Post, T., & Behr, M. (1987). Representations and translations among representations in mathematics learning and problem solving. In C. Janvier (Ed.), *Problems of representation in the teaching and learning of mathematics* (pp. 33–40). Lawrence Erlbaum.

Markus, M. L., & Silver, M. S. (2008). A foundation for the study of IT effects: A new look at DeSanctis and Poole's concepts of structural features and spirit. *Journal of the Association for Information Systems, 9*(10), 5.

Masterson, V. (2023, May). Future of jobs 2023: These are the most in-demand skills now—And beyond. In *World economic forum, Geneva, Switzerland* (Vol. 1).

McHale, S. M., Dotterer, A., & Kim, J. Y. (2009). An ecological perspective on the media and youth development. *American Behavioral Scientist, 52*(8), 1186–1203.

Mikeladze, T., Meijer, P. C., & Verhoeff, R. P. (2024a). A comprehensive exploration of artificial intelligence competence frameworks for educators: A critical review. *European Journal of Education, 59*(3), e12663.

Mikeladze, T., Meijer, P. C., & Verhoeff, R. P. (2024b). A comprehensive exploration of artificial intelligence competence frameworks for educators: A critical review. *European Journal of Education, 59*(1), 123–145. https://doi.org/10.1111/ejed.12663

Mitcham, C. (2004). Philosophy of information technology. In *The Blackwell guide to the philosophy of computing and information* (pp. 327–336).

Molenaar, I. (2022a). The concept of hybrid human-AI regulation: Exemplifying how to support young learners' self-regulated learning. *Computers and Education: Artificial Intelligence, 3*, 100070.

Molenaar, I. (2022b). Towards hybrid human-AI learning technologies. *European Journal of Education, 57*(4), 632–645.

Montalbano, V. (2014). Seeing and interacting with the invisible: A powerful tool for the learning of science. *arXiv*. https://arxiv.org/abs/1401.3047

nanoHUB. (n.d.). *nanoHUB: Making data and simulation pervasive*. Retrieved 13 June, 2025, from https://serc.carleton.edu/msu_nanotech/Nano_Hub.html

National Research Council. (2012). *A framework for K-12 science education: Practices, crosscutting concepts, and core ideas*. The National Academies Press. https://doi.org/10.17226/13165

Nikolic, S., Daniel, S., Haque, R., Belkina, M., Hassan, G. M., Grundy, S., et al. (2023). ChatGPT versus engineering education assessment: A multidisciplinary and multi-institutional benchmarking and analysis of this generative artificial intelligence tool to investigate assessment integrity. *European Journal of Engineering Education, 48*(4), 559–614.

Orlikowski, W. J., & Iacono, C. S. (2001). Research commentary: Desperately seeking the "IT" in IT research—A call to theorizing the IT artifact. *Information Systems Research, 12*(2), 121–134.

Ostmeyer, J., Schaerf, L., Buividovich, P., Charles, T., Postma, E., & Popovici, C. (2024). Synthetic images aid the recognition of human-made art forgeries. *PLoS One, 19*(2), e0295967.

Özgün-Koca, S. A., Meagher, M., & Edwards, M. T. (2010). Preservice teachers' emerging TPACK in a technology-rich methods class. *The Mathematics Educator, 19*(2), 10–20.

Partnership for 21st Century Skills. (2009a). *P21 framework definitions*. Retrieved 13 June, 2025, from http://static.battelleforkids.org/documents/p21/P21_Framework_DefinitionsBFK.pdf

Partnership for 21st Century Skills. (2009b). *P21 framework definitions*. Retrieved 13 June, 2025, from https://files.eric.ed.gov/fulltext/ED519462.pdf

Sunarti, T., & Septiana, D. A. (2019). The effect of problem-based learning with gallery walk strategy to creativity and communication skills. In *Mathematics, informatics, science, and education international conference (MISEIC 2019)* (pp. 144–148). Atlantis Press.

Tabak, M. A., Norouzzadeh, M. S., Wolfson, D. W., Sweeney, S. J., Vercauteren, K. C., Snow, N. P., Halseth, J. M., Di Salvo, P. A., Lewis, J. S., White, M. D., Teton, B., Beasley, J. C., Schlichting, P. E., Boughton, R. K., Wight, B., Newkirk, E. S., Ivan, J. S., Odell, E. A., Brook, R. K., et al. (2019). Machine learning to classify animal species in camera trap images: Applications in ecology. *Methods in Ecology and Evolution, 10*(4), 585–590. https://doi.org/10.1111/2041-210X.13120

UNESCO. (2010). *UN decade of education for sustainable development (2005–2014): Final report*. UNESCO. Retrieved 13 June, 2025, from https://unesdoc.unesco.org/ark:/48223/pf0000152453

UNESCO. (2017). *Education for sustainable development goals: Learning objectives*. UNESCO.

UNESCO. (2021). *Recommendation on the ethics of artificial intelligence*. UNESCO. Retrieved 13 June, 2025, from https://unesdoc.unesco.org/ark:/48223/pf0000380455

Valtonen, T., Hoang, N., Sointu, E., Näykki, P., Virtanen, A., Pöysä-Tarhonen, J., et al. (2021). How pre-service teachers perceive their 21st-century skills and dispositions: A longitudinal perspective. *Computers in Human Behavior, 116*, 106643.

Vaterlaus, J. M., Jones, R. M., & Tulane, S. (2015). Perceived differences in knowledge about interactive technology between young adults and their parents. *Cyberpsychology, 9*(4), 1.

Von Garrel, J., & Mayer, J. (2023). Artificial intelligence in studies—Use of ChatGPT and AI-based tools among students in Germany. *Humanities and social sciences communications, 10*(1), 1–9.

Wang, W., Schmidt-Crawford, D., & Jin, Y. (2018). Preservice teachers' TPACK development: A review of literature. *Journal of Digital Learning in Teacher Education, 34*(4), 234–258.

Weintrop, D., Beheshti, E., Horn, M., Orton, K., Jona, K., Trouille, L., & Wilensky, U. (2016). Defining computational thinking for mathematics and science classrooms. *Journal of Science Education and Technology, 25*, 127–147.

Wiek, A., Withycombe, L., & Redman, C. L. (2011). Key competencies in sustainability: A reference framework for academic program development. *Sustainability Science, 6*(2), 203–218.

Wiggins, G., & McTighe, J. (2005). *Understanding by design* (2nd ed.). Association for Supervision and Curriculum Development.

Zheng, L. (2021). A performative history of STEM crisis discourse: The co-constitution of crisis sensibility and systems analysis around 1970. *Discourse: Studies in the Cultural Politics of Education, 42*(3), 337–352.

CHAPTER 7

Assessment and Evaluation in STEM Education in the Age of AI

Ismail Çuhadar, Beyzanur Yalvaç, and Ayşe Akkır

Introduction

STEM education is vital for providing students with essential skills and expertise in science, technology, engineering, and mathematics when acquiring responsibility and making informed decisions for the future world (Zhai & Krajcik, 2024). Although STEM education has developed significantly over the past decade, there are some important issues in STEM education, such as pedagogical problems, program development difficulties, student concerns, and insufficient teacher support (Martín-Páez et al., 2019; Suhirman & Prayogi, 2023). Moreover, among these challenges, the assessment of STEM education emerges as a significant concern in the STEM education area (Fang & Hsu, 2019; Gao et al., 2020). For example, practitioners tend to implement a product-oriented

I. Çuhadar (✉)
Ministry of Education, Istanbul, Türkiye

B. Yalvaç
Boğaziçi University, Istanbul, Türkiye

A. Akkır
Turkish Maarif Foundation, Istanbul, Türkiye

R. Elmas, M. Akarsu (eds.), *STEM Education*, Maarif Global Education Series,
https://doi.org/10.1007/978-981-95-9023-0_7

assessment rather than a process-oriented assessment in STEM education (Gao et al., 2020; Akarsu et al., 2022). However, the quality of the final product depends on the quality of the process through which it is created. For this reason, the process is as important as the product itself. This necessitates the assessment of not only the product but also the process.

On the other hand, the complexity and interdisciplinary nature of STEM learning make it challenging to assess not only the product but also the learning process. Specifically, it is challenging to create authentic STEM problems for assessing complex skills with interdisciplinary STEM tasks (Herro et al., 2017; Zhai et al., 2023). For this reason, although teachers can achieve to implement the STEM education in an interdisciplinary manner, they may fail to carry out its assessment with the same level of interdisciplinarity and end up with assessing only a single discipline (Fang & Hsu, 2019; Gao et al., 2020).

Scoring students' performance is another challenge teachers face when assessing the complex skills with authentic STEM tasks (e.g., collaboration in Herro et al., 2017; group work in Theobald et al., 2020). Because the targeted learning outcomes are complex in STEM education, developing rubrics for assessing these complex learning outcomes is difficult for teachers. Although the measurement tools need to be practical for teachers to use in the classroom, scoring student performances using these rubrics is also time-consuming when assessing whether students understand STEM concepts and can apply them to practical STEM problems by using their problem-solving skills (Gao et al., 2020). Teachers may struggle to balance the need for extensive assessment with their restricted time, which may lead to reliance on basic tests that ignore important elements of problem-solving skills.

Additionally, in STEM education, where the process is as important as the product, it is also crucial to monitor students' learning and performance by using the results of such assessments. In this way, teachers can track students' engagement with the course, their progress, and the areas where they struggle, which may help teachers identify the students who may be at risk (Jiao et al., 2022). However, as in many areas of education, monitoring each student's progress is another challenge teachers face in STEM education. This is due to the fact that STEM programs frequently entail complex tasks, which may make it challenging to assess each student's progress. This demonstrates the necessity for cutting-edge assessments that are appropriate for STEM education.

On the other hand, whether the product or the process is being assessed in STEM education, using assessment results as feedback for students, teachers, and curriculum is essential, as it is one of the primary goals of assessment. For example, the assessment of process may provide feedback to students to help them acquire the skills and expertise they need when developing a product in STEM education. Accordingly, timely scoring of classroom-assessment practices and providing students with prompt feedback are essential. Delayed feedback may lead to the persistence of potential misconceptions in students, which may negatively affect their subsequent learning. Providing timely and objective feedback to students is another challenge teachers face in STEM education (Tempelaar et al., 2015; Zengin et al., 2020; Zhai et al., 2023). The complexity of the skills assessed in tasks, along with students' diverse learning patterns, necessitates more individualized feedback, which makes delivering effective feedback in STEM education even more difficult.

Overall, implementing a process-oriented assessment, using interdisciplinary assessment, designing authentic STEM tasks, scoring student performance to assess complex skills, monitoring student progress, and providing instant feedback are some important issues for teachers in the assessment of STEM education. These challenges can be addressed by integrating artificial intelligence (AI) into STEM education, which can transform the assessment practices (Ouyang et al., 2023; Zhai et al., 2020). There is a growing emphasis on the significance of applications of AI technologies within STEM education (Jantakun et al., 2025). For example, AI technologies are currently being implemented in STEM education for many purposes, such as providing personalized learning, improving the efficiency of instruction, and enhancing student engagement (Joseph & Uzondu, 2024). Similarly, AI can automate many assessment applications in STEM education and help reduce teachers' workload related to assessment. Overall, the purpose of this chapter is to provide an extensive investigation of the possibilities of employing AI in the assessment of STEM education. First, the opportunities of AI in STEM education assessment with potential applications are presented—from authentic task generation to instant feedback—considering the current challenges in assessing STEM education. Then, the challenges of using AI in STEM education assessment are elaborated. This is followed by a discussion on the future directions of using AI in STEM education assessment.

Opportunities of AI in STEM Education Assessment

AI may be used for many purposes in STEM education assessment, such as providing personalized learning and resource suggestions, automating item generation, monitoring student behavior, and predicting achievement to identify at-risk students (Xu & Ouyang, 2022). Specifically, Machine Learning (ML) and Natural Language Processing (NLP) enable to assess higher-order skills, design authentic STEM problems, and provide appropriate and timely feedback to students on their educational progress (Debuse & Lawley, 2016; Saputra et al., 2024; Tempelaar et al., 2015; Theobald et al., 2020). In addition, ML and NLP play an important role when scoring students' answers to open-ended questions (Ali, 2024). Overall, these AI tools can be particularly helpful for educators by providing fast and efficient feedback to students and significantly reducing educators' workload when grading students' work. These kinds of opportunities for AI in STEM education assessment are elaborated in the following sections.

Designing Authentic STEM Problems

Generating authentic STEM-related tasks is one of the issues that teachers may confront in STEM education assessment. Authentic tasks are defined as those that include subtasks involving multiple disciplines, multistep challenging inquiries, unclear difficulties, and numerous solutions to a problem (Amory, 2014). Bertrand and Namukasa (2020) highlight the importance of using authentic tasks in a variety of disciplines, especially in STEM, by claiming that focusing on authentic tasks in STEM education may help students develop skills such as perseverance and adaptability. Although the complexity of authentic tasks may be a challenge for teachers, AI is a possible solution when generating these STEM tasks for assessment purposes.

Many studies have been conducted to determine how AI can help with item generation. Chan et al. (2025) investigated how successfully GPT-3.5 and GPT-4 create questions and multistep answers in STEM topics (i.e., chemistry, physics, and mathematics) for year 12 students, with an emphasis on conceptual accuracy and quality. Their study indicates that applying structured prompting approaches such as Chain-of-Thought (CoT) and CoT with coding language (CoT-CL) considerably enhances output quality, particularly with GPT-4. Their work yields encouraging results allowing teachers to easily apply AI for item generation. Specifically, their study shows that providing explicit contextual information, such as sample questions and answers, instructing the model with step-by-step reasoning and coding verification, verifying outputs with subject-matter specialists, and revising prompts using domain-specific factors are all crucial for producing high-quality questions with large-language models. As for automated item generation, Circi et al. (2023) summarize research, which indicates that most of the current studies are experimental, rather than operational. In other words, automatic item generation tools need to be employed in real settings to produce test items and tasks so that its efficiency and accuracy can be observed and analyzed in real settings rather than controlled ones.

Template- or rule-based procedures dominate automatic item generation, which provides good concept alignment but little creativity. In other words, templates or rules can generate items, but such items share the same qualities as the templates or rules and hence are not creative in nature. While template-based item generation can generate many items, items may lack the capacity to include innovative or unexpected components. This might lead to predictable tests that do not adequately challenge students. Additionally, even though AI tools transform item generation by quickly making a large sample of items, these items still require verification to ensure they are both psychometrically valid and pedagogically correct.

On the other hand, automatic item generation provides tailored exam questions using AI approaches such as deep learning and natural language processing. However, issues remain in syntax correctness, semantic relevance, and item modeling complexity (e.g., calibrating test items using statistical models). Accordingly, more research is needed to investigate the item quality produced by automatic item generation in terms of being psychometrically valid and pedagogically sound.

Automated Scoring

In the context of STEM, scoring students' work faces challenges due to its interdisciplinary nature, which requires scoring knowledge and skills across multiple disciplines. AI mitigates these challenges by automating the assessment process. Automated scoring extends back to the 1960s, when Page (1966) created the first automated scoring engine. Since then, automated scoring has advanced significantly, and current systems are capable of far more than they were in the past.

In recent years, automated scoring has evolved through natural language processing, large language models, and machine learning methods (Foltz et al., 2020; Jung et al., 2024). In addition, AI technologies can be leveraged to develop systems that automatically score assessments. For instance, natural language processing techniques have been utilized to evaluate students' performance in programming (Saito & Watanobe, 2020). Morris et al. (2024) created a system to automatically score students' NAEP math explanations by first cleaning the data to ensure accuracy, then adding more examples to improve the model's understanding, and finally training a language model called DeBERTa-v3-large. This system was able to match the performance of human graders on most questions and showed no bias toward any group. This example demonstrates that advanced techniques could be used to create AI models that are able to perform automated scoring on a large number of responses. However, a lack of technical expertise, a shortage of finances, or the complexity of implementation may make it impractical for teachers to use such innovative technologies. Fortunately, Latif and Zhai (2024) developed ChatGPT for automated scoring. ChatGPT is a natural language programming model developed by using texts from journals and many sources, such as Wikipedia. However, students do not use the same language found in journals, so ChatGPT could not effectively score students' responses (Latif & Zhai, 2024). To overcome this challenge, Latif and Zhai (2024) trained and fine-tuned ChatGPT, and their results indicate that fine-tuned ChatGPT shows accurate scoring results compared to Google's latest AI technology, BERT, which is a benchmarking technology because it allows machines to comprehend the context and meaning of language, considerably enhancing performance in a wide variety of natural language processing programs.

When assessing STEM education, students' work on performance tasks needs to be scored if their work has met the predefined criteria. Accordingly, the assessment process necessitates the establishment of criteria and observation of results, which can be very time-consuming. While automated scoring is one example of how AI may help with STEM education assessment, creating evaluation rubrics is another beneficial application of AI. Based on a large body of educational research, rubrics are assessment tools that may assist students to develop self-regulation skills and autonomy and help students understand the instructional expectations (Andrade

& Du, 2005; Panadero & Jonsson, 2013). Moreover, rubrics assist teachers to provide constructive feedback and ensure that assessment criteria aligned with instructional objectives increases validity and reliability (Brookhart, 2013; Moskal & Leydens, 2000). Accordingly, using AI to design rubrics or to evaluate students' performance tasks based on pre-existing rubrics can make the assessment process easier and more objective. For example, in Ariely et al.'s (2023) study, machine learning and natural language processing are used to automatically score students' biology explanations. Using almost 500 samples, it offers comprehensive analytical grading rubrics for formative assessment that show good agreement with human experts.

Personalized and Dynamic Assessment

Dynamic assessment is a type of assessment which differentiates from standardized tests in some dimensions such as goals, orientation, context, interpretation of results, and nature of tasks (Tzuriel, 2001). According to Tzuriel (2001), the goals of dynamic assessment are to assess change, mediation, deficient cognitive functions, and non-intellective factors. Dynamic assessments are effective for capturing learners' attention and facilitating change. This assessment technique helps students improve their learning processes rather than just evaluate them (Abrar et al., 2025). In dynamic assessment, the assessor plays an interactive role by providing guidance and assistance to students throughout the assessment process. A very important part of STEM education is the evaluation of the process and provision of the necessary assistance (Akarsu et al., 2020). Accordingly, when students engage in practicing their learning through solving problems independently, AI may function as an intelligent tutor by providing hints, scaffolding, and feedback tailored to students' individual learning needs.

Integrating AI technologies into STEM education can also create opportunities for innovation by offering learning materials that are tailored to individual student needs (Chen et al., 2020). For instance, Zhai and Wiebe (2023) indicate that automated scoring can be used for personalized learning by turning scores into actionable information. In another study by Gerard and Linn (2016), automatic scoring based on the knowledge integration framework made it possible to automate teaching and provide learning assistance for each individual student. For example, in a photosynthesis assignment in this study, a student's written response received a rubric score of 2 out of 5. In response, the system gave targeted learning support, such as emphasizing the student's current views, suggesting relevant evidence, and advising a modification. Their research shows that AI technologies have potential to personalize assessment to students' needs. Akhil et al. (2025) also developed an AI powered platform to personalize assessment to students' needs. Their platform SMART ED considerably improved student learning outcomes, with post-test performance rising by 24% and follow-up evaluation results increasing by 31.5%, owing to its adaptive learning approach that matched the presentation of material to individual cognitive abilities. Its predictive analytics and interactive features, such as real-time evaluations and dynamic quizzes, increased engagement and assisted in filling knowledge gaps, resulting in a 32% increase in time given for engaging with learning content (Akhil, et al., 2025). Moreover, some software is present that can be used for personalized assessments, such as Dreambox, Knewton, and Carnegie (Wang et al., 2025). Carnegie Learning's AI-powered MATHia platform, which is utilized in US middle and high schools, personalizes math education and has been demonstrated to enhance standardized test scores by 8–12% compared to the results in traditional educational settings (Wang et al., 2025).

Personalized learning using AI has greatly enhanced STEM teaching in schools so that students may study at their own speed and receive rapid feedback following the completion of activities and exercises, which increases learning efficiency and students' confidence in problem-solving (Zhang et al., 2024). Intelligent tutoring systems such as Khan Academy's Khanmigo and adaptive learning platforms such as Smart Sparrow can be used for personalized assessment and then providing additional resources to enhance personalized learning (Zhang et al., 2024). Computer adaptive testing may also help personalized learning by dynamically adjusting the items' difficulty levels based on each student's individual performance, which provides personalized tests identifying students' achievement levels and learning needs more accurately than fixed test designs. Overall, AI can be used to create assessment procedures tailored to each student, which provides an opportunity for the students to learn on their own path.

Sentiment Analysis

Sentiment analysis refers to a natural language processing method that detects the emotions, opinions, or attitudes communicated in written content such as feedback or reviews (Sharma & Dash, 2023). In educational settings, this method has received growing attention as it facilitates a more profound comprehension of students, not just regarding their academic achievements but also their emotional and psychological states. In addition to providing feedback based on measurable data, AI can also evaluate students' emotions, attitudes, and interests, thereby supporting a more holistic understanding of the learning process.

Recent research has demonstrated various applications of sentiment analysis and associated technologies in education. For example, Tseng et al. (2023) evaluated students' personality traits and helped them enhance their academic performance by employing machine learning and physiological signals. Similarly, Egger et al. (2019) used different systems based on deep learning and machine-human interaction to identify emotions and evaluate the usefulness of these systems. These initiatives demonstrate how emotional and behavioral data can be incorporated into learning analytics using AI-driven tools. Moreover, Sharma and Dash (2023) proposed a natural language system that serves as a tool to understand students' answers to open-ended questions and help evaluate the sentiments behind the answers. This provides an increasing focus on understanding the students holistically by examining their emotional states in their answers beyond their quantitative scores. Recently, Yang et al. (2024) emphasized that the use of ChatGPT in sentiment analysis has demonstrated favorable impacts on students' emotional experiences and affective aspects of learning. However, it is important to recognize the challenges that come with these tools, especially regarding ethics, misinformation, and misinterpretation, as highlighted by Islam and Islam (2024).

Through the integration of sentiment analysis into AI-assisted educational assessment, teachers can gain a deeper insight into students' learning conditions, especially within the context of STEM education (Ouyang et al., 2023). Thus, this evolving method offers new opportunities to support students not only cognitively but also emotionally and personally. Sentiment analysis can be used in STEM education assessment to help teachers better understand students' feelings about challenging concepts, individual or group projects. Teachers can modify their lessons by identifying frustration, confusion, or engagement levels through written reflections or discussion posts and provide timely emotional and academic support. This emotional insight enhances student well-being and might contribute to perseverance and deeper learning in STEM environments.

Providing Timely Feedback

Timely and specific feedback is pivotal in the learning process, especially in formative assessments where feedback serves not only for diagnosis but also for guidance. Nevertheless, traditional methods frequently struggle to provide feedback at the required speed and individualization because of the time limitations faced by educators. As highlighted in Chng et al. (2023), "it is almost impossible for teachers to deliver quality individual feedback to students" (p. 402). This disparity has resulted in the incorporation of AI systems within educational settings, where AI's ability to provide immediate, personalized, and data-driven feedback has demonstrated significant potential.

AI-based platforms, particularly those leveraging natural language processing, deep learning, and improved optimization algorithms like IPSO-BP, have demonstrated notable improvements in instructional methods and quality (Ouyang et al., 2023). For instance, He and Fu (2022) evaluated the instructional quality through the use of advanced algorithms. They focused on improving the evaluation of engineering cost teaching, which has poor instructional quality. They provided a systematic and data-driven assessment for instructional quality through a proposed deep learning-based approach. AI-based technologies facilitate real-time feedback systems, allowing students to instantly understand and correct their mistakes. For example, Corbett and Anderson (1994) developed a knowledge tracing model within intelligent tutoring systems that could accompany students in their problem-solving process. These systems provide fast and personalized feedback to help students overcome difficulties and continue learning effectively. By making feedback more accessible, students gain more control of their learning and become more active participants. In this context, AI presents a promising solution to one of the key challenges in formative assessment—namely, the delay in delivering feedback—by ensuring that students receive timely support that can immediately influence their learning process.

Timely feedback is even more important in STEM education, where problem-solving, critical analysis, and creative thinking are key components. Immediate responses enable students to improve their methods and thinking in real time, which is very beneficial when they are involved in scientific investigation, mathematical reasoning, or engineering design. Thus, AI-based real-time feedback might help students stay engaged and better handle complex STEM tasks. In this way, AI not only improves assessment but also supports a more effective and personalized STEM learning experience.

Monitoring Student Progress

Monitoring student progress is a practice for teachers to systematically track learners' development over time to ensure their teaching efficacy and make better instructional decisions (Safer & Fleischman, 2005). It is essential to monitor progress effectively in STEM education, for teachers to understand students' learning progress, detect misconceptions early, to personalize support, and to meet instructional goals.

AI-based assessment systems can enhance monitoring student progress and provide predictive analysis based on students' academic behaviors and performance data. For instance, the Open Academic Analytics Initiative (OAAI) is an example of how students' data, such as demographics, aptitude, grades of assignments and exams, and students' interaction with the system, can be used for early warning systems in education (Jayaprakash et al., 2014). Aljuaid and Said (2021) proposed a method that uses deep learning to assess students' performance and predict their success in software engineering teamwork. AI enhances predictive assessment in education by analyzing patterns in students' past and present performance to anticipate potential learning challenges and identify at-risk students (Arnold & Pistilli, 2012; Jayaprakash et al., 2014). Using novel assessment methods (i.e., concept maps) integrated with machine learning can support the identification of students who have a risk of dropping out of courses (e.g., biology in Bertolini et al., 2021). These examples demonstrate how AI can be used in different functions of STEM educational assessment, such as academic performance, learning state, and instructional quality assessments (Ouyang et al., 2023).

AI-based educational assessments have positive effects on students by helping them overcome academic challenges, providing effective learning methods, identifying and supporting learners who are at risk, and enhancing the validity and efficacy of assessment techniques (Ouyang et al., 2023). A study by Arfah Baharudin and Lajis (2021) highlighted which competencies instructors need to assist students struggling in computer programming. A key challenge faced by higher education institutions is the difficulty of accurately assessing student competencies due to their continued reliance on manual methods. Through this research, both students and lecturers can detect learning difficulties at an early stage, improving students' chances of successfully passing the course. To facilitate learning, Lin and Chen (2020) presented a deep-learning-based recommendation system enhanced with augmented reality (AR) technology aimed at helping students from various academic backgrounds, including those not majoring in programming. Their findings showed that students using AR combined with AI-based suggestions significantly improved their learning outcomes and computational thinking abilities, particularly in creativity, logical reasoning, critical thinking, and problem-solving, when compared to the control group.

AI assessment can support teachers in monitoring student progress in real-time, assessing their performances across different areas, and revealing complex performances (Holstein et al., 2018; Shi & Choi, 2024). Through detailed reports prepared by AI on students' progress, teachers can decide on future instructional strategies and assessments (Shi & Choi, 2024). For example, AI-generated data-driven insights can reveal trends in students' learning behaviors, enabling educators to identify trends, skill deficiencies, and misconceptions that may not be visible through traditional approaches. These insights can guide pedagogical adjustments and curriculum improvements. In many technology-enhanced classrooms, learning analytics dashboards—such as those integrated into Moodle LMS—might be used to monitor student involvement and academic progress, providing real-time indicators that support personalized learning approaches. Furthermore, AI based assessments can evaluate past performance data to predict future academic results and identify students may need additional support before they struggle. For instance, Stadlman et al. (2022) predicted students' performance in synchronous online classes by using a predictive analytics system. These predictive abilities not only enhance early intervention systems but also contribute equitable learning environments by ensuring that no student's needs go unnoticed.

These AI-driven monitoring systems might support STEM assessment from a static and standardized process to a dynamic and individualized one. In STEM classrooms, AI technologies can analyze student learning patterns, monitor progress in real time, and provide immediate and personalized feedback. These features enable educators to detect learning difficulties early and improve instructional strategies to better support student success. For instance, deep learning methods have been used in computer programming courses to identify students' difficulties and suggest personalized interventions (Arfah Baharudin & Lajis, 2021). In biology education, predictive analytics using concept maps has helped identify students at risk of dropping out (Bertolini et al., 2021). Similarly, in software engineering courses, AI-supported assessment models have been developed to evaluate teamwork performance and predict student success (Aljuaid & Said, 2021). These examples demonstrate how AI enhances the quality of STEM assessments by providing data-informed insights, supporting early interventions, and fostering key competencies such as problem-solving, logical reasoning, and critical thinking. Ultimately, AI technologies serve not only as tools for automation but also as strategic assets in shaping responsive, evidence-based, and student-centered assessment practices in STEM education.

Challenges of AI in STEM Education Assessment

Despite many opportunities of using AI in STEM education assessment, some challenges and limitations are associated with its use in classroom settings. Using AI cautiously with an awareness of these limitations is crucial to ensure the accuracy of assessment practices. Otherwise, the opportunities offered by AI may fail to effectively serve the purposes teachers aim to achieve through assessment and may carry the risk of misguiding educational processes. In this regard, employing AI for the STEM education assessment may lead to some important issues related to fairness, ethics, inclusiveness, reliability, and validity (Akgün & Krajcik, 2024; Luckin et al., 2016; Panjwani-Charania & Zhai, 2024; Xing & Li, 2024).

Data Representativeness

When developing a test for a particular group, the sample of participants should represent the target population so that the test can be used for all members in the target group later. Similarly, the success of AI-based assessment practices depends on the representativeness and quality of the data used to train the AI models (Akgün & Krajcik, 2024). Accordingly, training data should be sampled in a way that reflects a mini version of the population, taking into account key variables, such as age, gender, and socioeconomic status. If a model developed for assessment in STEM education is trained on a dataset that represents only a specific group and fails to reflect others, the assessment results tend to be biased when the AI model is applied to underrepresented groups in the dataset (Holmes et al., 2019; Owan et al., 2023; Xing & Li, 2024). For instance, when considering the use of data in digital environments to train AI models, educational institutions in underdeveloped regions with limited digital resources may be underrepresented in these training datasets. In this circumstance, the use of AI in STEM education assessment in socio-economically disadvantaged schools may lead to unfair assessment results. For groups with limited representation in the training data, AI-generated items or tasks may lack cultural sensitivity. Thus, measurement non-invariance, which means the assessment does not have the same value for the same construct across different groups, may become an issue when assessing STEM education, potentially disfavoring the students in these kinds of schools (Gunn et al., 2019).

Data Privacy

Processing various types of student data is required to conduct STEM education assessment with AI. For example, private data such as students' item responses, project details, prior knowledge, emotions, personality traits, or other background information may be needed to employ AI for STEM education assessment when giving feedback to students or predicting student achievement. This may lead to concerns regarding the privacy of student data (Holmes et al., 2019; Luckin et al., 2016). AI platforms that process such student data for STEM education assessment should clearly disclose the purposes for which the data will be used, how long it will be stored, and with whom it may be shared. It is also essential for these AI platforms to provide solutions that ensure the security of such private information, which must be kept confidential. Otherwise, educational institutions may face serious legal and ethical issues (Holmes et al., 2019). The data ownership regarding the information uploaded to AI platforms is another critical issue (Holmes et al., 2019; Owan et al., 2023).

Alignment with Curriculum

One of the main purposes of assessment in education is to determine and monitor the extent to which students have attained the learning objectives in the curriculum. In order to use AI in STEM education assessment, the AI platform should present an assessment that aligns with the target learning objectives defined in the current STEM education curriculum. Otherwise, a misalignment may occur between the intended learning outcomes of STEM education and the constructs measured by AI-driven assessment systems. For example, these systems might emphasize memorizing facts rather critical thinking, which may conflict with the purpose of developing problem-solving skills in STEM education. Therefore, the feedback provided to students and teachers based on such assessment results may misguide the learning process. Accordingly, it is important to be aware that every AI model developed for STEM education assessment may not be applicable across all curricula.

Issues with Automated Scoring

In STEM education assessment, complex skills such as problem-solving and analytical thinking may need to be measured through performance tasks (Gao et al., 2020; Herro et al., 2017; Theobald et al., 2020). AI-based automated scoring is often more suitable for scoring short-answer responses. When measuring complex skills through performance tasks, AI may be limited and raise concerns regarding reliability and validity. AI-based automated scoring may struggle to assess the depth and creativity required for complicated problem-solving activities, which are essential in STEM education.

In addition, in STEM tasks where the process matters, such as projects or engineering design, AI-based automated scoring may be overly product-oriented and limited in its ability to score the process through the information in the intermediate steps (Holmes et al., 2019). Such limitations may lead to concerns regarding the reliability and accuracy of results obtained with AI in STEM education assessment. When training AI models for automated scoring, it is also essential to ensure both data representativeness and alignment with human grading. For each new item and performance task, the AI model should be retrained using representative data. Throughout the AI-based automated scoring process, the level of agreement between AI and human scoring should be continuously monitored. Failing to address these issues may jeopardize the reliability of scores obtained by AI-based automated scoring systems. The absence of human interaction in scoring STEM tasks through AI-based automated systems may lead to certain validity issues. Specifically, when a STEM task is scored only by AI, some critical aspects of student performance that can only be observed through human interaction may be unnoticed (Owan et al., 2023).

Transparency Challenges in AI-Powered Assessment

Test development requires adherence to established assessment standards that must be followed to ensure that the results are reliable, and the test serves its intended purpose. Such details are explicitly defined and shared when the test is designed by a teacher or a human expert. In addition, once the test is administered, a clear and detailed explanation is provided regarding how student responses are scored. However, due to the black-box nature of many AI models, it may be unclear how these models define and operationalize this assessment process. Moreover, teachers may not fully understand how these technologies operate or how the assessment results are generated (Owan et al., 2023). Such a lack of transparency may lead to concerns about the trustworthiness of the assessment results and raise questions regarding their acceptability among teachers and students. When students are not happy with the scores they receive or the feedback generated by AI, it should be possible to explain how and why those assessment results are produced. When teachers are not familiar with the underlying mechanisms of the AI model, they may find it challenging to evaluate the legitimacy of student grade appeals, which may lead to a lack of trust in AI-based assessment systems among students. Accordingly, it is important that the systems selected for AI-based assessment in STEM education are transparent, and the details of assessment process in these systems are explainable for accountability purposes (Luckin et al., 2016).

Role of Teachers

The integration of AI into assessment practices has resulted in a reconsideration and redefinition of the teachers' roles. In traditional classroom-based assessment practices, teachers are responsible for designing STEM-related tasks and scoring student performance. However, in AI-driven assessments, teachers have become primarily responsible for guiding the system and supervising the interpretation of the results (Luckin et al., 2016). As a result, teacher training on AI has become important. In order for teachers to use AI systems effectively and purposefully, they need to understand how these systems operate, how they are trained with data, and what algorithms drive their analyses. Accordingly, it is important for AI systems to be user-friendly and understandable for teachers so that teachers can easily integrate these systems into classroom assessment practices in STEM education (Pellegrino, 2024). Otherwise, teachers may face some obstacles when using AI systems in the STEM education assessment, due to being still a new tool for teachers. A lack of adequate AI training for teachers may lead to the emergence of an achievement gap in STEM education between schools (Chiu & Li, 2023). Teachers' preference for current conventional assessment practices and their resistance to change may lead to another challenge to the integration of AI in STEM education assessment. Additionally, developers of AI systems may sometimes overlook teachers' expectations of the system (Luckin & Cukurova, 2019). This may represent another challenge to the successful implementation of AI systems in schools. It is also important to avoid relying on only AI by ignoring the role of teachers when using AI for assessment practices (Luckin et al., 2016). Accordingly, rather than focusing just on how AI can be used in STEM education assessment, it is important to consider how teachers can use AI models for this purpose.

Conclusion

Assessment remains a cornerstone of effective STEM education, serving not only as a measure of learning outcomes but also as a vital mechanism for feedback, instructional planning, and student support. However, the interdisciplinary nature of STEM learning poses considerable challenges to traditional assessment methods. The incorporation of AI in the assessment of STEM education presents transformative opportunities for improving educational methodologies. AI technologies have the potential to improve the assessment of STEM education by providing many opportunities, such as designing authentic STEM problems, scoring student's work automatically, offering personalized and dynamic assessment, identifying the overall perspective and opinion expressed in a text, tracking student development, and offering immediate feedback. AI-based assessment systems that generate items or score students' work automatically have shown promising outcomes in facilitating educators' responsibilities while providing timely and tailored feedback, vital for student learning and engagement in STEM fields. Additionally, innovations such as Machine Learning (ML) and Natural Language Processing (NLP) have broadened the assessment scope to encompass higher-order cognitive skills, process-centered learning, and even emotional and behavioral insights through sentiment analysis. Each of these opportunities offers the potential to replace traditional, one-size-fits-all strategies with more adaptive, efficient, and student-centered models.

On the other hand, AI should be incorporated into the STEM assessment with caution, as AI comes with its own set of challenges. Concerns related to the data representativeness, data privacy, alignment with the curriculum, automated scoring, transparency, and the role of educators present significant challenges to the successful implementation of AI technologies in STEM education assessment, which may lead to some important issues related to fairness, ethics, inclusiveness, reliability, and validity when employing AI for the STEM education assessment. Accordingly, AI-based STEM education holds great potential but requires further research to ensure fairness, ethics, inclusiveness, reliability, and validity in the implementation of AI for STEM assessment in classrooms. To address these challenges, future work and regulations should consider the following directions:

1. Improving Data Quality and Representativeness: It is important to ensure that AI systems are trained on inclusive and diverse datasets that reflect a variety of demographics and educational situations. This may help reduce biases and foster equity in assessment outcomes. More research is needed to investigate the quality and representativeness of data used for training the models in the current AI based systems.
2. Reinforcing Data Privacy and Ethical Standards: Establishing comprehensive data management policies that secure personal information and adhere to global data protection regulations is required. Clear policies regarding data usage and storage must be developed to build confidence among educators, students, and parents when using AI for educational purposes.
3. Aligning Curriculum and Adaptive Assessment Models: AI-driven assessments should closely align with curriculum standards to maintain validity in STEM education assessment. Personalized adaptive learning systems can be designed to dynamically respond to students' learning journeys, and offer real-time support and evaluations that align with curriculum goals.
4. Human and AI Collaboration: Teachers have important roles in interpreting AI-generated assessments and supplying the human insights that machines cannot provide. Thus, training programs for enhancing teachers' knowledge and comfort with AI technologies are essential to bridge the gap between technology and classroom use. Incorporating human expert evaluations into AI-driven assess-

ment processes can help ensure the accuracy and reliability of feedback on students' STEM learning.

5. Ensuring Transparency and Explainability in AI Models: AI-driven assessment tools need to be transparent and explainable. Both educators and students need to have access to the descriptions of how assessment results are generated to build trust and enhance learning outcomes.
6. Encouraging Collaboration When Developing AI Tools: Collaboration among AI researchers, teachers, curriculum developers, and policymakers is essential for designing AI tools that are pedagogically effective and ethically suitable for classroom use.

In summary, the future of assessment in STEM education depends on the deliberate use of AI technologies that not only improve efficiency but also support fairness, transparency, and inclusivity. By addressing current challenges and considering the requirements for accurate assessment results, AI-driven assessment systems can contribute to more equitable and impactful STEM learning experiences, preparing students for the complex needs of today's world.

Disclaimer The authors gratefully acknowledge the assistance of OpenAI's ChatGPT in the preparation of this chapter, particularly for its assistance with preliminary literature exploration, structural organization of the content, and language-related tasks such as translation and semantic refinement.

References

Abrar, M., Aboraya, W., Khaliq, R. A., Subramanian, K. P., Al Husaini, Y., & Al Hussaini, M. (2025). AI-powered learning pathways: Personalized learning and dynamic assessments. *International Journal of Advanced Computer Science & Applications, 16*(1), 454–462.

Akarsu, M., Okur Akçay, N., & Elmas, R. (2020). STEM eğitimi yaklaşımının özellikleri ve değerlendirilmesi [Characteristics and evaluation of STEM education approach]. *Bogazici University Journal of Education, 37*, 155–175.

Akarsu, M., Okur Akçay, N., & Elmas, R. (2022). *STEM eğitimi yaklaşımı [STEM education approach]*. Pegem Akademi Yayıncılık.

Akgün, S., & Krajcik, J. (2024). Artificial intelligence (AI) as the growing actor in education: Raising critical consciousness towards power and ethics of AI in K–12 STEM classrooms. In X. Zhai & J. Krajcik (Eds.), *Uses of artificial intelligence in STEM education* (pp. 494–521). Oxford University Press.

Akhil, K., Kumar, M. K. S., & Vanathi, M. (2025). SMART ED: An AI-powered educational platform for personalized learning. *Journal Publication of International Research for Engineering and Management, 5*(3), 1–8.

Ali, Q. I. (2024). Towards more effective summative assessment in OBE: A new framework integrating direct measurements and technology. *Discover Education, 3*(1), 107. https://doi.org/10.1007/s44217-024-00208-5

Aljuaid, L. Z., & Said, M. Y. (2021). Deep learning-based method for prediction of software engineering project teamwork assessment in higher education. *Journal of Theoretical and Applied Information Technology, 99*(9), 2012–2030.

Amory, A. (2014). Tool-mediated authentic learning in an educational technology course: A designed-based innovation. *Interactive Learning Environments, 22*(4), 497–513.

Andrade, H., & Du, Y. (2005). Student perspectives on rubric-referenced assessment. *Practical Assessment, Research & Evaluation, 10*(3). https://doi.org/10.7275/g367-ye94

Arfah Baharudin, S., & Lajis, A. (2021). Deep learning approach for cognitive competency assessment in computer programming subject. *International Journal of Electrical and Computer Engineering Systems, 12*(Special Issue), 51.

Ariely, M., Nazaretsky, T., & Alexandron, G. (2023). Machine learning and Hebrew NLP for automated assessment of open-ended questions in biology. *International Journal of Artificial Intelligence in Education, 33*(1), 1–34. https://doi.org/10.1007/s40593-021-00283-x

Arnold, K. E., & Pistilli, M. D. (2012, April). Course signals at Purdue: Using learning analytics to increase student success. In *Proceedings of the 2nd international conference on learning analytics and knowledge* (pp. 267–270).

Bertolini, R., Finch, S. J., & Nehm, R. H. (2021). Testing the impact of novel assessment sources and machine learning methods on predictive outcome modeling in undergraduate biology. *Journal of Science Education and Technology, 30*, 193–209.

Bertrand, M. G., & Namukasa, I. K. (2020). STEAM education: Student learning and transferable skills. *Journal of Research in Innovative Teaching & Learning, 13*(1), 43–56.

Brookhart, S. M. (2013). *How to create and use rubrics for formative assessment and grading*. ASCD.

Chan, K. W., Ali, F., Park, J., Sham, K. S. B., Tan, E. Y. T., Chong, F. W. C., Qian, K., & Sze, G. K. (2025). Automatic item generation in various STEM subjects using large language model prompting. *Computers and Education: Artificial Intelligence, 8*, 100344.

Chen, L., Chen, P., & Lin, Z. (2020). Artificial intelligence in education: A review. *IEEE Access, 8*, 75264–75278. https://doi.org/10.1109/ACCESS.2020.2988510

Chiu, T. K. F., & Li, Y. (2023). How can emerging technologies impact STEM education? *Journal for STEM Education Research, 6*(1), 375–384. https://doi.org/10.1007/s41979-023-00113-w

Chng, E., Tan, A. L., & Tan, S. C. (2023). Examining the use of emerging technologies in schools: A review of artificial intelligence and immersive technologies in STEM education. *Journal for STEM Education Research, 6*(3), 385–407.

Circi, R., Hicks, J., & Sikali, E. (2023). Automatic item generation: Foundations and machine learning-based approaches for assessments. *Frontiers in Education, 8*, 858273.

Corbett, A. T., & Anderson, J. R. (1994). Knowledge tracing: Modeling the acquisition of procedural knowledge. *User Modeling and User-Adapted Interaction, 4*(4), 253–278. https://doi.org/10.1007/BF01099821

Debuse, J. C., & Lawley, M. (2016). Benefits and drawbacks of computer-based assessment and feedback systems: Student and educator perspectives. *British Journal of Educational Technology, 47*(2), 294–301. https://doi.org/10.1111/bjet.12232

Egger, M., Ley, M., & Hanke, S. (2019). Emotion recognition from physiological signal analysis: A review. *Electronic Notes in Theoretical Computer Science, 343*, 35–55.

Fang, S. C., & Hsu, Y. S. (2019). Assessment challenges in STEM reforms and innovations. In Y. S. Hsu & Y. F. Yeh (Eds.), *Asia-Pacific STEM teaching practices: From theoretical frameworks to practices* (pp. 191–203). Springer.

Foltz, P. W., Yan, D., & Rupp, A. A. (2020). The past, present, and future of automated scoring. In D. Yan, A. A. Rupp, & P. W. Foltz (Eds.), *Handbook of automated scoring: Theory into practice* (pp. 1–10). Chapman and Hall/CRC.

Gao, X., Li, P., Shen, J., & Sun, H. (2020). Reviewing assessment of student learning in interdisciplinary STEM education. *International Journal of STEM Education, 7*, 1–14. https://doi.org/10.1186/s40594-020-00225-4

Gerard, L. F., & Linn, M. C. (2016). Using automated scores of student essays to support teacher guidance in classroom inquiry. *Journal of Science Teacher Education, 27*(1), 111–129.

Gunn, H. J., Grimm, K. J., & Edwards, M. C. (2019). Evaluation of six effect size measures of measurement non-invariance for continuous outcomes. *Structural Equation Modeling: A Multidisciplinary Journal, 27*(4), 503–514. https://doi.org/10.1080/10705511.2019.1689507

He, X., & Fu, S. (2022). [Retracted] Data analysis and processing application of deep learning in engineering cost teaching evaluation. *Journal of Mathematics, 2022*(1), 8944570. https://doi.org/10.1155/2022/8944570

Herro, D., Quigley, C., Andrews, J., & Delacruz, G. (2017). Co-measure: Developing an assessment for student collaboration in STEAM activities. *International Journal of STEM Education, 4*(1), 26–37. https://doi.org/10.1186/s40594-017-0094-z

Holmes, W., Bialik, M., & Fadel, C. (2019). *Artificial intelligence in education: Promises and implications for teaching and learning*. Center for Curriculum Redesign.

Holstein, K., McLaren, B. M., & Aleven, V. (2018). Student learning benefits of a mixed-reality teacher awareness tool in AI-enhanced classrooms. In *Artificial intelligence in education: 19th international conference, AIED 2018, London, UK, June 27–30, 2018, proceedings, Part I 19* (pp. 154–168). Springer International Publishing.

Islam, I., & Islam, M. N. (2024). Exploring the opportunities and challenges of ChatGPT in academia. *Discover Education, 3*(1), 31.

Jantakun, T., Jantakun, K., & Jantakoon, T. (2025). Bibliometric analysis of artificial intelligence in STEM education. *Higher Education Studies, 15*(1), 69–81. https://doi.org/10.5539/hes.v15n1p69

Jayaprakash, S. M., Moody, E. W., Lauría, E. J., Regan, J. R., & Baron, J. D. (2014). Early alert of academically at-risk students: An open-source analytics initiative. *Journal of Learning Analytics, 1*(1), 6–47. https://doi.org/10.18608/jla.2014.11.3

Jiao, P., Ouyang, F., Zhang, Q., & Alavi, A. H. (2022). Artificial intelligence-enabled prediction model of student academic performance in online engineering education. *Artificial Intelligence Review, 55*(8), 6321–6344. https://doi.org/10.1007/s10462-022-10155-y

Joseph, O. B., & Uzondu, N. C. (2024). Integrating AI and machine learning in STEM education: Challenges and opportunities. *Computer Science & IT Research Journal, 5*(8), 1732–1750. https://doi.org/10.51594/csitrj.v5i8.1379

Jung, J. Y., Tyack, L., & von Davier, M. (2024). Combining machine translation and automated scoring in international large-scale assessments. *Large-Scale Assessments in Education, 12*(1), 1–18.

Latif, E., & Zhai, X. (2024). Fine-tuning ChatGPT for automatic scoring. *Computers and Education: Artificial Intelligence, 6*, 1–10.

Lin, P. H., & Chen, S. Y. (2020). Design and evaluation of a deep learning recommendation based augmented reality system for teaching programming and computational thinking. *IEEE Access, 8*, 45689–45699.

Luckin, R., & Cukurova, M. (2019). Designing educational technologies in the age of AI: A learning sciences-driven approach. *British Journal of Educational Technology, 50*(6), 2824–2838. https://doi.org/10.1111/bjet.12861

Luckin, R., Holmes, W., Griffiths, M., & Forcier, L. B. (2016). *Intelligence unleashed: An argument for AI in education*. Pearson.

Martín-Páez, T., Aguilera, D., Perales-Palacios, F. J., & Vílchez-González, J. M. (2019). What are we talking about when we talk about STEM education? A review of literature. *Science Education, 103*(4), 799–822. https://doi.org/10.1002/sce.21522

Morris, W., Holmes, L., Choi, J. S., & Crossley, S. (2024). Automated scoring of constructed response items in math assessment using large language models. *International Journal of Artificial Intelligence in Education*. Advance online publication.

Moskal, B. M., & Leydens, J. A. (2000). Scoring rubric development: Validity and reliability. *Practical Assessment, Research & Evaluation, 7*(10). https://doi.org/10.7275/q7rm-gg74

Ouyang, F., Dinh, T. A., & Xu, W. (2023). A systematic review of AI-driven educational assessment in STEM education. *Journal for STEM Education Research, 6*(3), 408–426. https://doi.org/10.1007/s41979-023-00112-x

Owan, V. J., Abang, K. B., Idika, D. O., Etta, E. O., & Bassey, B. A. (2023). Exploring the potential of artificial intelligence tools in educational measurement and assessment. *EURASIA Journal of Mathematics, Science and Technology Education, 19*(8), 1–15. https://doi.org/10.29333/ejmste/13428

Page, E. B. (1966). The imminence of grading essays by computer. *The Phi Delta Kappan, 47*(5), 238–243.

Panadero, E., & Jonsson, A. (2013). The use of scoring rubrics for formative assessment purposes revisited: A review. *Educational Research Review, 9*, 129–144. https://doi.org/10.1016/j.edurev.2013.01.002

Panjwani-Charania, S., & Zhai, X. (2024). AI for students with learning disabilities: A systematic review. In X. Zhai & J. Krajcik (Eds.), *Uses of artificial intelligence in STEM education* (pp. 469–493). Oxford University Press.

Pellegrino, J. W. (2024). A new era for STEM assessment: Considerations of assessment, technology, and artificial intelligence. In X. Zhai & J. Krajcik (Eds.), *Uses of artificial intelligence in STEM education* (pp. 17–37). Oxford University Press.

Safer, N., & Fleischman, S. (2005). Research matters: How student progress monitoring improves instruction. *Educational Leadership, 62*(5), 81–83.

Saito, T., & Watanobe, Y. (2020). Learning path recommendation system for programming education based on neural networks. *International Journal of Distance Education Technologies (IJDET), 18*(1), 36–64.

Saputra, I., Kurniawan, A., Yanita, M., Putri, E. Y., & Mahniza, M. (2024). The evolution of educational assessment: How artificial intelligence is shaping the trends and future of learning evaluation. *The Indonesian Journal of Computer Science, 13*(6), 9056–9074. https://doi.org/10.33022/ijcs.v13i6.4465

Sharma, P., & Dash, B. (2023). AI and VR enabled modern LMS for students with special needs. *Journal of Foreign Language Education and Technology, 8*(1), 1–14.

Shi, L., & Choi, I. (2024). Application types, pedagogical roles, and technological characteristics. In *Uses of Artificial Intelligence in STEM Education* (p. 321). https://doi.org/10.1093/oso/9780198882077.003.0015

Stadlman, M., Salili, S. M., Borgaonkar, A. D., & Miri, A. K. (2022). Artificial intelligence based model for prediction of students' performance: A case study of synchronous online courses during the COVID-19 pandemic. *Journal of STEM Education: Innovations and Research, 23*(2), 39–46.

Suhirman, S., & Prayogi, S. (2023). Overcoming challenges in STEM education: A literature review that leads to effective pedagogy in STEM learning. *Jurnal Penelitian Pendidikan IPA, 9*(8), 432–443. https://doi.org/10.29303/jppipa.v9i8.4715

Tempelaar, D. T., Rienties, B., & Giesbers, B. (2015). In search for the most informative data for feedback generation: Learning analytics in a data-rich context. *Computers in Human Behavior, 47*(1), 157–167. https://doi.org/10.1016/j.chb.2014.05.038

Theobald, E. J., Hill, M. J., Tran, E., Agrawal, S., Arroyo, E. N., Behling, S., Chambwe, N., Cintrón, D. L., Cooper, J. D., Dunster, G., Grummer, J. A., Hennessey, K., Hsiao, J., Iranon, N., Jones, L., II, Jordt, H., Keller, M., Lacey, M. E., Littlefield, C. E., Lowe, A., Newman, S., Okolo, V., Olroyd, S., Peecook, B. R., Pickett, S. B., Slager, D. L., Caviedes-Solis, I. W., Stanchak, K. E., Sundaravardan, V., Valdebenito, C., Williams, C. R., Zinsli, K., & Freeman, S. (2020). Active learning narrows achievement gaps for underrepresented students in undergraduate science, technology, engineering, and math. *Proceedings of the National Academy of Sciences, 117*(12), 6476–6483. https://doi.org/10.1073/pnas.1916903117

Tseng, C. H., Lin, H. C. K., Huang, A. C. W., & Lin, J. R. (2023). Personalized programming education: Using machine learning to boost learning performance based on students' personality traits. *Cogent Education, 10*(2), 2245637. https://doi.org/10.1080/2331186X.2023.2245637

Tzuriel, D. (2001). *Dynamic assessment of young children*. Kluwer Academic/Plenum Publishers.

Wang, J., Hussain, Y., & Mao, C. (2025). Artificial intelligence-driven personalized learning: Psychological implications and educational outcomes. *International Journal of Education, Humanities and Social Science, 2*(1), 1–16.

Xing, W., & Li, C. (2024). Fair artificial intelligence to support STEM education. In X. Zhai & J. Krajcik (Eds.), *Uses of artificial intelligence in STEM education* (pp. 522–546). Oxford University Press.

Xu, W., & Ouyang, F. (2022). The application of AI technologies in STEM education: A systematic review from 2011 to 2021. *International Journal of STEM Education, 9*(1), 59. https://doi.org/10.1186/s40594-022-00377-5

Yang, S., Dong, Y., & Yu, Z. G. (2024). ChatGPT in education: Ethical considerations and sentiment analysis. *International Journal of Information and Communication Technology Education, 20*(1), 1–19.

Zengin, N., Kaya, G., & Pektaş, M. (2020). Investigation of measurement and assessment methods used in STEM-based research. *Gazi University Journal of Gazi Educational Faculty, 40*(2), 329–355.
Zhai, X., & Krajcik, J. (2024). Artificial intelligence-based STEM education. In X. Zhai & J. Krajcik (Eds.), *Uses of artificial intelligence in STEM education* (pp. 3–14). Oxford University Press.
Zhai, X., & Wiebe, E. (2023). Technology-based innovative assessment. In C. J. Harris, E. Wiebe, S. Grover, & J. W. Pellegrino (Eds.), *Classroom-based STEM assessment: Contemporary issued and perspectives* (pp. 99–125). Springer.
Zhai, X., Yin, Y., Pellegrino, J. W., Haudek, K. C., & Shi, L. (2020). Applying machine learning in science assessment: A systematic review. *Studies in Science Education, 56*(1), 111–151. https://doi.org/10.1080/03057267.2020.1735757
Zhai, X., Neumann, K., & Krajcik, J. (2023). Editorial: AI for tackling STEM education challenges. *Frontiers in Education, 8*, 1–3. https://doi.org/10.3389/feduc.2023.1183030
Zhang, L., Qu, J., & Zhang, Z. (2024). The impact of personalized learning driven by artificial intelligence on stem education in primary and secondary schools. *International Journal of New Developments in Education, 6*(10), 29–34.

CHAPTER 8

Design Decision-Making in the Age of Artificial Intelligence

Tuğba Yüksel and Şenay Purzer

Introduction

Design has served as an integrative element of STEM education pedagogies (Purzer & Quintana-Cifuentes, 2019). The teaching approaches that build on design practices promote student engagement, creativity, and decision-making skills. According to Atman et al. (2007), the seemingly linear design processes encountered in the early stages of engineering education lay the groundwork for the more complex and nuanced decision-making that characterizes expert practices.

Through the integration of generative AI in educational settings, students are now engaging in co-design processes, where artificial intelligence (AI) assists in idea generation, analysis, or optimization. These AI-supported design activities lessen the cognitive load needed for ideation and bring opportunities for ethical and societal dimensions to the

T. Yüksel (✉)
Recep Tayyip Erdoğan University, Çayeli/Rize, Türkiye
e-mail: tugba.yuksel@erdogan.edu.tr

Ş. Purzer
Purdue University, West Lafayette, IN, USA
e-mail: purzer@purdue.edu

R. Elmas, M. Akarsu (eds.), *STEM Education*, Maarif Global Education Series,
https://doi.org/10.1007/978-981-95-9023-0_8

forefront, prompting even young learners to ask critical questions such as: Who benefits from this design? Who might be excluded or harmed? Engineering design thus becomes a powerful context not only for developing technical and creative skills but also for nurturing ethical awareness and responsible innovation from an early age. For this reason, the chapter situates engineering design as a focal point for understanding how decision-making is developed in AI-mediated learning environments.

Design is not only about generating solutions—it is fundamentally about making informed decisions that satisfy multiple and competing criteria (Fosmire, 2017). Developing decision-making competence among students requires a multifaceted approach—one that integrates cognitive strategies, authentic experiences, reflective practice, and inclusive methods. A human-centered integration of AI into education is essential for preparing students to make ethically responsible, technically sound, and socially informed decisions in increasingly complex design environments.

With the wide and mainstream use of AI tools, new perspectives are needed for re-defining design expertise, judgment, and practices. This chapter aims to inform engineering design education at the K–12 level because design decision-making, especially in the age of artificial intelligence, offers a unique lens through which to explore opportunities. While STEM education broadly emphasizes the integration of science, technology, engineering, and mathematics, engineering disciplines, it is important to focus on decision-making practices to help guide students compare different pieces of information, evaluate trade-offs, and examine the broader impacts of their solutions.

Pedagogies That Support Design Decisions

Scholars emphasize the importance of decision-making in equipping future generations to navigate intricate technical and social challenges (Householder & Hailey, 2012; Jonassen et al., 2006; Van den Beemt et al., 2020). Decision-making in this context involves evaluating, selecting, and implementing solutions that align with specific goals, constraints, and user needs. It requires a delicate balancing of trade-offs such as functionality, cost, aesthetics, sustainability, usability, and accessibility (Alexander, 1982; Beitz et al., 1996; Spirochkin, 2023).

To support students in developing competencies in design, researchers emphasize the value of scaffolding, authentic design experiences, and reflective practice. For example, Mentzer et al. (2015) argue that embedding real-world design challenges in curricula allows students to confront ambiguity and conflicting goals, thereby strengthening students' engineering judgment. Similarly, Dym et al. (2005) highlight the pedagogical benefits of early exposure to iterative design thinking, where learners engage in continuous evaluation and refinement. According to Dym and colleagues project-based learning and hands-on experiences contribute to the cultivation of foundational design skills among engineering students.

Educators can further enhance students' decision-making abilities by introducing structured frameworks such as decision matrices and multi-criteria analysis tools, which help make complex evaluations more transparent and deliberate (Crismond & Adams, 2012). Reflection is also essential; when students are encouraged to articulate the rationale behind their design decisions, they develop metacognitive awareness and improved capacity for informed judgment (Adams et al., 2003). Wendell et al. (2017) demonstrate that reflective decision-making strategies can significantly enhance the quality of students' design processes by helping them consider diverse perspectives and understand trade-offs more deeply. These practices are particularly effective when integrated into a broader culture of inquiry and collaboration, while fostering a deeper appreciation for the ethical and societal consequences of technical decisions.

Decision-making through design promotes both personal agency and social empowerment. As Cech (2014) emphasizes, inclusive learning environments further encourage students to engage with the social dimensions of engineering. Moreover, considering cultural perspectives allows educators to acknowledge and leverage the diverse experiences students bring into the classroom. Samuelson and Litzler (2016) argue that valuing students' unique cultural assets can empower underrepresented groups, fostering persistence and encouraging advocacy for systemic improvements. In this way, decision-making education becomes a vehicle for both technical and social empowerment. Instructional strategies such as simulation-based learning further enrich this process. Feijoo-Garcia et al. (2024) illustrate how simulations incorporating argumentation provide dynamic, real-time opportunities for students to integrate scientific reasoning with practical considerations, thereby refining their ability to assess design trade-offs collaboratively. Finally, recognizing that students approach decision-making through various lenses is vital. While Dringenberg et al. (2021) found that engineering students often default to rationalistic approaches, they underscored the need for educational interventions that also cultivate intuitive and empathetic modes of reasoning. These prior studies collectively highlight that developing decision-making competence among students is a multifaceted task that requires experiential learning and opportunities for reflective practice. As advanced technological elements become more embedded in design processes, there is a growing need for targeted strategies that support ethical judgment and social awareness, along with technical acumen, in navigating increasingly complex design environments.

Evolution of Decision Methods in Design

The historical evolution of design education reflects a transition from traditional technical approaches to more dynamic and context-sensitive methodologies. Central to this transformation is the incorporation of fuzzy decision-making models and simulation-based experimentation and learning environments. Fuzzy logic provides a structured, yet flexible, means of addressing the uncertainties inherent in complex engineering problems, enabling more refined and adaptive decisions (Arastoopour Irgens et al., 2017; Surya et al., 2016). Simulation-based platforms, especially when augmented by artificial intelligence (AI), offer students with interactive, real-world learning experiences that build their critical thinking, problem-solving, and decision-making competencies (Chaturvedi et al., 2023).

This evolution also reflects a growing recognition of the need for holistic decision-making in engineering—one that combines technical proficiency with ethical awareness and societal context. Educational paradigms have increasingly prioritized interdisciplinary learning, prompting the integration of ethical considerations, environmental sustainability, and cultural dimensions into engineering solutions (Cech, 2014; Van de Poel & Royakkers, 2023). AI technologies can support this shift by providing tools that analyze complex systems, integrate diverse data sources, and enable inclusive and equitable design outcomes (Shao et al., 2025).

However, the extent to which students engage with this approach in practice remains underexplored. This limited engagement can be attributed to several factors, including a general lack of familiarity among educators, the abstract and mathematically nuanced nature of fuzzy logic, and its minimal integration into traditional engineering curricula (Dixit et al., 2025). Furthermore, the scarcity of accessible educational resources and real-world case studies hampers students' ability to connect fuzzy decision-making with practical engineering problems (Aliev et al., 2012; Dixit et al., 2025). Assessment challenges also persist, as evaluating student performance in fuzzy logic-based tasks requires context-sensitive judgments rather than standardized answers, making it difficult to implement broadly in instructional settings (Dixit et al., 2025; Zadeh, 1996).

Furthermore, effective design decision-making requires not only technical proficiency but also the integration of ethical and human-centered considerations. As Bras and Mistree (1991) and Olewnik and Svihla (2025) argue, structured decision processes can foster creativity, minimize resource waste, and support the inclusion of diverse perspectives. Ethical dimensions—such as responsibility, foresight, and harm mitigation—are equally vital to ensure that technological innovations serve the public good (Corple et al., 2020; Van de Poel & Royakkers, 2023).

In design education, the convergence of technology and human-centered perspectives has long been recognized. Building on decades of research, AI has evolved in a mainstream role as a transformative tool in the design space, powering the design process through natural language processing, image recognition, and optimization algorithms (Biondi et al., 2023; Liao et al., 2023; Osasona et al., 2024). Generative AI tools act as co-creators in the design process, offering designers with a broad spectrum of creative design alternatives and design analysis options (Chandrasekera et al., 2025; Figoli et al., 2022). Effective design decisions demand a careful balancing of technical knowledge, critical thinking, and ethical judgment (Dym et al., 2005; Purzer & Arigye, 2024). In other words, design is a cognitively loaded process requiring creative, technical, and ethical judgment. Human–machine collaboration through AI can reduce some of this load enabling the designer to more intentionally account for design implications in terms of equity, sustainability, and long-term societal benefits and burdens (Becker & Mentzer, 2015; Olewnik & Svihla, 2025).

AI and Education

The integration of AI into education has been predominantly through the large language modeling tools such as ChatGPT. As AI becomes a mainstream tool in instructional delivery, its integration into education demands careful consideration of its transformative potential as well as risks. AI's ability to collaborate with the learner is well-documented but often viewed as a challenge to academic integrity. More intentional uses of AI offer promising opportunities to personalize learning, automate routine tasks, and enhance student engagement through real-time feedback. For example, a recent study in Villegas-Ch et al. (2025) implemented deep-learning–based intelligent tutoring in STEM subjects, including engineering-adjacent content, for students. The system offered real-time personalized guidance and feedback, resulting in measurable improvements in concept mastery (e.g., 85% in programming). The authors emphasize that infrastructure and student digital literacy play crucial roles in effectively deploying such systems. Despite the advantages of AI in engineering education, studies have shown that practical implementation remains uneven, often hindered by instructor preparedness, curriculum development barriers, and institutional inertia (Mah & Groß, 2024; Ren & Wu, 2025).

The AI implementation challenges of higher education are mirrored in K–12 STEM education. According to Khanolkar et al. (2021), while the potential for AI to transform engineering design education is widely recognized, most curricula have yet to fully embrace AI. This gap is attributed to the lack of curricular frameworks and absence of instructor training. The efforts and discourse are heavily focused on challenges. For example, the integration of AI raises concerns about diminishing students' foundational problem-solving and critical thinking skills. As AI systems begin to provide optimized solutions or guide decision-making processes, students may become overly reliant on these tools, thereby bypassing the deeper engagement traditionally fostered through iterative, hands-on design experiences (C. Zhai et al., 2024).

Compounding these pedagogical concerns is the opacity of many AI systems. The "black box" nature of AI decision-making can inhibit students' ability to critically evaluate design choices or understand the rationale behind specific outputs. This undermines the reflective practices central to engineering education and challenges educators to not only teach how to use AI tools but also to promote algorithmic transparency, bias detection, and ethical reasoning (Slimi & Villarejo-Carballido, 2023). Equity and access further complicate the landscape of AI integration. Schools in under-resourced or historically marginalized communities often lack the infrastructure, teacher training, and technological access to implement AI effectively. Without deliberate strategies to close these gaps, there is a real risk of exacerbating existing inequalities in STEM education (Holstein & Doroudi, 2021). Quinde et al. (2024) highlight that student engagement with AI tools varies widely, pointing to disparities in exposure that could affect readiness for future careers in a tech-driven economy.

Nonetheless, AI's potential to revolutionize engineering education remains significant if integrated thoughtfully and ethically. Miao and Shiohira (2024) emphasize the need for educational systems to embed AI literacy within curricula, equipping students with the technical and ethical competencies necessary to interact meaningfully with AI systems. The *UNESCO's* guidance for generative AI in education argues for a human-centered approach. The *AI Competency Framework* of UNESCO further elaborates on the importance of protecting human agency when using AI tools (Miao & Holmes, 2023), with guidelines on skill acquisition and the cultivation of critical awareness about AI's role in society. These publications also discuss the beneficial uses of AI as a coaching tool for project-based learning and as a Socratic challenger, along with risks due to learner's limited abilities to evaluate accuracy due to limited prior knowledge and AI-infused echo-chambers preventing diverse viewpoints.

Decision-making involves continuously assessing different options tied to a specific objective, and ultimately selecting the one that best aligns with the expected results of a chosen action (Harrison, 1999). The decision-making process in engineering design has undergone a profound transformation over the decades, evolving from deterministic, linear models to more nuanced, iterative, and adaptive methodologies (Parra et al., 2023). Traditionally, engineering design was shaped by prescriptive approaches grounded in mathematical optimization and rule-based logic, where solutions were derived from fixed parameters and objective criteria. This technical rationality, long embedded in engineering education and practice, often overlooked the complexity, ambiguity, and contextual variability inherent in real-world challenges (Dym et al., 2005). Early curricula primarily emphasized technical proficiency, treating design as a sequence of logical steps while neglecting creative, social, and iterative dimensions (Guzey et al., 2014).

A significant shift began in the 1970s with the introduction of fuzzy logic, which allowed for more flexible reasoning in the face of uncertainty and imprecision (Zadeh, 1975). This marked a pivotal move toward modeling human-like decision-making, laying the foundation for more human-centric design thinking. As societal needs and expectations evolved, so did the understanding of engineering design—prompting a growing emphasis on incorporating diverse inputs, social implications, and user-centered perspectives into the design process (Cech, 2014; Murray et al., 2019).

Researchers and educators increasingly recognize that effective engineering solutions must reflect ethical considerations, cultural awareness, and stakeholder perspectives (Van de Poel & Royakkers, 2023; Walther et al., 2017).

In the late twentieth and early twenty-first centuries, there was a rise in simulation-based learning environments and constructivist pedagogies, which enabled students and professionals to engage with dynamic, real-world systems and explore multiple design paths (Arastoopour Irgens et al., 2017). These educational innovations were bolstered by computational tools that supported iterative prototyping, collaborative learning, and scenario-based problem-solving (Dym et al., 2005).

Building on this transformation, the emergence of AI and machine learning has introduced new paradigms in design decision-making. AI technologies, such as predictive analytics, generative models, and real-time data processing, have begun to complement traditional methodologies by enhancing designers' ability to analyze complex trade-offs and adapt to evolving requirements (Figoli et al., 2022). These tools increase efficiency and enable engagement with multifaceted problems, promoting a shift toward context-aware and ethically informed practices.

Research on AI and Design Education

The growing presence of artificial intelligence (AI) in everyday life is beginning to ripple through the foundations of engineering education, particularly in K–12 learning environments. What was once the domain of experts and high-level research is now entering classrooms, bringing promise and complexity. As AI tools—such as generative design platforms, machine learning algorithms, and natural language processing—become more accessible, they reshape how students learn to approach engineering challenges. These technologies are not merely tools for automation; they are becoming co-participants in the learning process, helping students move beyond rote tasks to engage in strategic thinking, conceptual ideation, and ethical deliberation.

One of the most transformative aspects of AI in education is its capacity to shift the cognitive load. For example, in a study with eight grade students, Zhao et al. (2025) investigated how generative AI helped reduce extraneous cognitive load, allowing students to focus more on germane (deep-learning) processing. In the cognitive tool group, generative AI (GAI) automated low-level tasks such as material generation, allowing students to redirect their mental resources toward deeper understanding. In the thinking tool group, students critically analyzed and revised AI-generated content, fostering higher-order thinking. This resulted in better in-depth learning performance when students engaged critically with the AI-generated content. In another study, Edwards et al. (2024) conducted a quasi-experimental study with 120 middle school students to evaluate the effect of their AI chatbot, iLearnTech, on cognitive load and learning outcomes. They found that students in the chatbot group experienced substantial reduction in cognitive load, especially extraneous load, compared to the control group. This reduction was accompanied by significantly higher achievement scores, improved attitudes toward learning, and better retention rates. By automating repetitive elements such as drafting, data processing, or design evaluation, AI enables students to redirect their mental energy toward higher-order problem-solving and critical thinking (Luckin et al., 2016). This pedagogical shift aligns with the vision articulated by Clive Dym, a pioneer in integrating artificial intelligence into engineering design. Dym's engagement with AI began in the mid-1980s through his work with expert systems—early knowledge-based tools that laid the foundation for modern AI applications in engineering. He consistently emphasized the importance of design thinking and reflective decision-making as core components of engineering education (Dym et al., 2005). In this evolving context, AI-powered tools do not replace human cognition but instead restructure it, guiding learners toward deeper, more meaningful engagement with complex, real-world engineering challenges.

The integration of AI also supports more authentic, dynamic learning experiences. Tools like intelligent tutoring systems and adaptive simulation platforms create environments where students can make decisions, receive immediate feedback, and iterate in response to changing variables (Arastoopour Irgens et al., 2017; Zhai et al., 2021). In a study, Tseng and Yadav (2023) developed ActiveAI, an interactive platform designed to help middle school students build AI literacy through authentic, goal-based scenarios. The system allowed students to experiment with real-world AI problems, receive immediate feedback, and iteratively adjust their decisions. Their findings showed that ActiveAI increased students' understanding of AI concepts, boosted engagement, and promoted reflective, adaptive learning by simulating authentic problem-solving experiences. This type of iterative learning is crucial in developing judgment and resilience. In another study, Holstein and Aleven (2022) explored how AI can effectively complement human teachers in K–12 classrooms by designing tools that provide real-time, actionable insights to support instructional decisions. Their study focused on the Lumilo smart glasses system, which delivers adaptive feedback to teachers based on student data, enabling dynamic scaffolding tailored to individual learners. They found that this human–AI partnership enhanced teachers' ability to respond promptly to students' needs, fostering more personalized and engaging learning experiences. As the studies indicate, these tools can personalize the learning journey, adapting content difficulty and pacing to individual needs—a pedagogical strength that supports more equitable learning outcomes, especially for students from underrepresented or underserved backgrounds (Popenici & Kerr, 2017).

Just as important as personalization is collaboration. Engineering is rarely a solitary endeavor, and the educational use of AI increasingly mirrors the collaborative nature of professional practice. AI-supported platforms now enable students to engage in joint problem-solving, co-design, and data-sharing tasks across disciplines and locations. These digital ecosystems simulate the interdisciplinary nature of real-world engineering projects, where designers, engineers, stakeholders, and even end-users must communicate and negotiate design choices. As Khanolkar et al. (2023) suggest, AI-enhanced collaboration tools not only increase design performance but also cultivate teamwork skills and creative fluency—competencies often under-emphasized in traditional engineering curricula. As artificial intelligence (AI) becomes increasingly integrated into educational settings, several promising initiatives have emerged that explore its role in enhancing collaboration, motivation, and conceptual understanding among K–12 learners. For instance, the FLECKS[1] project (Fostering Learning and Collaboration with Knowledgeable and Supportive Agents) was developed to examine how intelligent virtual agents can support collaborative learning during pair programming tasks in computer science education. By embedding AI companions that prompt discussion, offer strategic feedback, and model cooperative behaviors, FLECKS has demonstrated meaningful impacts on students' engagement, problem-solving abilities, and conceptual grasp. Building on this foundation, Earle-Randell et al. (2024) investigated the influence of *near peer virtual learning companions*—AI agents designed to resemble slightly older student peers—on elementary-aged students working within a block-based coding

[1] For detailed information about the project: http://flecksproject.org/

environment. Their study involved fifteen student pairs who either collaborated with these AI agents or worked without them in a control condition. The findings revealed that students partnered with AI agents asked significantly more high-level questions and reported more positive attitudes toward computer science. Qualitative feedback further indicated that children found the agents helpful in improving peer interaction and expressed enthusiasm about continuing to work with them. These results highlight the potential of AI agents to foster collaborative, inquiry-driven learning environments that could extend to interdisciplinary STEM domains, including engineering.

Similarly, in a three-year pilot study, Lišnić et al. (2025) explored how AI tools could be integrated into K–12 programming education through an extracurricular module titled RIWA—"Development of Intelligent Web Applications." Conducted across twelve schools in Croatia's Split-Dalmatia region, the program combined theoretical instruction with hands-on, project-based learning, in which students developed AI-enhanced web applications using JavaScript and the ml5.js machine learning library. Over the course of the study, 112 student-led projects were completed. Analysis of these projects, along with student interviews, revealed that the AI-supported approach not only sustained high levels of student motivation, but also promoted meaningful development in programming skills—without requiring additional instructional time. Students demonstrated proficient use of computational concepts such as variables, loops, and functions, and reported increased confidence in their design choices. Notably, they also began to grapple with essential principles of machine learning, including dataset diversity and model accuracy. Together, these studies underscore the growing potential of AI-powered platforms to support joint problem-solving, co-design, and cross-disciplinary collaboration in K–12 STEM education.

However, the emergence of AI also demands a broader understanding of what it means to be "literate" in engineering today. Technical proficiency is no longer sufficient. Students must be equipped with the capacity to interrogate the design and decision-making processes embedded in AI tools. This includes understanding algorithmic bias, limitations of data-driven models, and the ethical implications of AI-generated outcomes (Jobin et al., 2019; Morley et al., 2020). AI literacy, then, must include a critical dimension: teaching students not just how to use AI, but when, why, and under what conditions its use is appropriate (Walter, 2024).

Educators, too, face a significant learning curve. While AI can assist teachers through automated grading and student progress tracking (Zawacki-Richter et al., 2019), it also introduces new responsibilities: selecting appropriate technologies, ensuring equity in access, and guiding students in ethical and responsible AI usage. The role of the teacher is increasingly that of a facilitator and mentor—someone who can bridge the gap between technical content and human-centered design values (Luckin, 2018). As Ferrer et al. (2021) pointed out, without deliberate implementation, AI can easily reinforce existing biases or obscure transparency in educational decision-making.

This ethical dimension becomes even more critical as engineering problems themselves grow in social, cultural, and environmental complexity. Incorporating AI into engineering education is not merely a matter of efficiency or engagement—it's a moral imperative to prepare students for the far-reaching impacts of their future work. Scholars such as Cech (2014) and Walther et al. (2017) argue that engineering education must foreground human values, empathy, and social justice alongside technical skill. A compelling example comes from Virginia Tech's IDEEAS Lab[2] introduced a case-study module in which civil engineering students used generative AI to design a bridge for a developing region. As part of the exercise, they were prompted to reflect on who benefits from the design, who might be marginalized, and how to validate AI-generated recommendations against community needs (i.e., raising questions such as whether environmental impacts were equitably distributed, or which stakeholder voices might be missing).

Such AI-mediated learning environments, the key is not simply to teach students to use powerful tools, but to engage them actively in questioning whose problems are being addressed, who benefits, and at whose expense. Stahl et al. (2023) offer a collection of real-world case studies that highlight pressing ethical concerns in AI, such as algorithmic bias in hiring practices, privacy violations, manipulation of user behavior, and threats to human dignity. Each vignette serves as a powerful pedagogical tool, encouraging students to interrogate whose perspectives are prioritized and whose are marginalized. Importantly, the authors provide concrete strategies—such as AI impact assessments, stakeholder analysis, and ethics-by-design approaches—that can be readily embedded into project-based

[2] For more detail please visit https://ideeaslab.com/resources/teaching/engineering-ethics-ai/

engineering courses. Through these integrative practices, students are not only building AI systems, but also learning to critically evaluate their social consequences, asking who benefits, who may be harmed, and how systems might be redesigned in the service of justice.

Ultimately, the emergence of AI in engineering education is not about replacing traditional methods, but about reimagining what effective, inclusive, and future-ready engineering learning looks like. If implemented thoughtfully, AI can serve as a catalyst for more engaging, reflective, and socially responsive educational experiences. But this transformation requires more than technology—it demands a pedagogical commitment to critical thinking, equity, and ethical action at every level of the educational system.

Research at the Intersection of AI, Engineering Design, and Engineering Education

A bibliometric analysis was conducted to explore the current trajectory and focal areas of research at the intersection of AI, engineering design, and engineering education. This analysis aimed to uncover prevailing themes, frequently used keywords, and emerging research trends. Data was collected from two leading academic databases—Scopus and Web of Science (WoS)—using the keywords *engineering education*, *design*, and *artificial intel*. The search was limited to publications categorized under the social sciences and education disciplines. This process yielded a dataset comprising 465 studies from Scopus and 184 from WoS, which served as the foundation for identifying key patterns and directions in the scholarly discourse surrounding AI's integration into engineering education and design practices.

A keyword co-occurrence analysis was conducted across the included studies. In order to ensure comparability, a consistent number of keywords was reported to enable cross-study examination of conceptual and thematic developments. Within the subdomain of AI in engineering education, a total of 3763 keywords were initially extracted from the reviewed studies. Applying a minimum threshold of five occurrences per keyword, 191 keywords met the inclusion criteria for the cluster analysis. Using a clustering algorithm, these keywords were grouped into four distinct thematic clusters, each representing a coherent research focus (Fig. 8.1):

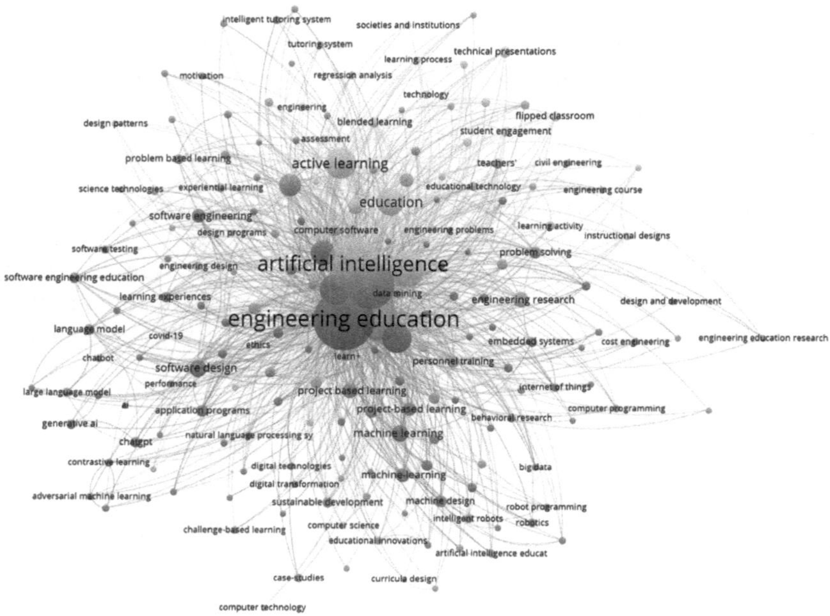

Fig. 8.1 Scopus bibliometric analysis

The red cluster encompasses keywords such as *engineering education*, *design support system*, *e-learning*, and *human–computer interaction*. These terms indicate a focus on the technological enhancement of engineering education, particularly through systems designed to support design processes and digital learning environments. Including *human–computer interaction* suggests a parallel interest in user experience and interaction design within educational settings. Key terms in the yellow cluster include *active learning*, *design*, *assessment*, *design education*, *blended learning*, and *engineering*. This grouping reflects pedagogical approaches within engineering education, with an emphasis on integrating active and blended learning strategies into design-based curricula. The presence of *assessment* and *design education* also highlights an interest in evaluating learning outcomes and instructional effectiveness.

Blue cluster is centered on terms such as *artificial intelligence*, *chatbot*, *conceptual framework*, *generative AI*, *personalized learning*, and *learning experience*. These keywords signify a strong research orientation toward developing and integrating AI tools in education. The inclusion of *chatbots* and *generative AI* points to the application of emerging technologies for enhancing personalization and engagement in learning environments, while the *conceptual framework* indicates ongoing theoretical work in structuring and understanding these applications. The Green Cluster comprises keywords such as *challenge-based learning*, *computational thinking*, *design and development*, and *educational innovation*. This cluster suggests a focus on innovative instructional models that promote problem-solving, creativity, and computational skills. These themes indicate a broader educational reform movement integrating real-world challenges and technology-enhanced learning.

Additionally, the keyword *ethics* appeared with relatively limited co-occurrence, linking primarily to terms such as *engineering education*, *artificial intelligence*, *ethical technology*, and *students*. This limited connectivity may reflect an emerging but still underdeveloped discourse around the ethical dimensions of AI integration in engineering education. The presence of *ethics* in connection with both *AI* and *students* underscores the growing awareness of the need to address ethical considerations in educational design and technological development. In general, the keyword co-occurrence and clustering analysis reveals four main thematic areas in the current literature on AI in engineering education: (1) technology-enhanced learning systems, (2) pedagogical design and assessment, (3) AI-driven personalized learning tools, and (4) innovative, challenge-based educational models. The relatively marginal presence of ethics suggests a potential area for further exploration and integration in future research.

Future Directions in AI-Driven Design Decisions

The infusion of artificial intelligence into design decision-making processes within K–12 education, particularly in engineering, presents a compelling and evolving narrative of promise, transformation, and complexity. As educational institutions increasingly embrace AI-driven approaches to both personalize learning and innovate curricular models, the landscape of engineering education is undergoing a significant shift. Emerging research underscores how AI is not merely a technical add-on but a powerful force capable of reshaping pedagogical and administrative paradigms alike. In the engineering classroom, students use AI in diverse ways: as a co-designer, assistant/intern, tutor, and writing coach. They use AI for technical purposes such as engineering analysis, optimization, and communication (Martin & Roth, 2025).

In a recent study, Chang et al. (2024) explore how middle and high school students can engage as co-design partners in imagining ideal educational technologies, including AI-powered tools. Rather than positioning students as passive users, the study emphasizes their role as transformative learners who critically reflect on how AI might assist with learning, tutoring, and creative expression. Through speculative design workshops, students proposed AI systems that function not only as academic tutors or assistants—offering personalized help in writing and math—but also as empathetic collaborators that support emotional well-being and social justice in school environments. The study highlights how involving students in co-designing AI fosters deeper engagement, critical thinking, and agency, while also surfacing nuanced expectations about the ethical and social roles AI should play in educational spaces. These outcomes provide examples aligned with UNESCO's guidance for generative AI, where human agency is maintained (Miao & Holmes, 2023).

One of the most notable trends in this domain is the capacity of AI to support individualized learning pathways. Studies consistently demonstrate that AI-driven systems can adapt content delivery to suit individual learning styles and paces, thereby improving engagement and educational outcomes. Zhang (2024) and Lin et al. (2024) have shown how adaptive educational management systems—empowered by AI—enable real-time feedback and formative assessment, creating dynamic learning environments that continuously respond to student progress. These findings resonate with Lin and Chen (2024), whose work in engineering technology education highlights AI's role in automating routine tasks and personalizing instruction, allowing learners to focus more on creativity and problem-solving. Furthermore, Harry (2023) argues that through effective data analysis, AI can empower educators to make pedagogical decisions that are more responsive and tailored, optimizing instructional alignment with learner needs. However, as Igbokwe (2024) and Osasona et al. (2024) caution, educational leaders must navigate concerns related to data privacy, algorithmic bias, and equitable access. These concerns are magnified within K–12 settings, where inclusivity is paramount and the risk of deepening existing disparities through technology misuse must be carefully mitigated.

The evolving role of AI in curricular development is equally consequential. Researchers such as Mosly (2024), and Adeyeye and Akanbi (2024), emphasize the necessity of integrating AI literacy into engineering curricula. They argue that understanding AI's capacity to solve complex systems-level problems is increasingly vital in preparing students for a technologically saturated workforce. Jauregui-Correa and Sen (2024) extend this perspective by advocating for a fundamental redesign of engineering education, proposing curriculum models that move away from traditional classes, courses, and grading systems. In this model, the development of AI literacy becomes embedded within knowledge nodes, modular, interconnected learning units that integrate theoretical content, practical activities, and digital tools. These nodes serve as flexible entry points into core topics, guiding students in acquiring both theoretical and applied engineering competencies. For example, a knowledge node on sensor systems might combine interactive AI simulations, lab-based data collection tasks, and formative assessments. These nodes leverage AI tools not only for content delivery and assessment but also to support students in self-directed learning, project development, and laboratory practices. Professors, shifting into roles as facilitators and curators, help students navigate AI-enhanced resources, ensuring they develop critical AI competencies alongside core engineering skills.

The promise of AI cannot be realized without addressing the practical and ethical challenges accompanying its implementation. A significant body of work underscores the urgent need for comprehensive professional development to support educators in navigating this new terrain. Xie et al. (2024) call for teacher co-design in AI curricula development, emphasizing that educators need content knowledge and pedagogical strategies to leverage AI tools effectively. This insight is highlighted by Aljemely (2024) and Luckin et al. (2022), who stress the importance of equipping teachers with practical training and ongoing support to overcome implementation barriers. Ethical concerns also remain central, particularly in ensuring that AI tools used in education do not perpetuate bias or exclude marginalized student populations. Porayska-Pomsta et al. (2023) and Osasona et al. (2024) advocate for embedding ethics explicitly within AI curricula and governance frameworks to safeguard equitable access and responsible use.

Another emerging focus is the collaborative dynamic between human educators and AI systems. As AI tools become more sophisticated, their role in design decision-making is increasingly seen not as a replacement for human judgment, but as a complement. Baltà-Salvador et al. (2025) explore how AI can enhance student ideation and foster holistic, interdisciplinary thinking aligned with the principles of Education 5.0 when employed as a co-creative partner in engineering design. This aligns with Li et al. (2023), who underscore the potential of human–AI teaming, particularly in contexts where critical thinking and empathy are required alongside analytical precision. This partnership model envisions educators using AI-generated insights to inform, rather than dictate, pedagogical decisions—thereby enriching the learning experience while retaining the central role of human agency.

The integration of artificial intelligence into K–12 STEM education is reshaping both pedagogy and curriculum in profound ways. AI is emerging not just as a tool for efficiency, but as a co-creative partner—supporting students in roles such as co-designer, tutor, assistant, and writing coach. Research across K–12 and higher education highlights AI's potential to foster personalized learning, enhance student agency, and promote interdisciplinary, project-based engagement. At the same time, educators and students are increasingly involved in the co-design of AI tools, ensuring that ethical, social, and emotional dimensions—such as equity, empathy, and well-being—are not sidelined.

While AI can empower both learners and teachers through adaptive feedback, automation, and enhanced access to learning resources, it also raises critical challenges around bias, data privacy, and equitable implementation. As the field moves forward, successful integration of AI in design education will depend not only on technological innovation but also on reimagining pedagogical roles, embedding AI literacy across curricula, and fostering collaborative human–AI partnerships grounded in ethical responsibility and inclusive design.

Conclusion

The emergence of artificial intelligence in education—particularly in design decision-making across K–12 levels—represents a pivotal transformation in how future generations are prepared to engage with complex, real-world challenges. Far beyond a technological enhancement, AI introduces a foundational shift in both pedagogical strategy and STEM practice. It redefines the role of educators and learners alike, positioning AI not as a replacement for human ingenuity, but as a co-creative partner that augments human judgment, fosters innovation, and enhances strategic thinking.

AI technologies, such as generative design, predictive modeling, and natural language processing, are already reshaping the contours of the design process. Tools that autonomously explore vast design spaces, optimize competing criteria, and provide real-time feedback allow for more informed, agile, and sustainable decision-making. These capabilities offer the promise of early-stage design optimization, reduced late-stage revisions, and an overall increase in design efficiency and creativity. Moreover, the integration of AI with big data analytics further enables intelligent, data-driven workflows that extend beyond traditional engineering boundaries into domains such as urban planning, manufacturing, and environmental systems. For instance, in urban planning, AI algorithms are used to analyze traffic patterns and simulate pedestrian flow to optimize city infrastructure and reduce congestion (Zhu et al., 2024).

However, these technical advances must be met with a parallel evolution in educational frameworks. Preparing students to thrive in a world increasingly mediated by AI requires more than technical proficiency. It necessitates the cultivation of interdisciplinary literacy, ethical awareness, and critical reflection. As AI systems become more capable, so too must students become adept at understanding the implications of their design choices—ensuring that algorithmic efficiency does not come at the cost of transparency, equity, or human-centered values.

Institutions, therefore, face a dual imperative: to integrate AI meaningfully into curricula and to do so with strategic foresight and ethical sensitivity. This includes aligning AI competencies with existing educational standards, redesigning courses to incorporate human–AI collaboration, and investing in robust professional development for educators. Faculty preparedness remains a critical concern, as does the development of curricula that balance technical depth with ethical inquiry and social responsibility.

Ultimately, the trajectory of AI in K–12 education points toward a future in which design thinking, creativity, and decision-making are deeply intertwined with intelligent technologies. As the design process evolves from deterministic methods to adaptive, collaborative ecosystems, the goal is not to automate creativity, but to expand its boundaries—enabling students to become more thoughtful, capable, and socially responsible designers. By embedding AI thoughtfully into education, we can prepare learners not just to use technology, but to shape it—building a future in which human potential and machine intelligence work together in service of inclusive innovation.

Disclaimer This research made partial use of generative AI tools to support the development of academic content. ChatGPT, developed by OpenAI, was employed to assist in shaping and articulating scholarly arguments. For literature discovery and citation tracking, Scite played a key role in streamlining the review process. Additionally, Perplexity was used during editing of text written by the authors. These tools provided alternative terminology, assisted in grammar, and helped refine the clarity of language. It is important to note that while these tools enhanced the overall workflow, all core ideas, interpretations, and conclusions are solely those of the authors. The AI tools served as supplementary aids and did not produce original content or influence the analytical outcomes. Any remaining inaccuracies or shortcomings in the work are entirely the responsibility of the author.

References

Adams, R. S., Turns, J., & Atman, C. J. (2003). Educating effective engineering designers: The role of reflective practice. *Design Studies, 24*(3), 275–294.

Adeyeye, O. J., & Akanbi, I. (2024). Artificial intelligence for systems engineering complexity: A review on the use of AI and machine learning algorithms. *Computer Science & IT Research Journal, 5*(4), 787–808.

Alexander, E. R. (1982). Design in the decision-making process. *Policy Sciences, 14*(3), 279–292.

Aliev, R. A., Fazlollahi, B., & Aliev, R. R. (2012). *Soft computing and its applications in business and economics* (Vol. 157). Springer.

Aljemely, Y. (2024). Challenges and best practices in training teachers to utilize artificial intelligence: A systematic review. *Frontiers in Education, 9*, 1470853. Frontiers Media SA.

Arastoopour Irgens, G., Chesler, N. C., Linderoth, J. T., & Williamson Shaffer, D. (2017). Data-enabled cognitive modeling: Validating student engineers' fuzzy design-based decision-making in a virtual design problem. *Computer Applications in Engineering Education, 25*(6), 1001–1017.

Atman, C. J., Adams, R. S., Cardella, M. E., Turns, J., Mosborg, S., & Saleem, J. (2007). Engineering design processes: A comparison of students and expert practitioners. *Journal of Engineering Education, 96*(4), 359–379.

Baltà-Salvador, R., El-Madafri, I., Brasó-Vives, E., & Peña, M. (2025). Empowering engineering students through artificial intelligence (AI): Blended human–AI creative ideation processes with ChatGPT. *Computer Applications in Engineering Education, 33*(1), e22817.

Becker, K., & Mentzer, N. (2015). Engineering design thinking: High school students' performance and knowledge. In *2015 international conference on interactive collaborative learning (ICL)* (pp. 5–12). IEEE.

Beitz, W., Pahl, G., & Grote, K. (1996). Engineering design: A systematic approach. *MRS Bulletin, 71*, 30.

Biondi, G., Cagnoni, S., Capobianco, R., Franzoni, V., Lisi, F. A., Milani, A., & Vallverdú, J. (2023). Ethical design of artificial intelligence-based systems for decision making. *Frontiers in Artificial Intelligence, 6*, 1250209.

Bras, B., & Mistree, F. (1991). Designing design processes in decision-based concurrent engineering. *SAE Technical Paper Series, 1*, 1019–1040.

Cech, E. (2014). Culture of disengagement in engineering education? *Science Technology & Human Values, 39*(1), 42–72. https://doi.org/10.1177/0162243913504305

Chandrasekera, T., Hosseini, Z., & Perera, U. (2025). Can artificial intelligence support creativity in early design processes? *International Journal of Architectural Computing, 23*(1), 122–136.

Chang, M. A., Wong, R. Y., Breideband, T., Philip, T. M., McKoy, A., Cortez, A., & D'Mello, S. K. (2024, May). Co-design partners as transformative learners: Imagining ideal technology for schools by centering speculative relationships. In *Proceedings of the 2024 CHI conference on human factors in computing systems* (pp. 1–15).

Chaturvedi, I., Cambria, E., & Welsch, R. E. (2023). Teaching simulations supported by artificial intelligence in the real world. *Education Sciences, 13*(2), 187.

Corple, D. J., Zoltowski, C. B., Kenny Feister, M., & Buzzanell, P. M. (2020). Understanding ethical decision-making in design. *Journal of Engineering Education, 109*(2), 262–280.

Crismond, D. P., & Adams, R. S. (2012). The informed design teaching & learning matrix. *Journal of Engineering Education-Washington, 101*(4), 738.

Dixit, A. C., Prakasha, K. N., Harshavardhan, B., Anand, A., & Taranum, A. (2025). Barriers to integrating AI in curriculum for enhanced engineering education: A fuzzy ISM approach. *Journal of Engineering Education Transformations*, 137–147.

Dringenberg, E., Guanes, G., & Leonard, A. (2021). Student and faculty beliefs about diverse approaches to engineering design decisions. *Studies in Engineering Education, 2*(2).

Dym, C. L., Agogino, A. M., Eris, O., Frey, D. D., & Leifer, L. J. (2005). Engineering design thinking, teaching, and learning. *Journal of Engineering Education, 94*(1), 103–120.

Earle-Randell, T. V., Wiggins, J. B., Ma, Y., Celepkolu, M., Bounajim, D., Gao, Z., et al. (2024). The impact of near-peer virtual agents on computer science attitudes and collaborative dialogue. *International Journal of Child-Computer Interaction, 40*, 100646.

Edwards, B. I., Olugbade, D., & Ojo, O. A. (2024). Facilitating cognitive load management and improved learning outcomes and attitudes in middle school technology and vocational education through AI Chatbot. *Journal of Technical*

Education and Training, 16(3), 114–131. https://doi.org/10.30880/jtet.2024.16.03.009

Feijoo-Garcia, M. A., Holstrom, M. S., Magana, A. J., & Newell, B. A. (2024). Simulation-based learning and argumentation to promote informed design decision-making processes within a first-year engineering technology course. *Sustainability, 16*(7), 2633.

Ferrer, X., Van Nuenen, T., Such, J. M., Coté, M., & Criado, N. (2021). Bias and discrimination in AI: A cross-disciplinary perspective. *IEEE Technology and Society Magazine, 40*(2), 72–80. https://doi.org/10.48550/arXiv.2008.07309

Figoli, F. A., Rampino, L., & Mattioli, F. (2022). AI in the design process: Training the human-AI collaboration. In *Proceedings of the 24th international conference on engineering and product design education (E&PDE 2022)* (pp. 1–6).

Fosmire, M. (2017). Making informed decisions: The role of information literacy in ethical and effective engineering design. *Theory Into Practice, 56*(4), 308–317.

Guzey, S. S., Tank, K., Wang, H. H., Roehrig, G., & Moore, T. (2014). A high-quality professional development for teachers of grades 3–6 for implementing engineering into classrooms. *School Science and Mathematics, 114*(3), 139–149. https://doi.org/10.1111/ssm.12061

Harrison, E. F. (1999). *The managerial decision-making process.* Houghton Mifflin.

Harry, A. (2023). Role of ai in education. *Interdiciplinary Journal and Hummanity (Injurity), 2*(3), 260–268. https://doi.org/10.58631/injurity.v2i3.52

Holstein, K., & Aleven, V. (2022). Designing for human–AI complementarity in K-12 education. *AI Magazine, 43*(2), 239–248.

Holstein, K., & Doroudi, S. (2021). *Equity and artificial intelligence in education: will "aied" amplify or alleviate inequities in education?* Cornell University. https://doi.org/10.48550/arXiv.2104.12920

Householder, D. L., & Hailey, C. E. (Eds.). (2012). *Incorporating engineering design challenges into STEM courses.* Retrieved on May 2025 from http://ncete.org/flash/pdfs/NCETECaucusReport.pdf

Igbokwe, I. C. (2024). Artificial intelligence in educational leadership: Risks and responsibilities. *European Journal of Arts, Humanities and Social Sciences, 1*(6), 3–10.

Jauregui-Correa, J. C., & Sen, M. (2024). Revolutionizing engineering education: Adapting curricula to address artificial intelligence challenges and opportunities. *Asean Journal of Engineering Education, 8*(1), 64–69.

Jobin, A., Ienca, M., & Vayena, E. (2019). The global landscape of AI ethics guidelines. *Nature Machine Intelligence, 1*(9), 389–399. https://doi.org/10.1038/s42256-019-0088-2

Jonassen, D., Strobel, J., & Lee, C. B. (2006). Everyday problem solving in engineering: Lessons for engineering educators. *Journal of Engineering Education, 95*(2), 139–151. https://doi.org/10.1002/j.2168-9830.2006.tb00885.x

Khanolkar, P. M., Gad, M., Liao, J., Hurst, A., & Olechowski, A. (2021). A pilot study on the prevalence of artificial intelligence in Canadian engineering design curricula. *Proceedings of the Canadian Engineering Education Association (CEEA)*.

Khanolkar, P. M., Vrolijk, A., & Olechowski, A. (2023). Mapping artificial intelligence-based methods to engineering design stages: A focused literature review. *AI EDAM, 37*, e25.

Li, Z., Lu, Z., & Yin, M. (2023, June). Modeling human trust and reliance in ai-assisted decision making: A markovian approach. *Proceedings of the AAAI Conference on Artificial Intelligence, 37*(5), 6056–6064.

Liao, H., He, Y., Wu, X., Wu, Z., & Bausys, R. (2023). Reimagining multi-criterion decision making by data-driven methods based on machine learning: A literature review. *Information Fusion, 100*, 101970.

Lin, H., & Chen, Q. (2024). Artificial intelligence (AI)-integrated educational applications and college students' creativity and academic emotions: Students and teachers' perceptions and attitudes. *BMC Psychology, 12*(1), 487.

Lin, M. P. C., Liu, A. L., Poitras, E., Chang, M., & Chang, D. H. (2024). An exploratory study on the efficacy and inclusivity of AI technologies in diverse learning environments. *Sustainability, 16*(20), 8992.

Lišnić, B., Zaharija, G., & Mladenović, S. (2025). Integration of artificial intelligence in K-12: Analysis of a three-year pilot study. *AI, 6*(3), 49. https://doi.org/10.3390/ai6030049

Luckin, R. (2018). *Machine learning and human intelligence. The future of education for the 21st century*. UCL institute of education press.

Luckin, R., Holmes, W., Griffiths, M., & Forcier, L. B. (2016). *Intelligence unleashed - an argument for AI in education*. Retrieved in May 2025 from http://discovery.ucl.ac.uk/1475756/

Luckin, R., Cukurova, M., Kent, C., & Du Boulay, B. (2022). Empowering educators to be AI-ready. *Computers and Education: Artificial Intelligence, 3*, 100076.

Mah, D. K., & Groß, N. (2024). Artificial intelligence in higher education: Exploring faculty use, self-efficacy, distinct profiles, and professional development needs. *International Journal of Educational Technology in Higher Education, 21*(1), 58.

Martin, K., & Roth, B. (2025, May). *How can AI help design? Lessons learned from engineering students and faculty*. Clive L. Dym Mudd Design Workshop XIV. Claremont, CA.

Mentzer, N., Becker, K., & Sutton, M. (2015). Engineering design thinking: High school students' performance and knowledge. *Journal of Engineering Education, 104*, 417–432. https://doi.org/10.1002/jee.20105

Miao, F., & Holmes, W. (2023). *Guidance for generative AI in education and research*. UNESCO Publishing. https://doi.org/10.54675/EWZM9535

Miao, F., & Shiohira, K. (2024). *AI competency framework for students.* UNESCO Publishing.

Morley, J., Floridi, L., Kinsey, L., & Elhalal, A. (2020). From what to how: An initial review of publicly available AI ethics tools, methods and research to translate principles into practices. *Science and Engineering Ethics, 26*(4), 2141–2168.

Mosly, I. (2024). Artificial intelligence's opportunities and challenges in engineering curricular design: A combined review and focus group study. *Societies, 14*(6), 89.

Murray, J. K., Studer, J. A., Daly, S. R., McKilligan, S., & Seifert, C. M. (2019). Design by taking perspectives: How engineers explore problems. *Journal of Engineering Education, 108*(2), 248–275.

Olewnik, A., & Svihla, V. (2025). Engineering requirements and their role in engineering undergraduates' design decision making. *International Journal of Technology and Design Education, 35*(2), 673–698.

Osasona, F., Amoo, O. O., Atadoga, A., Abrahams, T. O., Farayola, O. A., & Ayinla, B. S. (2024). Reviewing the ethical implications of AI in decision making processes. *International Journal of Management & Entrepreneurship Research, 6*(2), 322–335.

Parra, X., Tort-Martorell, X., Alvarez-Gomez, F., & Ruiz-Viñals, C. (2023). Chronological evolution of the information-driven decision-making process (1950–2020). *Journal of the Knowledge Economy, 14*(3), 2363–2394.

Popenici, S. A., & Kerr, S. (2017). Exploring the impact of artificial intelligence on teaching and learning in higher education. *Research and Practice in Technology Enhanced Learning, 12*(1), 22.

Porayska-Pomsta, K., Holmes, W., & Nemorin, S. (2023). The ethics of AI in education. In *Handbook of artificial intelligence in education* (pp. 571–604). Edward Elgar Publishing.

Purzer, S., & Arigye, J. (2024). Engineering design reasoning for conscientious decision-making. In *Ways of thinking in STEM-based problem solving* (pp. 178–191). Routledge.

Purzer, S., & Quintana-Cifuentes, J. P. (2019). Integrating engineering in K-12 science education: Spelling out the pedagogical, epistemological, and methodological arguments. *Disciplinary and Interdisciplinary Science Education Research, 1*(1), 13. https://doi.org/10.1186/s43031-019-0010-0

Quinde, G. A. L., Muñoz, M. Y. T., Suárez, J. M. R., Villarreal, R. E. P., Vélez, W. A. Z., & Laínez, A. A. D. P. (2024). Perception of university students on the use of artificial intelligence (AI) tools for the development of autonomous learning. *Revista de Gestão Social e Ambiental, 18*(2), 1–20.

Ren, X., & Wu, M. L. (2025). Examining teaching competencies and challenges while integrating artificial intelligence in higher education. *Techtrends, 69*, 1–20.

Samuelson, C. C., & Litzler, E. (2016). Community cultural wealth: An assets-based approach to persistence of engineering students of color. *Journal of Engineering Education, 105*(1), 93–117.

Shao, K. Z., Wilson, D., Cho, E. K. M., Tang, S., Ma, H., & Makhsous, S. (2025). Bridging educational equity gaps: A systematic review of AI-driven and new technologies for students living with disabilities in STEM education. *2025 Collaborative Network for Engineering & Computing Diversity (CoNECD).*

Slimi, Z., & Villarejo-Carballido, B. V. (2023). Navigating the ethical challenges of artificial intelligence in higher education: An analysis of seven global AI ethics policies. *Tem Journal, 12*(2).

Spirochkin, Y. (2023). *Human factors and design* (1st ed., pp. 1–195). Springer. https://doi.org/10.1007/978-981-19-8832-5

Stahl, B. C., Schroeder, D., & Rodrigues, R. (2023). *Ethics of artificial intelligence: Case studies and options for addressing ethical challenges* (p. 116). Springer Nature.

Surya, A. A., Kurian, M. K., & Varghese, S. M. (2016). Overall performance evaluation of engineering students using fuzzy logic. *International Journal on Cybernetics & Informatics (IJCI), 5*(2), 71–78.

Tseng, Y. J., & Yadav, G. (2023). ActiveAI: Introducing AI literacy for middle school learners with goal-based scenario learning. https://doi.org/10.48550/ARXIV.2309.12337

Van de Poel, I., & Royakkers, L. (2023). *Ethics, technology, and engineering: An introduction.* John Wiley & Sons.

Van den Beemt, A., MacLeod, M., Van der Veen, J., Van de Ven, A., Van Baalen, S., Klaassen, R., & Boon, M. (2020). Interdisciplinary engineering education: A review of vision, teaching, and support. *Journal of Engineering Education, 109*(3), 508–555.

Villegas-Ch, W., Buenano-Fernandez, D., Navarro, A. M., & Mera-Navarrete, A. (2025). Adaptive intelligent tutoring systems for STEM education: Analysis of the learning impact and effectiveness of personalized feedback. *Smart Learning Environment, 12*, 41. https://doi.org/10.1186/s40561-025-00389-y

Walter, Y. (2024). Embracing the future of artificial intelligence in the classroom: The relevance of AI literacy, prompt engineering, and critical thinking in modern education. *International Journal of Educational Technology in Higher Education, 21*(1), 15.

Walther, J., Miller, S. E., & Sochacka, N. W. (2017). A model of empathy in engineering as a core skill, practice orientation, and professional way of being. *Journal of Engineering Education, 106*(1), 123–148.

Wendell, K. B., Wright, C. G., & Paugh, P. (2017). Reflective decision-making in elementary students' engineering design. *Journal of Engineering Education, 106*(3), 356–397.

Xie, B., Sarin, P., Wolf, J., Garcia, R. C., Delaney, V., Sieh, I., et al. (2024). Co-designing AI education curriculum with cross-disciplinary high school teachers. *Proceedings of the AAAI Conference on Artificial Intelligence, 38*(21), 23146–23154.

Zadeh, L. A. (1975). Fuzzy logic and approximate reasoning: In memory of Grigore Moisil. *Synthese, 30*, 407–428.

Zadeh, L. A. (1996). Fuzzy logic = computing with words. *IEEE Transactions on Fuzzy Systems, 4*(2), 103–111.

Zawacki-Richter, O., Marín, V. I., Bond, M., & Gouverneur, F. (2019). Systematic review of research on artificial intelligence applications in higher education–where are the educators? *International Journal of Educational Technology in Higher Education, 16*(1), 1–27.

Zhai, X., Krajcik, J., & Pellegrino, J. W. (2021). On the validity of machine learning-based next generation science assessments: A validity inferential network. *Journal of Science Education and Technology, 30*, 298–312.

Zhai, C., Wibowo, S., & Li, L. D. (2024). The effects of over-reliance on AI dialogue systems on students' cognitive abilities: A systematic review. *Smart Learning Environments, 11*(1), 28.

Zhang, Q. (2024). Harnessing artificial intelligence for personalized learning pathways: A framework for adaptive education management systems. *Applied and Computational Engineering, 82*, 167–172.

Zhao, G., Sheng, H., Wang, Y., Cai, X., & Long, T. (2025). Generative artificial intelligence amplifies the role of critical thinking skills and reduces reliance on prior knowledge while promoting in-depth learning. *Education Sciences, 15*(5), 554. https://doi.org/10.3390/educsci15050554

Zhu, W., He, W., & Li, Q. (2024). Hybrid AI and big data solutions for dynamic urban planning and smart city optimization. *IEEE Access*. Retrieved in June 2025fromhttps://ieeexplore.ieee.org/stamp/stamp.jsp?arnumber=10794774

CHAPTER 9

Preparing Teachers for STEM Education in the Age of AI: A Literature and Projects Review Within the Slovenian Educational Context

Mateja Ploj Virtic and Kosta Dolenc

Introduction

Generative artificial intelligence (GenAI)—most prominently represented by ChatGPT—has rapidly emerged as a transformative force in various fields, including Slovenian education (Aristovnik et al., 2023; Dolenc, 2024; Gartner & Krašna, 2024). Its ability to produce coherent, human-like responses and creative content has fuelled both optimism about new teaching and learning opportunities and concern about ethical use, academic integrity, and the preservation of critical thinking skills (Lipovec & Flogie, 2023). In Slovenia, GenAI is increasingly being used for tasks such as information retrieval, text refinement, and the creation of teaching materials. These applications emphasize the growing need to prepare teachers—not only to understand AI at a technical level, but also to

M. Ploj Virtic (✉) • K. Dolenc
University of Maribor, Maribor, Slovenia
e-mail: mateja.plojvirtic@um.si; kosta.dolenc@um.si

R. Elmas, M. Akarsu (eds.), *STEM Education*, Maarif Global Education Series,
https://doi.org/10.1007/978-981-95-9023-0_9

integrate it into the classroom meaningfully. Over the past two years, groundbreaking research efforts and pilot projects have underscored the promise of GenAI, but also revealed the lack of comprehensive guidelines and highlighted the risk of using AI-driven "shortcuts" that can undermine genuine learning processes (Dolenc & Brumen, 2024; Licardo et al., 2025).

The chapter situates the discussion within the contemporary framework of STEM education and takes an integrated view in which science, technology, engineering, and mathematics are linked through common practices of modeling, evidence-based reasoning, and iterative design. GenAI intersects with this agenda in three key ways. First, it lowers the barriers to entry for large-scale data investigations and enables students to test hypotheses on authentic data sets within a single lesson. Second, it accelerates the generation and critique of multiple solution prototypes (code snippets, construction schemes, or statistical simulations) extending the established cycles of engineering design and problem-based learning. Third, it requires new layers of algorithmic and ethical competence, forcing teachers to guide students in assessing the validity of AI explanations and defusing embedded biases. Slovenia's pilot projects are already capitalizing on these opportunities—for example, SETCOM's AI-powered simulation tasks that combine digital content creation with reflective recordings, and GEN UI's forthcoming cross-faculty STEM labs—while also highlighting the challenges of disciplinary rigor and equal access. By foregrounding these tensions, the chapter positions GenAI not as an autonomous solution, but as a catalyst that must be embedded in disciplined enquiry and sound pedagogical guidance.

Despite the growing interest in GenAI, there is still a critical knowledge gap on how to introduce this technology systematically and responsibly in Slovenian primary and secondary schools and in the faculties that train future educators. Empirical studies that provide long-term evaluations, best practices, and a deeper understanding of the impact of GenAI at different educational levels are still scarce. To address this gap, this chapter aims to:

1. Review the literature and ongoing projects (2022–2025) that address GenAI in the Slovenian educational context,
2. Analyze the main challenges and opportunities in using GenAI for educational purposes, including ethical considerations, and
3. To formulate recommendations for future research and practice, emphasizing how GenAI can enhance rather than replace the development of basic skills and critical thinking.

The chapter begins with a theoretical overview of GenAI, followed by a detailed explanation of the methodology used to review the existing literature and relevant projects. We then summarize the main findings and conclude with a discussion of the critical issues and the prospects for further development of research on GenAI in Slovenian education.

Definition and Characteristics of Generative Artificial Intelligence (GenAI)

GenAI explicitly refers to AI systems that are designed to autonomously generate new, coherent and contextualized content that simulates or mimics human-like creativity and originality (Dolenc, 2024; Aristovnik et al., 2023). This category of AI technology utilizes complex algorithmic models such as deep learning and neural networks to generate outputs in various formats, including text data, images, audio, and multimedia content (Idil, 2023; Sun et al., 2025).

Key features of GenAI include its ability to produce new novel content without direct human intervention independently and autonomously, adapt to different contexts and learning environments, and respond interactively and dynamically to user input (Sun et al., 2025; Parviz, 2024). In addition, GenAI systems demonstrate an ability to personalize educational applications by adapting teaching methods and materials to individual learner profiles, improving engagement and learning outcomes (Sun et al., 2025). Furthermore, GenAI technologies are characterized by scalability and indefatigability. They are able to continuously engage with users without fatigue, expanding access to high-quality educational interactions (Parviz, 2024). However, alongside these benefits, GenAI also brings specific challenges, including the risk of generating misleading or biased information (due to biases in the training datasets), ethical concerns regarding data privacy, and the potential to foster student dependency on AI-generated solutions (Kohnke & Zaugg, 2025). Understanding these defining characteristics and complexities is critical for educators, policymakers, and stakeholders who want to responsibly harness the educational potential of GenAI while mitigating the associated risks.

GenAI offers significant benefits in education, including personalized learning experiences, instant feedback, content creation support, and increased student creativity and engagement.

- Firstly, personalized learning experiences are greatly enhanced by AI technologies. GenAI analyze large amounts of data to tailor teaching content to each student's learning pace and style systems, thus catering to individual learner needs (Sun et al., 2025). AI-driven personalized learning environments have been shown to improve motivation, engagement and overall learning outcomes by providing adaptive learning pathways that cater to individual academic strengths and weaknesses (Idil, 2023).
- Secondly, GenAI facilitates immediate feedback, which is essential for formative assessment methods. AI tools can assess complex student tasks, such as scientific arguments or explanations, and provide immediate feedback that leads students to deeper understanding and higher-order thinking skills (Zhai et al., 2023). Immediate feedback mechanisms are critical for supporting continuous learning progress and promoting effective teaching practices.
- Third, GenAI supports content creation and significantly helps teachers by automating the creation of high-quality teaching materials. For example, AI technologies can develop customized quizzes, simulations, interactive activities, and multimedia resources, which reduce teachers' workload and allows them more time for direct interaction with students and one-on-one mentoring (Chiu & Li, 2023).
- Finally, GenAI encourages student creativity and engagement. AI-driven interactive tools such as virtual reality environments and intelligent tutoring systems have been shown to be effective in promoting immersive and interactive STEM learning experiences (Chiu & Li, 2023). Students who interact with these tools report higher levels of engagement, increased self-efficacy and more positive attitudes toward STEM subjects (Sun et al., 2025).

Despite these significant benefits, integrating GenAI into education poses critical risks and challenges that need to be addressed. One primary concern is the threat to academic integrity. The ability of GenAI to create high-quality, human-like content has led to fears of an increase in plagiarism and academic dishonesty. Educators and institutions face the challenge of developing robust detection systems and clear guidelines for maintaining academic honesty (Žerovnik & Zapušek, 2024). Misinformation also poses a serious risk. GenAI models can occasionally produce false or misleading information that can mislead students. There is an urgent need for constant vigilance and critical oversight to ensure the accuracy and appropriateness of AI-generated content in an educational context (Gartner & Krašna, 2024). Ethical dilemmas further complicate the use of GenAI, particularly in relation to privacy, algorithmic bias, and transparency. The collection and use of student data to train AI systems raises significant privacy concerns, while biases embedded in algorithms can exacerbate existing educational inequalities (Kohnke & Zaugg, 2025). Ethical guidelines and transparent AI practices are essential to effectively mitigate these risks. Dependence on technology is another critical problem. Over-reliance on GenAI tools could undermine the development of essential critical thinking and problem-solving skills, turning students into passive recipients rather than active constructors of knowledge (Parviz, 2024).

Finally, the digital divide remains a significant barrier. Unequal access to technology and resources exacerbates existing educational inequalities. Schools with limited resources may struggle to utilize advanced GenAI tools, leading to inequalities in educational opportunities and outcomes between affluent and disadvantaged communities (Kohnke & Zaugg, 2025).

Methodological Approach

Between January 2022 and January 2025, we conducted a mixed-methods synthesis to investigate how generative artificial intelligence (GenAI) is finding its way into Slovenian classrooms and teacher education programs. The research was determined by a three-part filter: temporal (publications and project achievements from 2022–2025), geographical (the Republic of Slovenia), and thematic (GenAI in primary, secondary and higher education, with a focus on teacher education). Only peer-reviewed journal articles, conference papers, and project reports that fully presented their empirical or design research methods were considered; texts that only referred to AI without substantive pedagogical analysis were excluded.

The literature component began with a structured search in Scopus, Web of Science, and Google Scholar, where combinations such as "generative artificial intelligence" AND "education" AND "Slovenia" were paired with more targeted strings ("teacher training," "ChatGPT," "assessment feedback"). Because the GenAI research cycle exceeds conventional indexing, we supplemented these queries with AI-powered deep search tools to find emerging conference proceedings and technical papers. Two researchers independently reviewed the titles and abstracts against the inclusion criteria, resolved discrepancies through discussion, and then conducted a full-text review. This double-blind process strengthened reliability and mitigated selection bias.

As formal publications often lag behind practice, we mapped the current national landscape of AI projects in education in parallel. A project was eligible if it (a) involved a faculty of education, (b) provided empirical or design-based evidence with direct relevance to classroom practice, and (c) promised tangible impact—for example, curriculum pilots or large-scale professional development workshops. Publicly available findings from flagship initiatives such as SETCOM and GEN UI therefore formed a second corpus. No additional interviews or surveys were commissioned; instead, project artifacts were treated as naturalistic data, supplementing peer-reviewed records.

The resulting material combined statistical indicators (e.g., adoption rates, gains in competence) with narrative evidence (e.g., ethical dilemmas expressed by teachers) and therefore required an integrated analysis strategy. Quantitative findings were tabulated to identify scope and trend lines, while qualitative passages were thematically coded to capture constructs such as teacher readiness, opportunities for formative feedback and risks of overuse across sources. In a final comparative reading, the project-based findings were cross-referenced with the academic literature to identify convergences—such as widespread enthusiasm for AI-supported feedback loops—and tensions—such as ongoing concerns about the erosion of critical thinking skills.

Three limitations inform the following interpretations. First, the scientific corpus on GenAI in Slovenian education remains modest, so certain niches—especially the long-term effects on teaching—have yet to be explored. Second, the 2022–2025 timeframe captures an experimental phase rather than a mature, institutionalized practice. Thirdly, independent verification of some claims was not possible as we relied on published texts and project documentation. These limitations are acknowledged so that readers can weigh the evidence appropriately.

The Most Important Results of the Latest Studies and Reports in Slovenia

The literature review shows a rapidly developing picture of generative AI in Slovenian education. Overall, the studies show that AI tools are increasingly being experimented with in both K–12 and higher education. For the sake of clarity, we organize the results into four thematic categories: (1) student use and perceptions of GenAI, (2) teacher attitudes and readiness for GenAI, (3) applications of GenAI in educational practice, and (4) comparative analysis of perceptions and use of AI tools at different levels of education. Within each category, we summarize the purpose of each study, the type of GenAI application investigated, the main findings, and a critical evaluation of the methodology and limitations of the study.

Student Use and Perceptions of GenAI

A significant body of research in Slovenia has examined how students—primarily at the university level—perceive and use GenAI tools such as ChatGPT. This focus is not surprising given how widespread ChatGPT is among students globally and in Slovenia. A pivotal study by Aristovnik et al. (2023) investigated the perception of ChatGPT by Slovenian university students shortly after the tool was released. The study was conducted as part of an international conference on technology in education and aimed to investigate public administration students' attitudes toward ChatGPT and its perceived impact on their academic activities. The authors surveyed university students enrolled in a public administration degree program in Slovenia on various study-related topics associated with the use of ChatGPT—likely including perceived usefulness for learning tasks, ease of use, and impact on various skills.

The findings of Aristovnik et al. (2023) suggest that students generally view ChatGPT favorably as an educational tool, describing it as "an attractive platform for public administration students." Respondents reported that ChatGPT enriches learning, particularly by facilitating access to information and the generation of ideas—an observation that is consistent with global trends showing that students value the help of AI in tasks such as research, writing, and problem-solving. Crucially, however, the Slovenian students also pointed out the limitations they saw. They felt that the tool was less effective when it came to fostering higher order skills such as critical thinking, creativity, and personal communication skills. The students expressed concern that using ChatGPT could hinder the development of deeper cognitive processes. The limitations of the study include its focus on a specific group (public administration students) at a single institution and its relatively small sample size. In addition, the study was conducted in early 2023, shortly after the introduction of ChatGPT. Student attitudes may change as they gain more experience and universities introduce more straightforward guidelines. Nonetheless, this study provides valuable initial evidence that Slovenian students appreciate the benefits of GenAI but are also aware of the potential drawbacks in training deeper skills.

Another early study by Dolenc and Brumen (2024) investigated how students from social science and computer science programs perceive the integration of AI into foreign language teaching. The study was conducted shortly after GenAI tools had taken center stage at universities and examined students' attitudes, reported experiences, and perceived impact on learning outcomes. The authors looked at perceived usefulness, ease of use, and the impact of AI on key language skills. According to the study results, participants generally welcomed AI-based tools for their speed, convenience, and ability to provide diverse learning materials. The immediate feedback on grammar, pronunciation, and vocabulary motivated students and complemented traditional methods. At the same time, many interviewees expressed uncertainty about whether such tools actually foster deeper cognitive skills—particularly critical thinking, creativity, and sophisticated communication skills. They also expressed concerns about academic integrity, cautioning that AI could unintentionally promote shortcuts and undermine authentic language skills. The article highlights both the promise and pitfalls of AI-enhanced language learning: while AI can certainly enrich educational experiences, it cannot replace the human interaction and carefully designed pedagogy necessary for sound language development. The authors point out that their sample included students in social science and computer science programs at certain institutions and warn that the results may not be fully transferable to other disciplines or contexts.

A large-scale global study by Ravšelj et al. (2025), in which 109 countries, including Slovenia, were analyzed, underlines these findings. This project, the most comprehensive study of university students' use of ChatGPT to date (with over 23,000 participants worldwide), found that Slovenian students' responses were very close to global trends. Students appreciated ChatGPT's ability to simplify complex topics, provide structured guidance, and serve as a dialogue partner for learning. Nevertheless, they were also concerned about misinformation from AI and emphasized the need for proper guidance on the effective use of such tools. Concerns about academic integrity were also frequently expressed: The global survey found that many institutions, presumably including those in Slovenia, did not initially have clear guidelines for the use of AI. The inclusion of Slovenia in a rigorous large-N mixed methods study underpins the positive but cautious attitude observed by Aristovnik et al. (2023) and Dolenc and Brumen (2024). It also highlights a gap in policy: At the time of these surveys, formal regulations and support for AI-based courses were still emerging, leaving students largely to navigate the ethical dimensions on their own. In addition to these formal research efforts, informal data provide further insight into student use in Slovenia. A short survey conducted by the Faculty of Social Sciences (Ljubljana) in 2023 and reported by the Slovenian Press Agency (2023) found that over 90% of students had tried ChatGPT. Although this survey was not peer-reviewed, it confirms academic studies as it shows widespread use. Notably, about half of the respondents admitted to using ChatGPT to “relieve boredom or have the chatbot write their own research papers,” indicating concerns about cheating or circumventing meaningful studies. While these figures are self-reported and may be exaggerated, they do reflect the concerns about integrity that are often discussed in scientific papers. Overall, the new literature paints a picture of enthusiastic acceptance of generative AI among Slovenian students, largely due to the convenience and efficiency of the tools, tempered by an awareness of the need for self-regulation and the maintenance of critical thinking skills. Although many of these studies are methodologically sound (several are peer-reviewed and data-driven), further research is needed to assess the actual impact on learning outcomes—not just perceptions—and to go beyond the context of early adopters of ICT and social sciences.

The recent study by Licardo et al. (2025) on students’ use of GenAI has broadened the scope to investigate how GenAI is accepted and perceived at different levels of the Slovenian education system, from primary school pupils to secondary school students and university students. Using survey data collected at each level, the researchers found that familiarity with GenAI increases with age, although most respondents rated their AI knowledge as only moderate. Students between the ages of 13 and 15 generally stated that they had interacted the least directly with AI-based tools. They often first encountered ChatGPT or similar platforms informally—via social media or out of curiosity—rather than in the classroom. While they described the technology as ‘interesting’ or ‘fun’, they also expressed concerns about reliability and the possible spread of false information. Teachers also noted that student engagement with AI was sporadic and largely unstructured, emphasizing the need for formal guidance to ensure responsible use. Secondary school students are slightly more familiar with and use it more frequently. Around half of secondary school students surveyed said that they had used texting tools such as ChatGPT to help with homework, brainstorm, or check grammar and spelling. They acknowledged that such tools are often a valuable time-saver when deadlines are tight. Nevertheless, these students also expressed uncertainty about where to draw the line between legitimate help and academic dishonesty, suggesting that uncritical reliance on AI could undermine genuine learning or critical thinking. University students, for their part, indicated that they were most comfortable using and being comfortable with the technology ($M = 2.57$ for self-reported familiarity, SD = 0.91). Although few described themselves as "very knowledgeable," many regularly used AI-based text generators or similar platforms ($M = 3.40$, SD = 1.20 for frequency of use). They cited convenience, efficiency, and instant feedback as key benefits—particularly for tasks such as summarizing research findings and clarifying written tasks. However, there were still ethical concerns. Students worried about content accuracy and the possibility that overuse of AI could weaken personal analytical skills. Comparing the three groups, the study found that students were least likely to engage with GenAI in a structured way, often exploring it casually or out of curiosity. Secondary school students are more likely to use AI for coursework, but are unsure where the line between legitimate assistance and plagiarism lies. University students are more confident but emphasize that AI is not a substitute for in-depth knowledge or sound critical thinking. Licardo et al. (2025) conclude that a clearly defined framework—that takes into account reliability, ethical considerations and the development of key skills can enable learners to benefit from GenAI without jeopardizing their learning progress.

Teacher Attitudes and Preparedness for GenAI

As GenAI tools become more common in Slovenian classrooms and universities, the attitude and willingness of teachers to integrate these tools are of increasing importance. Several recent studies have investigated how teachers in Slovenia perceive the role of AI in education and how prepared they feel to use this technology.

One notable example is the study by Gartner and Krašna (2024), "The Oxymoron of AI in Education," in which 73 Slovenian primary and secondary school teachers (49 from primary schools and 24 from secondary schools) were interviewed about their use of information and communication technology (ICT) and AI. The study made a generational comparison, as many teachers belong to Generation X or Y, while students tend to be Generation Z "digital natives" who are often fluent with AI tools (e.g., ChatGPT). Gartner and Krašna found that more than 70% of teachers surveyed have used AI services such as ChatGPT—an unexpectedly high number that signals the willingness of these educators to explore the potential of AI. More than half planned to incorporate AI into their lesson plans (e.g., creating quizzes, providing explanations or supporting students with group projects). Despite this enthusiasm, less than 30% of respondents felt confident about assessing AI-generated outcomes. In other words, they were concerned about how to recognize errors or bias in ChatGPT-generated materials and ensure that the use of AI did not interfere with students' own analytical skills. About half of the teachers said they would use the AI feedback to improve their teaching methods, suggesting that the other half were unsure how to put these findings into practice. The authors describe this as a paradox: on the one hand, teachers feel the need to adapt to AI, especially as students are already experimenting with it; on the other hand, they experience a significant "confidence gap" due to limited training or unclear guidelines. Although the study by Gartner and Krašna (2024) was part of a monograph that was editorially reviewed (and therefore considered peer-reviewed to some extent), the small sample size and possible self-selection bias (teachers interested in AI were more likely to respond) limit its generalizability. Nonetheless, the study emphasizes the need for structured AI training—which includes not only the use of AI in the classroom, but also the critical evaluation of AI results in terms of accuracy, reliability, and bias.

Lipovec and Flogie (2023) conducted a study at the MECO 2023 conference with 82 prospective teachers enrolled at the Faculty of Education at the University of Maribor. Their survey assessed participants' self-reported AI knowledge, their attitudes toward AI in society and education, and their views on ethical issues related to AI. Overall, the prospective teachers viewed AI positively and recognized its potential to differentiate instruction and support social-emotional learning (SEL). Many believed that GenAI could save them time and allow for more individualized interaction with students. Interestingly, the study showed a correlation between teachers' openness to AI and their concern for students' emotional and social needs. In other words, teachers predisposed to an empathetic, student-centered pedagogy appeared to be more open to the introduction of novel technologies, possibly because they see AI as a complement rather than a threat to their core pedagogical values. Nevertheless, some participants expressed caution, particularly with regard to the possibility that AI could inadvertently undermine basic skills (e.g., writing) if students rely too heavily on automated support. While Lipovec and Flogie (2023) offered a rigorous, peer-reviewed conference report, their sample was from a single institution with a predominantly female student population (reflecting the general trend in elementary education programs). The short format (four pages) also limited the depth of the analysis. Nevertheless, the findings are consistent with the broader conclusions of Gartner and Krašna (2024): Future teachers are open to the benefits of AI, but need clearer guidelines, effective training, and ongoing support to use AI responsibly.

Finally, the most recent study by Licardo et al. (2025) expanded the scope and investigated how Slovenian primary, secondary, and higher education teachers use and perceive GenAI. The authors analyzed self-reported familiarity, classroom integration practices, and perceived challenges in these three groups of teachers. Despite differences in usage behavior and familiarity, all groups saw significant potential for AI to enrich student learning and automate certain classroom tasks.

- Primary school teachers ($N = 89$). Most were generally open to digital tools, but only a minority regularly used GenAI in lesson planning or classroom activities. Although 76.4% had experience with technology-oriented projects, few had received specific AI training. Many acknowledged that AI-generated exercises or quizzes could make routine work easier, but they were concerned that students might rely too much on automated support.
- Secondary school teachers ($N = 190$) are slightly more familiar with technology than primary school teachers, and almost 60% have participated in technology integration initiatives. They recognized the potential of AI to save time and diversify learning materials (e.g., practice questions, essay drafts). Respondents were concerned that students might submit AI-generated work without adequate supervision or self-reflection, undermining authentic learning.
- Teaching staff at universities ($N = 115$). Overall, this group reported being most familiar and comfortable with AI, although less than half used it regularly for teaching. Those that did often emphasized the value of efficiency gains, timely feedback for students, and new research-based teaching methods. However, they remained wary of

the risk of outsourcing too much cognitive work to AI—potentially preventing deep expertise.

Comparing these three groups, it appears that primary school teachers rarely use generative AI, partly due to limited experience and training, while secondary school teachers show a slightly higher level of acceptance but struggle with the balance between utilizing the benefits of AI and maintaining academic integrity. Higher education faculty seem most comfortable using generative AI tools, but share ongoing ethical concerns about maintaining high quality analytical skills and authentic student engagement. There is consensus at all levels that AI can simplify faculty work and expand student learning opportunities. However, better faculty training, institutional policies and continued reflection on when, why, and how AI can be used effectively remain urgent priorities.

Applications of GenAI in Educational Practice

In addition to investigating the perception of AI by teachers and students, several Slovenian studies have ventured into the practical integration of GenAI tools in real educational environments. These early implementations provide insights into how GenAI can be used in the classroom, the benefits it brings, and the challenges teachers and students face along the way.

An important example is a case study by Dolenc (2024), presented in a chapter entitled "Educational Transformation: Exploring the Benefits of GenAI Tools." This work, conducted at the Faculty of Education at the University of Maribor, is one of the first detailed accounts of how GenAI can be woven into university education in Slovenia. The study, which spanned the academic years 2022/2023 and 2023/2024, looked at students on a degree program in primary education—future primary school teachers—who used AI tools as part of their regular academic tasks. The overall aim was to explore how GenAI could enhance student learning, identifying both best practice and potential pitfalls. Prior to the interventions, all participating students attended a structured lecture/workshop covering the basics of GenAI (e.g., how large language models work), key capabilities, and ethical considerations. This preparatory step reflects the study's focus on 'informed use'. Three AI chatbots—ChatGPT, Bing Chat (AI), and Google Bard—were introduced because they are easily accessible and relatively simple to use in Slovenia. The students then applied these tools to various subject-specific tasks:

- Third-year students ($n \approx 63$): They used AI to support the preparation and defense of compulsory seminar papers. For example, they designed a science experiment for primary school students and then consulted ChatGPT to refine the instructions and content of the experiment.
- Fourth-year students ($n = 7$): They used AI tools to collect ideas for their bachelor thesis, especially on scientific and technological topics. The AI provided examples and clarifications that the students evaluated before incorporating them into their proposals.
- Fifth-year students ($n \approx 40$): Integrated AI into a research-focused course by using ChatGPT or Google Bard to improve the clarity of written summaries of scientific articles, receiving AI-generated feedback on organization, terminology, and possible additions.

According to Dolenc (2024), the qualitative observations of these applications were very encouraging. Students reported that AI-based feedback promoted a deeper understanding of course content and often prompted them to ask more focused questions or consider multiple perspectives on a topic. AI also facilitated the development of creative teaching ideas—such as alternative classroom activities—and helped students refine their written work through iterative feedback. Most importantly, students discovered the importance of formulating precise prompts. Better structured questions or instructions led to more useful, contextually relevant responses from the AI. Despite these benefits, the case study emphasized that "guided interaction and human supervision" are crucial. Researchers and faculty members monitored AI use, reminded students of academic integrity standards, and pointed out the limits of AI output. Rather than blindly accepting AI suggestions, students were encouraged to evaluate these ideas critically. This way, risks such as inaccurate information, biased content, or verbatim copying of text were avoided. Students' assignments also included a defense or presentation component to prevent them from relying solely on AI-generated work. Finally, the study pointed to ethical and logistical considerations: maintaining privacy in the use of personal accounts, recognizing algorithmic bias, and verifying the accuracy of 'corrections' provided by the AI Overall, these measures underlined that the successful integration of AI depends on robust pedagogical strategies and teacher engagement.

From a methodological point of view, Dolenc (2024) chose a qualitative approach with mixed methods: direct observation in the classroom, focus groups with students, and joint analyses with teachers. While this triangulation increases the credibility of the results, there was no control group for direct comparison of performance, and some subtasks involved small cohorts (e.g., only seven fourth-year students). Therefore, although the study provides meaningful proof-of-concept data, it cannot make a definitive statement about the learning gains attributable to AI or the optimal way to use different AI tools. However, as one of Slovenia's first peer-reviewed studies, it provides an important foundation for later, more targeted research.

Overall, these early implementations show that Slovenia is moving from theoretical discussions about AI to concrete, practical experiments. Although the database remains predominantly qualitative and thus limited in scope, the results consistently suggest that GenAI can promote efficiency, personalized support, and creative inspiration. At the same time, pre-deployment training, structured monitoring, and critical scrutiny of AI outcomes remain of paramount importance. As these initial pilots evolve, more robust studies are needed to quantify learning gains, examine the long-term impact on skill development, and identify best practices for scaling AI-based interventions in broader student populations.

Comparative Analysis of the Perception and Use of AI Tools Across Different Levels of Education

Studies on integrating AI in Slovenian primary, secondary, and tertiary education show different perceptions, patterns of use, and challenges. For the primary level, Kermavc (2024) and Žerovnik and Zapušek (2024) describe teaching experiments with interactive, AI-controlled didactic games. Their results suggest that these tools can increase student engagement and enable differentiated instruction by adapting to individual learning styles in real-time. Despite these promising results, the use of AI at the primary level is somewhat sporadic, partly due to limited teacher training and uncertainty about how best to integrate AI into standard curricula. At the secondary level, several studies suggest that the potential of AI to improve administrative efficiency and learning outcomes is increasingly being recognized. For example, Šprah (2024) examined how secondary school students use GenAI—particularly ChatGPT—for assignments and research projects. Students generally welcomed the technology, citing its benefits in organizing information, generating ideas and improving the overall quality of their work. However, they also expressed concerns about potential misuse, including the ease with which AI-generated texts can be passed off as their own. Teachers at this level often share these reservations and call for clearer guidelines to prevent plagiarism and maintain academic integrity. At the same time, they recognize that GenAI can be a valuable addition to traditional pedagogical methods.

A recent study by Licardo et al. (2025) provides a comprehensive comparative analysis of the use of GenAI in primary, secondary, and university contexts. It highlights crucial differences in terms of familiarity, adoption rates, and main concerns among pupils, secondary school and university students, and their respective teachers:

- Pupils and teachers in primary school. Pupils are the least familiar with AI and generally see AI-based activities as something new. While teachers appreciate the potential to automate routine tasks (e.g., creating quizzes or worksheets), few consistently integrate AI into lessons. Many state a lack of formal preparation and guidelines for AI's safe, age-appropriate use.
- Secondary school students and teachers. Students here increasingly rely on AI tools, often for homework, brainstorming or planning essays. They appreciate the efficiency of AI, but are concerned about misuse (e.g., creating AI-generated essays that could be submitted as original work). Teachers at this level are only moderately familiar with AI and may have participated in technology integration initiatives, but they too would like to see clearer institutional policies to balance the benefits of AI with the need for academic honesty and sound skills development.
- University students and lecturers. Undergraduate and graduate students report that they use GenAI the most. They use tools such as ChatGPT or Bard to streamline research tasks, refine texts, or create drafts for complex projects. Yet they remain wary of potential drawbacks: from factual inaccuracies to over-reliance on automated results. University teachers, who are generally more familiar with AI than their primary or secondary counterparts, nevertheless share the same concerns about academic integrity and maintaining analytical rigor.

Overall, these comparative results show that the acceptance and use of AI increases from primary school to tertiary education, while concerns about ethical use, reliability, and the development of critical thinking skills grow in parallel. While curiosity but little structured engagement can be observed in primary school, a more pronounced—but nevertheless more cautious—approach to AI can be observed in secondary school, and acceptance is greatest, but still cautious, in higher education. At all levels, both teachers and students emphasize the importance of training, clear guidelines, and a nuanced understanding of how GenAI tools can support rather than undermine genuine learning and ethical academic practices.

Important Project Objectives and Results

The SETCOM (Supportive Environments to Enhance Transversal Competencies in Education) project, which was carried out at the University of Maribor from July 2022 to April 2024, aimed to strengthen the transversal competences of educators—in particular digital competences, AI competence, social-emotional skills and interdisciplinary collaboration (Licardo & Lipovec, 2024). These goals were closely related to the strategic educational goals and emphasized pedagogical innovation, ethical considerations, and cooperation between faculties, schools, kindergartens, and non-governmental organizations.

One of SETCOM's key achievements was the establishment of the SETCOM Competence Center, a sustainable platform that continues to provide resources, professional development, and support for education professionals (Competence center SETCOM, 2023). This center promotes ongoing interdisciplinary collaboration, research, and innovation—particularly in transversal competences. It is important to note that SETCOM was already active when GenAI tools (such as ChatGPT) were attracting much attention. Due to the existing focus on digital competences and AI skills, the project was well-positioned to respond quickly. The researchers were able to adapt the project's activities in real time and bring new insights and best practices to Slovenian educational processes at different levels.

Within the project, GenAI tools were central to improving the interdisciplinary learning experience. For example, teachers used ChatGPT to create customized didactic games—centered around Slovenian folk tales, promoting students' literacy, engagement, and teamwork skills (Kermavc, 2024). Systematic training on AI tools, ethical data handling, and critical analysis of AI-generated results further improved educators' digital competences, especially regarding data literacy and safe AI use. Evaluations found measurable improvements in competences aligned with the DigComp 2.1 framework, particularly in information and data literacy, digital content creation, and creative use of digital technology (Carretero et al., 2017; Lipovec et al., 2024). These results confirm the effectiveness of the SETCOM method, which combines theoretical teaching with practical exercises and reflective practice (Licardo & Lipovec, 2024), ensuring that the integration of AI is done in a balanced and ethically sound way.

Building on these findings, the GEN UI project has recently been launched and is supported by a broad consortium representing the wider education and computing community in Slovenia. The consortium includes:

- All three Faculties of Education: University of Primorska (UP, Faculty of Education), University of Maribor (UM, Faculty of Education) and University of Ljubljana (UL, Faculty of Education).
- Both major computer science faculties: University of Ljubljana, Faculty of Computer and Information Sciences (UL, FRI), and University of Maribor, Faculty of Electrical Engineering and Computer Science (UM, FERI).
- Other important partners: University of Ljubljana, Faculty of Arts (UL, Filozofska fakulteta), the National Institute of Education Slovenia (Zavod RS za šolstvo), the Jožef Stefan Institute (Inštitute Jožef Štefan), and the Oskar Kavičič Primary School Škofije (OŠ Oskarja Kavčiča Škofije).

This broad partnership emphasizes that GEN UI encompasses the entire education and computer sector in Slovenia and ensures a solid, interdisciplinary approach. A dedicated web portal—https://gen-ui.si/—facilitates project activities and the dissemination of results. The main focus of GEN UI is to investigate how GenAI can be used effectively in different educational settings by assessing current practices, pedagogical approaches and barriers to successful integration. In the initial phase of the project, extensive surveys and focus groups will be conducted with students, educators, and administrators to identify effective strategies, ethical considerations, and best practices for the use of GenAI. Key priorities include promoting student engagement, personalizing learning and addressing concerns such as privacy, algorithmic bias, and academic integrity.

Preliminary results suggest that students welcome AI tools to increase efficiency, but are wary of potential drawbacks—particularly the erosion of critical thinking or creative skills if AI is used without proper guidance (Flogie, 2025). Accordingly, GEN UI aims to develop structured frameworks and guidelines to ensure responsible and ethical use of AI, with educators playing a central role in supporting AI-driven learning. As the project progresses, the upcoming pilot programs will empirically evaluate how GenAI can be integrated into different educational contexts to ultimately enable evidence-based policy and curriculum development.

Conclusions

This study shows how GenAI in Slovenian education proves to be both a powerful pedagogical tool and a source of ethical and cognitive challenges. Our mixed-methods analysis shows that educators and students value GenAI for its potential to individualize learning, automate routine tasks, and stimulate creative thinking. At the same time, the research also shows concerns about over-reliance on AI-generated results, insufficient development of critical thinking, and limiting policy prescriptions.

As Dolenc K. wisely notes, "People see AI as a shortcut to achieving their own goals rather than as a tool to support those goals. As a result, bad practices often occur because AI is working instead of them, not with them." The results of our study confirm this warning: when GenAI is used merely as a time-saver or convenient alternative to student effort, the risk of undermining genuine cognitive engagement and academic integrity grows. To prevent such misuse, teachers need to be prepared and institutionally supported in the use of AI, with a focus on reflective practice and explicit instruction on the ethical boundaries of AI use.

Furthermore, our results show that GenAI cannot replace differentiated pedagogy, which is essential for cultivating higher-order thinking. Ploj Virič M. emphasizes this point: "If schools want to prepare young people for a world with artificial intelligence, the greatest impact can be achieved by teaching students critical thinking, analytical skills, research methods, how to adapt to a changing world, the habit of reading classic thick books, and ways to be alone with themselves while remaining curious and creative." These insights align with our own findings, which show that even the most sophisticated GenAI solutions depend on learners' ability to question, refine, and contextualize AI-generated content.

To summarize, an effective introduction of GenAI in Slovenia rests on three pillars: (1) a balanced integration of AI tools, (2) solid teacher training that promotes AI skills and pedagogical reflection, and (3) a clear policy framework to ensure ethical standards. If these conditions are met, GenAI can serve as a catalyst for meaningful innovation in education. It enriches student learning rather than offering shortcuts that circumvent the development of critical and creative skills.

Looking at the film through the lens of the STEM subjects, a coherent storyline emerges that spans the four disciplines rather than treating them as isolated silos. In science, GenAI opens the doors to rapid hypothesis testing with authentic data sets, allowing students to move from curiosity to empirical evidence in a single lesson. The same technology becomes both the object of study and how personalized feedback and formative assessment are delivered in real time, but its opaque inner workings force teachers to model algorithmic transparency and data-ethical dialogues. From a technical perspective, AI enhances iterative design: learners can create multiple prototypes, simulate loads, or debug code in seconds, but it is the pedagogical scaffolding that puts safety, sustainability, and reflective analysis at the center of any design. Finally, in mathematics, these same tools provide immediate feedback on modeling, statistics, and optimization tasks, strengthening numerical literacy while bringing critical evaluation of the solutions proposed by AI to the fore. Taken together, these perspectives show that GenAI transforms classrooms into living laboratories where scientific research, technological literacy, engineering creativity and mathematical rigor co-evolve, each discipline illuminating and strengthening the others when the three policy pillars of balanced integration, solid teacher preparation and clear ethical frameworks are respected.

Disclaimer This work was supported by the Slovenian Research Agency under the core projects: 'Computationally intensive complex systems', grant number P1–0403 (Ploj Virtič Mateja) and 'Information systems', grant number P2–0057 (Kosta Dolenc). During the preparation of this work the author(s) used [ChatGPT 4.0 and 4.5 / Open AI and InstaText] in order to [improve readability and language, as well as to assist in proofreading the final draft]. After using this tool/service, the author(s) reviewed and edited the content as needed and take(s) full responsibility for the content of the publication.

References

Aristovnik, A., Umek, L., Brezovar, N., Keržič, D., & Ravšelj, D. (2023). The role of ChatGPT in higher education: Some reflections from public administration students. Technology in Education. In *Innovative practices for the new normal* (pp. 254–263). https://doi.org/10.1007/978-981-99-8255-4_22

Carretero Gomez, S., Vuorikari, R., & Punie, Y. (2017). *DigComp 2.1: The digital competence framework for citizens with eight proficiency levels and examples of use.* Publications Office of the European Union. https://doi.org/10.2760/38842

Chiu, T. K. F., & Li, Y. (2023). How can emerging technologies impact STEM education? *Journal for STEM Education Research, 6*(3), 375–384. https://doi.org/10.1007/s41979-023-00113-w

Competence center SETCOM. (2023, December 24). *Kompetenčni center SETCOM* [eng. Competence center SETCOM]. https://setcom.pef.um.si/

Dolenc, K. (2024). Educational transformation: Exploring the benefits of generative AI tools. In M. Licardo & A. Lipovec (Eds.), *Artificial intelligence literacy and social-emotional skills as transversal competencies in education* (pp. 59–84). Verlag Dr. Kovač. https://www.verlagdrkovac.de/open-access/978-3-339-13814-9.htm

Dolenc, K., & Brumen, M. (2024). Exploring social and computer science students' perceptions of AI integration in (foreign) language instruction. *Computers and Education: Artificial Intelligence, 7*, 100285. https://doi.org/10.1016/j.caeai.2024.100285

Flogie, A. (2025, January 15). *Generativna umetna inteligenca* [eng. Generative artificial intelligence]. https://gen-ui.si/

Gartner, S., & Krašna, M. (2024). The oxymoron of AI in education. In M. Licardo & A. Lipovec (Eds.), *Artificial intelligence literacy and social-emotional skills as transversal competencies in education* (pp. 85–118). Verlag Dr. Kovač. https://www.verlagdrkovac.de/open-access/978-3-339-13814-9.htm

Idil, S. (2023). What could be the role of artificial intelligence in STEM education in the 21st century? In C. Dignam & M. T. Hebebci (Eds.), *Current academic studies in technology and education 2023* (pp. 157–174). ISRES Publishing.

Kermavc, M. (2024). *Umetna inteligenca kot inovativni pomočnik učitelju pri ustvarjanju didaktičnih iger na razredni stopnji poučevanja:* magistrsko delo [Master work, University of Maribor, Faculty of Education]. https://dk.um.si/IzpisGradiva.php?id=88835

Kohnke, S., & Zaugg, T. (2025). Artificial intelligence: An untapped opportunity for equity and access in STEM education. *Education Sciences, 15*(1), 68. https://doi.org/10.3390/educsci15010068

Licardo, M., & Lipovec, A. (2024). *Artificial intelligence literacy and social-emotional skills as transversal competencies in education.* Verlag Dr. Kovač. https://www.verlagdrkovac.de/open-access/978-3-339-13814-9.htm

Licardo, M., Kranjec, E., Lipovec, A., Dolenc, K., Arcet, B., Flogie, A., Plavčak, D., Ivanuš-Grmek, M., Bednjički Rošer, B., Sraka Petek, B., & Laure, M. (2025). *Generativna umetna inteligenca v izobraževanju: analiza stanja v primarnem, sekundarnem in terciarnem izobraževanju: [Gen-UI]* (1st ed.). University of Maribor Press. https://press.um.si/index.php/ump/catalog/book/950

Lipovec, A., & Flogie, A. (2023). Empowering future teachers: Unveiling their attitudes and knowledge about AI in Slovenian K-12 education. *2023 12th Mediterranean Conference on Embedded Computing (MECO)*, 1–4. https://doi.org/10.1109/meco58584.2023.10155010.

Lipovec, A., Kaučič, B., & Arcet, B. (2024). Intersecting evaluations: Digital competence and AI attitudes among students. In M. Licardo & A. Lipovec (Eds.), *Artificial intelligence literacy and social-emotional skills as transversal competencies in education* (pp. 31–57). Verlag Dr. Kovač. https://www.verlagdrkovac.de/open-access/978-3-339-13814-9.htm

Parviz, M. (2024). AI in education: Comparative perspectives from STEM and non-STEM instructors. *Computers and Education Open, 6*, 100190. https://doi.org/10.1016/j.caeo.2024.100190

Ravšelj, D., Keržič, D., Tomaževič, N., Umek, L., Brezovar, N., Iahad, N. A., Abdulla, A. A., Akopyan, A., Segura, M. W. A., AlHumaid, J., Allam, M. F., Alló, M., Andoh, R. P. K., Andronic, O., Arthur, Y. D., Aydin, F., Badran, A., Balbontin-Alvarado, R., Saad, H. B., et al. (2025). Higher education students' perceptions of ChatGPT: A global study of early reactions. *PLoS One, 20*(2), e0315011. https://doi.org/10.1371/journal.pone.0315011

Slovenian Press Agency (STA). (2023, May 28). *Over 90% of social sciences students use ChatGPT, survey suggests.* https://english.sta.si/3175736/over-90-of-social-sciences-students-use-chatgpt-survey-suggests

Sun, D., Cheng, G., Yu, P. L. H., Jia, J., Zheng, Z., & Chen, A. (2025). Personalized stem education empowered by artificial intelligence: A comprehensive review and content analysis. *Interactive Learning Environments*, 1–23. https://doi.org/10.1080/10494820.2025.2462156

Šprah, L. (2024). *Mnenja in stališča študentov razrednega pouka o generativni umetni inteligenci ChatGPT* [Master work, University of Maribor, Faculty of Education]. https://dk.um.si/IzpisGradiva.php?id=91284

Zhai, X., Neumann, K., & Krajcik, J. (2023). AI for tackling STEM education challenges. *Frontiers in Education, 8*, 1183030. https://doi.org/10.3389/feduc.2023.1183030

Žerovnik, A., & Zapušek, M. (2024). *Uporaba generativne umetne inteligence v izobraževanju (1st electronic edition)*. Faculty of Education. https://zalozba.pef.uni-lj.si/index.php/zalozba/catalog/book/226

CHAPTER 10

STEAM Teacher Education and Learning with Emerging Technologies in the Digital Age

Jan Lundell and Kristof Fenyvesi

Introduction

This chapter explores how innovative, research-informed STEM education initiatives can be leveraged to prepare educators for teaching in digitally rich, student-centered environments. Drawing from Finnish and international models, we examine three case studies from the University of Jyväskylä that highlight teacher training and student engagement with emerging technologies. The *Kitchen Chemistry Online* project demonstrates how experiential chemistry learning was adapted from teacher

J. Lundell (✉)
University of Helsinki, Helsinki, Finland
e-mail: jan.lundell@helsinki.fi

K. Fenyvesi
University of Jyväskylä, Finnish Institute for Educational Research, Jyväskylä, Finland

Global Research Institute for Finnish Education, The Education University of Hong Kong, Ting Kok, Hong Kong
e-mail: kristof.fenyvesi@jyu.fi

R. Elmas, M. Akarsu (eds.), *STEM Education*, Maarif Global Education Series,
https://doi.org/10.1007/978-981-95-9023-0_10

training into a scalable online format for broader audiences using everyday materials. The *Modeling for Understanding* course showcases how chemistry teacher education can integrate digital tools, such as visualization and computational modeling, to enhance students' conceptual understanding and higher-order thinking. Finally, a case study from the *World at Play* mathematical art exhibition illustrates how primary school students used generative AI to visualize sustainability-focused future scenarios, merging STEM content with artistic imagination, and employing a variety of digital methods and models. Together, these examples offer practical insights into how teacher education and classroom innovation can align with national and European STEM strategies, supporting digital agency, creativity, and inclusivity in the next generation of science education.

In recent decades, teaching STEM subjects has changed significantly in many ways. One of the most important changes has been from traditional teacher-led instruction toward student-centered and phenomenon-based learning methods. Alas, more actively involved in the learning process by investigating and solving problems themselves to develop their critical thinking and problem-solving skills. In addition, the integration of technology into teaching has increased significantly. Digital tools such as simulations, programming, and robotics have become part of STEM subjects, making learning more interactive and hands-on. Furthermore, connecting STEM subjects with other disciplines like art (as in STEAM, where A stands for Art) has been shown to diversify the possibilities for engaging students in science. By implementing innovative science curricula that leverage each individual's unique field of study, personal interests, or cultural background, educators can tap into original and creative approaches to learning Further, STEM subjects connected with other disciplines like art (STEAM, A=Art) has proven to diverse the possibilities of engaging in science education by innovative science curricula, making use of the skills and talents in each individual's personal field of study, personal interest, or cultural background originally and creatively (Lerman & Morton, 2009).

Another significant change is an increase in collaboration and networking among different stakeholders. Schools, universities, and companies now work together more closely, enabling more diverse learning environments and experiences for students. Such collaboration also helps teachers stay up to date with the latest scientific and technological developments. Additionally, there has been greater investment in the continuing education and professional development of teachers to provide high-quality and state-of-the-art teaching.

Beyond issues of digital access and teacher training, a number of pedagogical challenges must be addressed as schools integrate emerging technologies and AI. Adapting traditional curricula (i.e. conventional, content-centered programs) to include digital competencies and STEM skills is a pressing priority. Educators need to weave new literacies (such as coding or data analysis) into existing syllabi without sacrificing depth in core content. Developing appropriate assessment methods is equally crucial. Teachers are exploring digital assessment tools from online quiz platforms to AI-driven adaptive tests, that can evaluate students' digital literacy and problem-solving skills in real time. For example, game-based quiz systems and learning analytics dashboards can provide immediate feedback and track complex competencies, complementing traditional exams with more personalized metrics. However, teachers require training to effectively use and interpret these new assessment forms.

Student engagement and motivation remain a central focus, since digital tools can be double-edged: while they enable interactive, hands-on learning, they also risk distracting students if used without clear purpose. Educators are responding by adopting active learning strategies and carefully curated technologies. Gamified learning environments and inquiry-based projects have been used to channel students' enthusiasm for gadgets into genuine enthusiasm for STEM content. Ensuring that digital learning experiences are meaningful and aligned with learning goals helps to maintain engagement and motivation. Ensuring equity and inclusion in this digital shift is paramount. All students, regardless of gender, ethnicity, socioeconomic status, or disability, should have equal opportunities to benefit from digitally enriched STEM education. This means actively addressing the digital divide (for example, by providing devices and internet access to underserved communities) and designing content that is culturally responsive and accessible. It also means leveraging assistive technologies so that students with special needs can fully participate. AI-powered assistive tools like text-to-speech readers, speech-to-text dictation, and predictive text suggestions are already helping students with visual, hearing, or writing impairments engage with STEM content on a more level-playing field. Likewise, targeted outreach initiatives are tackling longstanding gaps in participation. Such efforts illustrate how inclusion in STEM can be improved by combining digital resources with supportive human networks.

Finally, digital ethics and safety have become critical considerations. With the increased use of online platforms, AI applications, and data in education, teachers must address cybersecurity and privacy concerns to protect students' data and ensure a secure learning environment. They are also grappling with new ethical questions arising from AI integration. This includes teaching students about algorithmic bias and the importance of fairness and transparency in AI-driven systems and clarifying issues of intellectual property when digital tools (like generative AI) create content. For instance, the European Commission has emphasized that educators should be mindful of biases or copyright issues in AI-generated content used for learning. In sum, fostering an ethical digital mindset alongside technical skills is now part of the educator's role, so that students learn to use technology responsibly and thoughtfully.

Addressing these multifaceted challenges requires a collaborative effort from educators, policymakers, and communities to create a supportive, future-ready STEM education ecosystem. Teacher preparation programs particularly need to equip future teachers with the skills and mindset to navigate this complexity.

Emphasizing contextualization and meaningful use of technology during teacher training can help new educators see digital tools not as ends in themselves, but as means to enhance learning. Notably, international policy frameworks have begun to echo and reinforce these priorities. For instance, the OECD's recent Digital Education Outlook report underscores the importance of leveraging teachers' digital competencies to harness AI's opportunities in the classroom, while also putting in place "guidelines and guardrails" to ensure AI use remains effective, equitable, and trustworthy (OECD, 2023). UNICEF's policy guidance on AI for children (UNICEF, 2021) likewise calls for principles of inclusion, fairness, transparency, and accountability in educational AI, to ensure that such tools uphold children's rights and do not exacerbate existing inequalities.

Meanwhile, the World Bank has highlighted that without investments in robust infrastructure and human capacity, even the most advanced EdTech or AI innovations will fail to benefit all learners (World Bank, 2024). It advocates comprehensive strategies for AI in education, including upgrading internet connectivity, providing teacher training, and instituting ethical governance so that countries can responsibly harness AI to make learning more personalized and accessible.

These convergent perspectives from major organizations send a clear message: successful integration of AI and emerging technologies in STEM education will rest on preparing teachers and students with new competencies, securing the necessary tools and infrastructure, and upholding strong ethical standards. By heeding this guidance, educational systems can better navigate the digital transformation while ensuring that its benefits are shared broadly and ethically.

Strategic Development of STEM Education

The Finnish STEM strategy and its implementation plan (Ministry of Education and Culture, 2023a, b) recognize teachers and teacher education as one of the cornerstones to develop evidence-based STEM education for the future. Multidisciplinary STEM activities in education and teaching promote enthusiasm for studying mathematics and natural sciences, for example, including interdisciplinary approaches such as integrating arts and design to improve learners' self-direction, interaction, and critical thinking skills.

The main goals of the Finnish STEM strategy are to ensure high-level competence in STEM across all levels of education and society. The strategy promotes socially, ecologically, and economically sustainable growth and well-being. Key objectives include enhancing the quality and accessibility of STEM education from early childhood to higher education. This is done by ensuring that students develop strong foundational skills and interest in these subjects, and as such, it is directing the science and teacher education in general (Ministry of Education and Culture, 2023a). Moreover, lifelong learning and teacher development are essential to support the competencies of current and future STEM teachers, ensuring they have the necessary skills and knowledge to inspire and educate students effectively in the changing educational landscape (Ministry of Education and Culture, 2023b). Strategic and systemic advocating for STEM throughout society is needed to encourage students to pursue further education and careers in STEM fields by highlighting the wide-ranging opportunities in and the societal impact of these professions (Ministry of Education and Culture, 2023a).

STEM education at all stages of education relies heavily on systemic and research-based collaboration between schools, universities, industries, and other stakeholders to create a cohesive and supportive STEM education ecosystem (Ministry of Education and Culture, 2023b). Simultaneously, raising awareness among parents and the public about the importance of STEM education and its role in societal development is pivotal for fostering positive attitudes toward STEM and reinforcing its relevance (Ministry of Education and Culture, 2023a).

All the goals above are designed to create a robust STEM education framework supporting Finland's long-term competitiveness and sustainability. This is evident throughout the educational system in curricula work, the development of teacher education, the implementation of digital diversity as tools for learning and teaching, as well as the strive for multidisciplinary understanding and capabilities for evidence-based decision-making.

The STEM Competence Framework

A key prerequisite for developing STEM competence at various educational levels, formal and non-formal education, is high-quality teacher education based on research, including the development of teachers' competence during their careers.

The digital revolution in education introduces an ever-evolving technological landscape in schools. This technological advancement transforms traditional methodologies into dynamic platforms of learning, with students no longer passively absorbing information; instead, they actively participate, manipulate digital content, and collaborate in real-time. Simultaneously, the pedagogical, content-related, and context-related digital skills and competences of science educators become increasingly diverse, dynamic, and demanding.

Recently, an international STEM network (Little Scientist Foundation, n.d.) outlined six important key issues in implementing the digital revolution and its tools into STEM education (Pasquinelli et al., 2024). These key areas are to integrate technology with purpose, prioritize pedagogy, enhance the use and not replace existing good practices, prepare for success, empower digital literacy, and use evidence-based practices. To effectively promote cognitive engagement in STEM education, several strategies can be employed to connect learning with digitalization and enhance student involvement. Active learning techniques, such as problem-based and inquiry-based learning, encourage students to solve real-world problems and explore questions through experiments, fostering deeper understanding and critical thinking (Pohl, 2020). Collaborative learning through group projects and peer-to-peer teaching helps students develop communication skills and learn from diverse perspectives (Barlow & Brown, 2020).

Connecting STEM concepts to real-world applications makes learning more relevant and engaging by demonstrating how these concepts apply to everyday life or current events (DeVille, 2024). Technology integration is crucial, as tools like virtual labs, simulations, and educational software create interactive and immersive learning experiences that enhance understanding and engagement (Barlow & Brown, 2020). Scaffolded instruction provides support by gradually increasing task complexity and helping students build skills and confidence while maintaining motivation.

Formative assessments offer ongoing feedback, allowing students to understand their progress and areas for improvement, fostering a growth mindset. Providing students with choices in projects or topics increases their investment and interest in the subject matter (DeVille, 2024). Teaching metacognitive strategies helps students plan, monitor, and evaluate their learning, encouraging self-reflection and goal setting (Pohl, 2020). By implementing these strategies, educators can create a more engaging and effective STEM learning environment that leverages digital tools to benefit all learners. Incorporating these strategies helps students develop self-awareness and control over their learning, leading to more effective and meaningful educational experiences. Simultaneously, these methods shift students' focus away from the digital tool itself and toward what the tool enables them to understand about a phenomenon or theoretical concept in STEM.

The main goals of the EU Digital Competence Framework (DigComp) are to enhance digital literacy and skills across Europe, ensuring that all citizens can engage confidently, critically, and safely with digital technologies. The framework outlines five key areas of digital competence (Vuorikari et al., 2022):

- *Information and data literacy:* the ability to locate, retrieve, evaluate, and manage digital information and data.
- *Communication and collaboration:* the skills to interact, communicate, and collaborate effectively using digital technologies, considering cultural and generational diversity.
- *Digital content creation:* the capability to create, edit, and manage digital content while understanding copyright and licensing issues.
- *Safety and well-being:* ensuring personal and organizational digital security, including data protection and understanding the impact of digital technologies on well-being.

- *Problem-solving:* identifying digital needs and problems, using digital tools to solve them, and staying updated with technological advancements.

Emphasizing contextualization, meaningfulness, and relevance can help generate interest in digital tools as supplementary assets that enhance teaching and learning.

These goals aim to support lifelong learning, employability, and active participation in society, contributing to the EU's broader objectives of digital transformation and economic growth (European Commission, 2024). In the digitalized world, STEM education encounters several significant challenges that need to be addressed to ensure effective learning. The digital divide remains a major issue, as unequal access to technology and high-speed internet can lead to disparities in learning opportunities, especially for students in rural or low-income areas. Teacher training and professional development are crucial, as many educators may lack the skills and confidence to integrate digital tools into their teaching. Continuous support and training are essential to keep teachers updated with technological advancements and pedagogical strategies.

Adapting curricula to include digital competencies and STEM skills is another challenge, along with developing appropriate assessment methods to evaluate students' digital literacy and STEM proficiency. While digital tools can enhance learning, they can also be distracting, making student engagement and motivation a critical focus. Ensuring that digital learning environments are engaging requires careful planning and implementation. Equity and inclusion are persistent challenges, as all students, regardless of gender, ethnicity, or disability, should have equal opportunities to benefit from STEM education. Addressing biases and promoting diversity in STEM fields is essential for creating an inclusive learning environment. Additionally, with the increased use of digital tools, cybersecurity and privacy concerns must be addressed to protect students' data and ensure a secure learning environment.

Addressing these challenges requires a collaborative effort from educators, policymakers, and communities to create a supportive and equitable STEM education ecosystem. These issues should be addressed in teacher training to equip future teachers with the skills to effectively use digital tools in teaching. Simultaneously, artificial intelligence (AI) holds immense potential to revolutionize STEM education in several keyways, and transform STEM education, making it more personalized, engaging, and accessible for all students (Holmes & Tuomi, 2022). AI can facilitate personalized learning by analyzing student performance data to create customized learning paths that cater to individual strengths and weaknesses, ensuring each student receives the necessary support to succeed (The Princeton Review, n.d.). Additionally, AI-powered tools such as virtual labs and simulations can make learning more interactive and engaging, allowing students to experiment with complex concepts safely, thereby enhancing their understanding and retention. Moreover, AI can automate grading assignments and assessments, providing immediate feedback to students and saving educators' valuable time. This helps students quickly identify and address their learning gaps (Zhai & Krajcik, 2024).

Intelligent tutoring systems driven by AI can provide real-time feedback and guidance, adapting to each student's learning pace and style for a more personalized educational experience. AI also enhances accessibility by offering tools like text-to-speech, speech-to-text, and image recognition, making STEM education more inclusive for students with disabilities. AI can also assist in curriculum development by analyzing educational data to identify trends and gaps, enabling educators to design more effective and relevant STEM programs (Tucci, 2025).

By leveraging AI's capabilities, STEM education can be transformed to become more personalized, engaging, and accessible for all students. Fredricks, Blumenfeld, and Paris (2004) define engagement in learning as a multifaceted construct that includes behavioral, emotional, and cognitive dimensions. Behavioral engagement involves students' participation in academic, social, and extracurricular activities, such as attendance, involvement in class activities, and completing assignments. Emotional engagement refers to students' emotional reactions in the classroom, including their interest, boredom, happiness, sadness, and anxiety, reflecting their emotional connection to their school and learning environment (Alrashidi et al., 2016). Cognitive engagement involves investing in learning and the willingness to exert the effort necessary to comprehend complex ideas and master difficult skills, including being strategic and self-regulated in learning. These interconnected dimensions collectively contribute to a comprehensive understanding of student engagement. Educators can create more effective and engaging learning environments by addressing all three aspects.

Three examples from Finland

Here we aim to demonstrate how educational research, collaborative and co-creative practices, and educational STEM innovations can be combined and adopted to prepare teachers for student-centered STEM pedagogy for diverse contexts and purposes. At the same time, it is important to study examples of how students use emerging technologies to engage in creative or self-directed learning processes, paving the way for the digital transformation of education (Aksela, 2019).

The following sections address effective and engaging STEM education in three different contexts. The first example focuses on engaging students in hands-on learning through online education for the public. Second, we introduce a course implementation of molecular modeling and scientific visualization tools in chemistry teacher training for STEM education. Lastly, we showcase students' engagement with emerging technologies for personalized STEAM education.

Example 1: Kitchen Chemistry Online

Inquiry-based learning has become a central theme in science education globally, emphasizing active learning and reflective thinking over merely presenting scientific facts and theories. In chemistry, this approach is linked to the three levels of the chemistry triangle (Jegstad, 2023). However, there is concern that too much focus is placed on the sub-micro and symbolic levels, making it difficult for students to connect to the macro level. Current inquiry-based chemistry education often prioritizes topic knowledge and pre-packaged science over twenty-first century skills like creativity, critical thinking, and collaboration. To address this, some educators have expanded the chemistry triangle to include models that emphasize the purpose and real-life relevance of chemistry (Sjöström et al., 2020).

Experiments play a crucial role in chemistry education, as they allow students to engage in hands-on activities, operate scientific instruments, and observe processes firsthand. By actively experimenting, theoretical knowledge is transformed into practical understanding. Experiential learning fosters critical thinking, problem-solving skills, and a deeper appreciation for the subject. Moreover, experiments can ignite curiosity and enthusiasm, making complex concepts more accessible and engaging. By actively participating in experiments, students develop a better grasp of safety protocols and laboratory techniques, preparing them for future scientific endeavors. Overall, experimenting bridges the gap between classroom learning and real-world applications, making chemistry both exciting and relevant (Gendjova, 2007; Abrahams, 2017; Oliveira & Bonito, 2023).

Chemistry in the kitchen has been denoted as a life-relevant learning (LRL) environment that engages learners in science through personally meaningful goals (Yip et al., 2012). These LRL environments can be both formal and informal, with the kitchen serving as an everyday context for science learning. They allow participants to explore scientific concepts and practices through cooking, enabling learners to apply their science knowledge to personal investigations in the kitchen. Highlighting the importance of making science relevant to learners' lives, Clegg et al. (2010) created afterschool and summer camp programs that teach science through cooking. These programs aim to engage learners, regardless of their initial interest in science, by starting with their interest in cooking and connecting it to scientific principles. This approach fosters authentic scientific practices and meaningful learning experiences. Jacobsen (2011) also suggests that using non-laboratory spaces, like home economics classrooms, can make chemistry more accessible and relatable to students' everyday lives.

Cooking-related contexts typically attract a larger audience since all people have a relationship with food. In addition to being inexpensive to implement, the context of kitchen chemistry can increase interest and enthusiasm in studying chemistry (Grosser, 1984). It presents science applications in familiar and safe contexts. It uses techniques familiar to most learners, giving them a sense of control and success. Learners realize when they prepare food, they are essentially acting as chemists, and they can explain many cooking recipes with a few scientific principles. It is well-suited for self-directed experimentation at home with instructions, which makes it also an ideal approach to popularize science to the public. In chemistry education, it provides practical means to engage context and phenomenon-based learning in an easy and cost-effective way besides introducing the importance of chemistry as science for the learners.

Online learning presents a unique set of strengths and challenges. While it offers flexibility and accessibility, it also demands significant self-regulation and technological competence. Addressing these challenges requires thoughtful implementation and continuous adaptation to ensure that online education can meet the diverse needs of all learners. One of the primary challenges for educators is adapting successful in-person teaching methods to the online environment. The debate continues whether online learning can achieve the same outcomes as traditional face-to-face instruction (Cavinato et al., 2021; Hong et al., 2021). As the landscape of education evolves, so too must our approaches to teaching and learning in the digital age.

Connecting Online and Hands-on STEM Education for Public

At the University of Jyväskylä, chemistry in the kitchen has been used for a long time as the basis for contextual learning in chemistry teacher education and professional development courses (Nuora & Välisaari, 2020). Chemistry in the kitchen is a life-relevant learning environment that engages learners in science by pursuing personally relevant and meaningful goals. As a form of interdisciplinary learning, chemistry in the kitchen aims to develop boundary-crossing skills and to support the development of pupils' scientific thinking.

Connecting chemistry to everyday things like cooking can make the learning experience more engaging, especially for non-science majors (Nguyen & Keuseman, 2020). Adapting the experiential learning approach from chemistry teacher training courses to stand-alone general audience chemistry courses, particularly in the context of kitchen chemistry, offers unique opportunities for enhancing scientific understanding. This approach leverages everyday materials and environments to make chemistry accessible and engaging for a broader audience. However, implementing hands-on inquiry in online education presents several challenges that need to be addressed to maximize its effectiveness (Roponen et al., 2025).

The phenomenon-based learning approach (Akkas & Eker, 2021; Schaffar & Wolff, 2024) is the fundamental cross-disciplinary theme in the Finnish national curriculum for primary to upper secondary schools. A self-paced online course has been developed at the University of Jyväskylä Open University continuous learning program. The aim was to teach chemistry in an easily understandable context and link it to everyday phenomena. Moreover, as chemistry heavily relies on experimentation, the interest was how to initiate personal experimentation in an online education setting, that is, allowing students to conduct scientific experiments at home.

The course content was built into four modules, and the online learning platform, Moodle, was chosen for its reliability and familiarity among students. The first module was an introduction to the course and its content, as well as practical studying issues. The three following modules explore the chemical phenomena behind food preparation and examine the composition and properties of raw materials and the changes that occur in them during cooking. Food chemistry is a natural science that occurs in everyday life, and understanding it opens new perspectives on cooking and food ingredients. Through independently studied theory and self-directed experimental work, students learn the basics of both favorable and undesirable molecular changes in the kitchen. The learning outcomes include recognizing the most common food molecules, awareness of the importance of cooking, understanding the science behind browning reactions, and recognizing the importance of acids and bases in the kitchen.

Experience and learning theories indicate that students in asynchronous online courses often require more guidance. For instance, research highlights that students in asynchronous learning environments may feel socially isolated and require more structured guidance to stay engaged and motivated (Lin & Gao, 2020). Additionally, studies have shown that while asynchronous learning offers flexibility, it also necessitates more self-directed learning, which can be challenging without adequate support and feedback from instructors (Ang et al., 2024). These findings underscore the importance of providing clear instructions, regular feedback, and opportunities for interaction to help students succeed in asynchronous online courses. Therefore, this necessitates certain pedagogical strategies in course design. Communication between teachers and students needs to be facilitated through discussion forums, where students could ask questions and teachers could post course-related messages. To ensure a logical progression, some subsections need to be locked until the previous ones are completed. Practical assignments need to be followed by surveys, which not only collect data but also encourage students to reflect on their work, making the learning experience more meaningful. Model solutions should be provided after experiments since this gives the student a reference point to compare with their own ideas (Ang et al., 2024).

Given that students in asynchronous online courses often feel isolated and need more guidance, certain pedagogical strategies were implemented in the course design (Lin & Gao, 2020; Ang et al., 2024). Communication between teachers and students was facilitated through a discussion forum, which students used to ask questions and by teachers to send messages about course-related matters. The logical progression of the studies was ensured by keeping some subsections locked until the completion of previous ones. Practical course assignments were followed by surveys, which served as a data collection method while also directing students to reflect on their own work, making the overall experience more meaningful than just completing the experiment as a chore. Model solutions after the experiments provided students a point of reference to compare with their own ideas and to understand the importance of chemistry in food preparation. The course was graded pass or fail, and to pass, students had to complete all questionnaires, report performing the experiments, and acceptably manage all exams at the end of each learning module (Roponen et al., 2025).

Adopting experiential learning from chemistry teacher training courses to stand-alone general audience chemistry courses, particularly in the context of kitchen chemistry, presents several challenges in online education. One significant issue is the lack of physical interaction, as hands-on inquiry often requires direct manipulation of materials and equipment, which is difficult to replicate online. This absence can hinder the development of practical skills. Additionally, students may face limited access to necessary tools and materials at home, restricting their ability to perform experiments or engage in practical activities. Keeping students engaged and motivated in a virtual environment is another challenge, as the lack of immediate feedback and interaction can make it harder for students to stay focused. Technical issues, such as unreliable internet access and inadequate technology, can disrupt the learning process and create barriers to effective participation (Bekteshi, Gollopeni & Avdiu, 2023). Providing timely and effective feedback on hands-on activities is challenging online, making it difficult for teachers to accurately assess students' practical skills without observing them in person. Educators can adopt a blended learning approach to overcome these challenges, combining online theoretical instruction with periodic in-person practical sessions (York, 2020). Automatic and personalized feedback can be employed to support the students tackling in their studies. Usually, this is managed by a simple Question and Answer (Q&A) approach, but learning challenges, as well as study and time management issues, can be tackled with more personalized support (Tewari, 2025). Additionally, virtual labs and simulations can help bridge the gap by providing interactive and immersive experiences. The simplest supportive measure is to provide video clips that clarify practical work or explain scientific concepts prone to misunderstanding. By integrating approaches like relatable context, active learning, regular feedback, personalization, interactive discussions, and virtual office hours, as well as using engaging visuals, students can develop a deeper appreciation for chemistry and its practical applications in everyday life (Delaney, 2023; van Heerden, 2024; Hilner, 2025).

Example 2: Modeling for Understanding

As a branch of science, chemistry is inherently abstract, dealing with concepts that are often invisible to the naked eye. Educators rely heavily on various representations and models to bridge the gap between these abstract concepts and students' understanding. Understanding how to use these tools pedagogically is crucial in making chemistry education effective and meaningful. Representations and models in chemistry serve as essential tools for visualizing and conceptualizing chemical phenomena. They include symbolic representations like chemical formulas and equations, particulate representations describing, for example, molecular structures and diagrams, and macroscopic representations representing laboratory experiments and real-world examples (Johnstone, 2000). Each type of representation offers a unique perspective, helping students to grasp complex ideas by linking the abstract with the concrete.

Effective pedagogy involves selecting and using representations that align with the learning objectives and the students' cognitive levels (Ainsworth et al., 2020). For instance, when teaching chemical bonding, educators might use ball-and-stick models to illustrate the spatial arrangement of atoms, Lewis structures to show electron sharing, and energy diagrams to explain bond formation and breaking. By integrating these different representations, teachers can provide a comprehensive understanding of the concept.

Misconceptions in chemistry often arise from students' difficulties translating between different representations (Permatasari et al., 2022). For example, students might struggle to connect the symbolic representation of a chemical equation with the particulate nature of the substances involved. Pedagogical strategies that emphasize the connections between these representations can help address and correct these misconceptions. Teachers need to be aware of common pitfalls and design their instruction to guide students through these challenging transitions.

Using diverse representations and models can make chemistry more engaging and relatable for students. Interactive simulations, virtual labs, and 3D models can bring abstract concepts to life, sparking curiosity and motivation when students see the relevance of chemistry to real-world problems and applications, their engagement increases. Effective pedagogy involves using these tools and creating a learning environment where students feel encouraged to explore and ask questions. The ability to interpret and use different representations is a critical higher-order thinking skill in chemistry. It involves analysis, synthesis, and evaluation—skills that are essential for scientific inquiry and problem-solving. Pedagogical approaches that challenge students to create their own models, predict outcomes, and explain their reasoning can foster these skills. Teachers play a key role in scaffolding these activities, providing the necessary support while gradually increasing the complexity of tasks (Gilbert & Treagust, 2009).

Different students have different learning styles and needs. Some may find visual representations more helpful, while others might prefer hands-on models or interactive simulations. Pedagogical strategies incorporating various representations can cater to these diverse preferences, making chemistry education more inclusive. Additionally, technology can provide accessible options for students with disabilities, ensuring that all students have an opportunity to succeed.

Teacher Education Course on Models and Visualization

Modern computer-based molecular modeling opens new possibilities for chemistry teaching at different levels. An early study sought insight into Finnish school teachers' use of computer-based molecular modeling in teaching chemistry, into the various working and teaching methods used, and their needs for necessary support (Aksela & Lundell, 2008). The study suggested that most of the teachers studied need to personally discover the benefits of molecular modeling in their own work, which illustrates on a practical level how molecular modeling can provide added value for teaching and understanding in school chemistry. Teachers were found to need more pedagogical and technical training in molecular modeling so they can use it more and effectively in their own teaching. Furthermore, there appeared to be a need for easily adaptable learning and teaching materials to be made available to teachers in their domestic teaching language. The study was based on professional development training in the summer seasons for voluntary chemistry teachers. The research and the collaboration with teachers at school, besides the rapid advancement of digitalization and use of digital tools in chemistry research, led to the conclusion that it is crucial to provide teacher students the practical skills and experience of these new possibilities to enhance chemistry education in schools (Aksela & Lundell, 2008).

Concomitantly, a training course on "Chemistry models and visualization" is a combination of chemical theory and hands-on activities for chemistry teacher students implemented for the last decade at the University of Jyväskylä. The course gives 5 ECTS and it is aimed to provide future chemistry teachers fundamental and self-experienced skills in using digital tools in chemistry learning and teaching. The course discusses the application of computational chemistry, molecular modeling and computer-aided chemistry in illustrating and explaining chemical concepts and phenomena, as well as the possibilities of computer-aided chemistry in chemistry teaching as part of high-quality multi-modal education. The course pays particular attention to various visualization and modeling programs suitable for school use to support the interconnectivity of the three levels of chemical information to understand chemical concepts and phenomena.

The learning outcomes of the course are outlined as follows: The student

- *knows* the basic principles of models and modeling related to chemistry teaching.
- *can convey* chemical information meaningfully using different chemistry presentation methods.
- *knows* the possibilities of modern information and communication technologies (ICT) applications in chemistry teaching.
- *masters* the basic use of modern ICT applications in teaching situations.
- *can integrate* the possibilities offered by ICT into their own teaching.

Moreover, for a future chemistry teacher, it is important to recognize the working life skills acquired during the training like

- *able to use* drawing and modeling programs suitable for teaching.
- *competent in* the basic knowledge and skills of modeling and visualization suitable for chemistry teaching.

- *able to evaluate* the benefits and challenges of digital tools in implementing teaching.
- *able to guide* learning in a digital and visual operating environment.

The course content addresses especially chemistry concepts and phenomena considered difficult and abstract in upper secondary school education. The focus of the content is on 2D/3D visualization of molecules, orbitals and electron density, chemical bonding, chemical isomerism and reactivity, and vibrational spectroscopy. Moreover, the course gives an overview of digital tools implemented in the Finnish national chemistry curriculum and especially tools that are used in the digital matriculation examination test.

The course is built on five separate modules (see Table 10.1) and is supported by the Moodle digital learning environment. The modules represent the progression of general knowledge toward more elaborated use of digital tools for clarifying, understanding and doing chemistry. The structure of the online material repository is the same for all the modules, and it is based on three exemplary scientific research papers connected with reflective questions, and one graded exercise combining theory and hands-on activity. The overall grade assessment is based on the combined results of the grades from each module.

Measuring the impact of molecular modeling tools on learning can be approached through several methods:

- *Pre- and post-tests:* Assess students' knowledge and skills before and after using the tools to measure improvement. This can include tests on specific content knowledge, problem-solving abilities, and application of concepts (Wu, Krajcik & Soloway, 2001).
- *Surveys and questionnaires:* Collect feedback from students about their experiences, engagement, and perceived learning gains. This can help gauge their attitudes toward the tools and their self-reported learning outcomes (van Enckevort, 2014)
- *Observations:* Monitor students during the learning process to see how they interact with the tools, collaborate with peers, and apply their knowledge. This can provide insights into their learning behaviors and the effectiveness of the tools in facilitating understanding (Taber, 2017).
- *Interviews and focus groups:* Conduct in-depth discussions with students and teachers to gather qualitative data on their experiences and

Table 10.1 Learning modules of the course

	Focus area	*Teaching material*	*Assessment*
Module 1	ICT tools in chemistry	Open access web resources. University learning material on ICT and AI ethics.	The material forms a handbook of the school chemistry applications in use.
Module 2	Models and modeling in chemistry	Open access web resources Research papers on models and modeling Online drawing and modeling tools MarvinSketch	Three scientific papers on models in chemistry, mental models and teachers' models. Graded exercise #1 on MarvinSketch modeling
Module 3	Drawing and visualization	Open access web resources Research papers on models and modeling ACD ChemSketch	Three scientific papers on models in chemistry, mental models and teachers' models. All sections involve pass/fail open questions. Graded exercise #2.
Module 4	Animations and simulations in chemistry	Open access web resources Research papers ChemSense	Three scientific papers on models in chemistry, mental models and teachers' models. All sections involve pass/fail open questions. *Graded exercise #3.*
Module 5	Molecular modeling in chemistry	Research papers Specific lecture and exercise materials on computational chemistry; video material for exercises Spartan Edu	Compulsory laboratory exercises. *Graded exercise #4.*

perceptions. This can help identify strengths and areas for improvement in the use of molecular modeling tools (van Enckevort, 2014)

- *Performance tasks:* Assign tasks that require students to use the molecular modeling tools to solve real-world problems or complete projects. Evaluate their performance to determine how well they can apply what they've learned (Taber, 2017).
- *Comparative studies:* Compare the learning outcomes of students who use molecular modeling tools with those who use traditional methods. This can help determine the added value of the tools in enhancing learning (Quayson et al., 2022).

By combining these methods, educators can obtain a comprehensive understanding of the impact of molecular modeling tools on student learning. However, for a pedagogically meaningful teacher training course, a more interactive approach needs to be adopted. First, to adapt to the personal needs of the students, the course has been run in three different formats:

Course model 1 The course is delivered fully on-site with weekly interactive and applying student-activated teaching methods combining both classroom and computer laboratory time. In this model, the Moodle learning platform is used only as a site to deliver the course material and a site for returning the graded exercises. All reflective exercises are discussed in peer-to-peer and group discussion sessions. The course involves a school group visit for which a pair of students develop and teach a modeling workshop for the visiting students. The numeric assessment is based on the graded exercises, and self-assessed and peer-assessed school visit.

Course model 2 The course is a distant learning course where all the activity happens on the Moodle learning platform. The assessment takes place only based on the activity and graded exercises online. This possibility has been employed only in cases where there are restrictions in using the computer laboratory facilities on-site.

Course model 3 This model combines the on-site and online teaching in a hybrid model. The course can be run in a distant mode but with three to four hours long laboratory-based workshops included with each module teaching weeks. For the computational chemistry module, it has been noted that it is very beneficial to have at least two to three seminars to introduce the students deeply in the modeling program and its possibilities.

Secondly, for the benefit of the teacher students, more interactive and reflective teaching is needed. Besides mastering the digital tools, the students need to have time to discuss and consider the connection with the chemistry topics and as well as the pedagogical means to apply the digital tools. The teachers need to experience working with the tools themselves, and also understand the pedagogical and subject-specific rationales for using them. This pedagogical content knowledge aspect is needed to transfer the skills learned during the course to actual school settings.

Finally, the assessment of the course is based on four graded exercises. These exercises are constructively built-on the same chemical topic—usually on chemical bonding—to teach the students how the different digital tools can give different pedagogical and scientific possibilities for teaching, learning and assessment practices. However, the overall learning environment should support discussions, comparisons and testing of pedagogical solutions to use digital tools in everyday chemistry education. An added bonus of the course modularity is also to use the same course and the activities in-built there for professional development programs for in-field teachers or combined in-field teacher and teacher student groups.

Benefits of Digital Tools for Learning Chemistry

Chemistry teacher course on pedagogical use of digital tools on models, visualization, and computational chemistry supports higher-order thinking and understanding in chemistry education. The necessity of various representations in learning STEM subjects like chemistry lies in their ability to make complex and abstract concepts more understandable and relatable for students. Chemistry involves understanding phenomena at different levels: the macroscopic (what we can see and measure), the submicroscopic (atoms and molecules), and the symbolic (chemical equations and formulas) (Johnstone, 2000; Jegstad, 2023).

Using multiple representations helps bridge these levels, allowing students to visualize and conceptualize the interactions and processes that occur (Gilbert et al., 2008; Treagust, 2018). Each representation provides a different perspective, enhancing students' overall comprehension and ability to apply their knowledge in various contexts.

Incorporating digital tools like structural and data visualizations, simulations and animations can further enrich the learning experience by providing dynamic and interactive ways to explore chemical concepts (Permatasari et al., 2022). This multifaceted approach not only aids in deeper understanding but also helps in reducing misconceptions and improving problem-solving skills (Kelly & Akaygun, 2019).

Molecular modeling and visualization play a crucial pedagogical role in chemistry learning by helping students understand and internalize complex concepts that are often abstract and difficult to grasp. These tools provide visual and interactive representations of molecular structures and reactions, making it easier for students to comprehend the spatial arrangements and dynamic behaviors of molecules (Jones et al., 2005). Moreover, teachers need to be trained to use these digital resources pedagogically.

The target for using modern, research-related digital tools enhance conceptual understanding since using molecular models and visualizations students can see the three-dimensional structures of molecules and how they interact. This helps them understand concepts such as molecular geometry, bonding, and reaction mechanisms more intuitively (Quayson et al., 2022). Teachers and students alike need to engage in learning experiences. Interactive simulations and visualizations make learning more engaging and enjoyable. Learners can manipulate models, observe molecular interactions in real-time, and conduct virtual experiments, which can enhance their interest and motivation in studying chemistry (Treagust, 2018).

From a pedagogical point of view, it is crucial to bridge different representational levels. Molecular modeling helps bridge the gap between macroscopic observations (what we see in the lab) and microscopic phenomena (atomic and molecular interactions). This multilevel approach aids in developing a deeper and more integrated understanding of chemical principles (Jones et al., 2005). Additionally, using and learning with research-based tools enhances problem-solving skills. Working with molecular models encourages students to develop critical thinking and problem-solving with authentic research tools. They learn to predict molecular behavior, analyze reaction pathways, and solve complex chemical problems using visual and computational tools (Quayson et al., 2022).

Overall, pedagogical use and understanding of different representations and models are fundamental for effective chemistry education. Molecular modeling and visualizations are invaluable in making chemistry more accessible, interactive, and comprehensible for students. They enhance conceptual understanding, address misconceptions, foster engagement, develop higher-order thinking skills, and promote inclusivity (Khine & Saleh, 2011). Educators who are skilled in using these tools can create a dynamic and supportive learning environment that not only makes chemistry accessible but also inspires a lifelong interest in the subject. As the field of education continues to evolve, the integration of innovative ways to manage, apply, and present representations and models will remain a cornerstone of effective chemistry teaching.

Example 3: Mathematical Art and Generative AI in Education

The interplay of mathematical art and generative AI offers a novel avenue for STEAM education, expanding analytical thinking through design-based approaches such as speculative design and creative pedagogies. These approaches are essential for engaging learners with complex, interconnected topics—such as sustainability or planetary challenges—by fostering both critical inquiry (integrating logic and abstraction with ethical values) and imaginative exploration and problem-solving. In Finland, the national STEM strategy explicitly encourages integrating arts and design into STEM teaching to spark enthusiasm and innovation in learning (Ministry of Education and Culture, 2023b). This aligns with broader objectives to ensure high-level competence in science and mathematics through multidisciplinary approaches. It also reflects Finland's commitment to educational innovation as outlined in the National Roadmap for Research, Development and Innovation 2030, which calls for new partnerships and creative pedagogies to keep the country at the forefront of knowledge and technology (Ministry of Education and Culture, 2021).

At the European level, the importance of such integrative competencies is underscored by the EU's Digital Competence Framework for Educators (European Commission, 2020), which highlights that educators should be able to leverage emerging digital tools to enrich learning and foster students' creative skills (Redecker & Punie, 2017). Within this context, mathematical art combined with generative AI (GenAI) can serve as a powerful case study of innovative STEAM pedagogy in teacher education and practice. It also supports the European Commission's guidelines on the responsible use of generative AI in education (European Commission, 2025), emphasizing learner agency, ethical engagement, and co-creation in AI-augmented learning environments.

One illustrative example is the "World at Play" mathematical art exhibition (2023–2024), an initiative in Finland that blended visual arts, sustainability education, and AI-driven creativity. In this museum exhibition, developed initially as part of an international math-art program, nearly 2000 6th-grade students from Central Finland engaged with mathematically-inspired artworks created by peers from South Africa and then used a text-to-image generative AI platform to visualize their own sustainability-related visions. In practice, Finnish students were inspired by math-based art pieces addressing issues from UNESCO's set of Sustainable Development Goals, like climate change or biodiversity (United Nations, 2015). The first artworks, serving as initial artistic 'prompts,' were created by South African youth through a math-art competition. The Finnish students were invited to respond to the South African works by realizing their ideas for a sustainable future in a hands-on and AI-augmented way, considering their own local or planetary context. The Finnish students' AI-generated artworks were collected along with the related prompts for research purposes.

From a policy perspective, this case demonstrates how grassroots pedagogical innovation can inform teacher education and curriculum development. It offers a practice-based example of how GenAI can be integrated in ways that enhance, rather than undermine, learner agency and teacher professionalism—a concern echoed in the latest EU AI Act, UNESCO's AI Competency Framework for Teachers (2024), and Finnish national discussions on digitalization in education.

The GenAI system enabled the Finnish children to transform their concepts into unique images, essentially co-creating art with the assistance of AI. This process combined artistic expression with STEM concepts, encouraging learners to explore the creative intersections of technology and imagination. Crucially, the activity did not require advanced technical skills; even young students could input descriptive prompts into the AI tool and see a visual representation of their idea emerge. Teachers and facilitators guided hands-on STEAM workshops and discussions around the generated images, linking them back to mathematical patterns, scientific concepts, or societal impacts embedded in the art.

The World at Play case exemplified how GenAI can lower the barrier to entry for creative or speculative design in the classroom, allowing students of various ages and backgrounds to participate in interdisciplinary problem-solving. It shifted the teacher's role toward being a facilitator and curator of an enriched learning experience—a role that Finland's teacher education programs increasingly emphasize (Ministry of Education and Culture, 2023b).

By experiencing how AI can augment creativity and expression, pre-service and in-service teachers involved in such projects gain first-hand insight into the pedagogical potential of generative technologies. Educational outcomes from the World at Play project were notably aligned with key twenty-first-century skills. Students engaged in this AI-supported art exploration demonstrated heightened critical thinking, creativity, and collaboration—competencies essential for addressing complex global challenges (United Nations, 2015). The exhibition's design fostered a planetary dialogue on sustainability: Finnish children learned about issues through the eyes of their South African counterparts and then reflected on those issues by creating both traditional and AI-assisted art, thereby practicing empathy and systems thinking.

Research on informal STEAM learning environments supports these observations. In settings like science centers, art exhibits, or makerspaces, students often exhibit deeper engagement and curiosity than in traditional classrooms, thanks to the multisensory and exploratory nature of the experience (Thuneberg, Salmi, & Fenyvesi, 2017). The World at Play exhibition took place in a museum environment and leveraged these advantages of informal learning.

Surrounded by vivid student-created artworks and experimenting with hands-on STEAM activities and the interactive AI station, learners were physically and intellectually immersed in the topic. Such hands-on, object-based learning has been shown to foster critical thinking by allowing students to directly interact with authentic materials and phenomena. Moreover, these kinds of interactive environments provided opportunities for experimentation and collaboration that complement formal instruction (Citrin & Kney, 2013). In the World at Play program, for example, students often worked in pairs or small groups at the AI station, discussing what keywords to input and interpreting the AI-generated images together. This collaborative problem-solving mirrors professional STEAM practices and encourages peer learning.

From a practical standpoint, the use of generative AI and art to teach sustainability proved highly engaging. By expanding learning beyond textbooks into an interactive, creative realm, the initiative demonstrated a scalable model for enriching STEM learning experiences. Notably, this model can be adapted to formal classroom settings. While World at Play was implemented in a museum as a special event, its core approach—combining art, technology, and collaborative problem-solving around STEM themes—can be implemented throughout the fields of science education to build upon the added value of modern technology for meaningful learning experience.

Each AI-generated artwork told a story about a possible future, effectively merging narrative and visual expression. Some students envisioned scenes of technology harmonizing with nature (like the solar-powered city in harmony with lush greenery), while others depicted solutions to environmental problems (e.g. AI-generated images of oceans teeming with restored marine life, based on prompts about cleaning plastic pollution). In every case, the ideas originated from the students and the AI served as one of the tools, in addition to the possibility to create hands-on models and engaging rapid prototypes, to amplify their creative expression and make their ideas visible. This co-creation approach encouraged learners to explore the intersections of technology and imagination, art and science. Teachers and facilitators guided follow-up discussions around the generated images, prompting students to explain the math or science concepts hidden in their art and to reflect on the societal implications. In this way, the activity linked back to underlying STEM content: for instance, a fantastical image of an engineered forest might lead to a conversation about real carbon sequestration, geometry in nature, or the biodiversity concepts that inspired it.

From a practical standpoint, the use of generative AI and art to teach sustainability proved to be engaging. By expanding learning beyond textbooks into interactive and creative learning environments, the initiative demonstrated a scalable model for enriching STEM learning experiences. For STEAM teacher education, the implications of this case are significant. First, it highlights how teachers can act as designers of interdisciplinary learning experiences that harness students' natural creativity.

Finnish teacher preparation has increasingly incorporated design-based and phenomenon-based learning approaches (Aksela, 2019), and the integration of mathematical art with AI is a natural extension of this trend. This case also contributes to understanding mathematical art as a form of computational thinking and creative problem-solving, a perspective championed in Finnish LUMA pedagogies and emerging EU-funded research on AI in education (Holmes & Tuomi, 2022). By positioning AI not merely as a tool but also as a co-agent in distributed creativity (Ness & Glaveanu, 2019), the World at Play activities encouraged educators and researchers to reimagine the design of inclusive, future-oriented learning environments where mathematical reasoning, artistic expression, and AI-supported imagination converge. Teacher trainees exposed to projects like World at Play learn to appreciate the value of connecting STEM content with artistic practices—a strategy shown to improve students' attitudes toward science and mathematics (Thuneberg et al., 2017). They also gain familiarity with cutting-edge digital tools. Generative AI is rapidly evolving; by encountering it during training, future educators develop digital competence and critical awareness of AI's capabilities and limitations.

The European DigCompEdu framework emphasizes that teachers should not only be competent in using digital tools themselves, but also adept at facilitating learners' creative use of digital technology (Redecker & Punie, 2017). In the math-art AI project, teachers had to guide students in translating abstract ideas into textual prompts and interpreting the algorithm's output. This required pedagogical sensitivity—ensuring that the technology was used with a clear purpose and that students remained the authors of their creative work. Such practice aligns with recommended guidelines for integrating AI in education, which stress using technology with pedagogical intent, augmenting (not replacing) human creativity, and preparing learners with digital literacies (Pasquinelli et al., 2024).

Indeed, the teachers involved in World at Play framed the AI as a supportive tool—an "assistant" to visualize ideas—rather than as the source of knowledge. This approach empowered students to drive the creative process, with the technology amplifying their imaginations. From a teacher training perspective, this serves as a model for how to introduce AI in the classroom: integrate with purpose, prioritize the learning goals (in this case, sustainability and cross-cultural understanding), and use the tool to enhance student agency and engagement (Pasquinelli et al., 2024).

Another important aspect for teacher education is the collaborative and cross-sector nature of such initiatives. The World at Play exhibition was the result of partnerships between universities (both in Finland and South Africa), schools, and informal learning institutions (an art museum). This exemplifies the kind of collaboration that national strategies are calling for to strengthen STEM education ecosystems (Ministry of Education and Culture, 2023a). Finnish policy, for example, explicitly supports deeper cooperation between science education actors and teacher education programs (Ministry of Education and Culture, 2021). By participating in joint projects with museums and international partners, teacher candidates and practicing teachers broaden their professional horizons. They learn how community resources and global networks can be leveraged to create rich learning experiences for students. In the exhibition project, teachers had to coordinate with exhibition curators and possibly even communicate with the South African organizers of the math-art competition. Such experiences build teachers' capacity in project management and intercultural communication skills that are increasingly valuable as classrooms become more connected and interdisciplinary.

For instance, a group of Finnish teachers involved in the exhibition later reported greater confidence in guiding project-based learning and using art and story to explain scientific concepts. This kind of reflective feedback is essential in teacher education, reinforcing theoretical knowledge with lived practice. It also resonates with the EU's vision of teachers as innovators and lifelong learners who continuously update their methods (Redecker & Punie, 2017). In short, engaging with mathematical art and GenAI in teacher training not only equips educators with specific new tools but also instills a mindset of creativity, experimentation, and collaboration.

Looking more broadly, the integration of AI into STEM/STEAM education is gaining momentum worldwide, and examples like this serve as valuable case studies for shaping future practices. Early evidence from such initiatives is encouraging researchers and policymakers to pay attention. For example, a recent collection of international studies highlights diverse ways AI can transform STEM learning, from intelligent tutoring systems to creative coding and art generation (Zhai & Krajcik, 2024). These developments underscore that generative AI is not just a technical topic for computer science classes but a cross-cutting tool that can make traditionally abstract subjects more tangible and student-centered.

In the context of mathematics education, generative AI can help visualize complex patterns or bring real-world data to life through art and simulation, thereby engaging learners who might not be reached through formulae alone. For teachers, the challenge is to integrate these tools thoughtfully and ethically. Finnish and EU frameworks both insist on the dual importance of innovation and responsibility: educators should feel encouraged to try new digital approaches but also be guided by pedagogical best practices and awareness of issues like bias or intellectual property in AI-generated content (European Commission, 2020; Zhai & Krajcik, 2024).

In the World at Play project, these considerations were evident: students, for instance, discussed the difference between human-made art and AI-made images, and who "owns" the creations, which opened conversations about the nature of creativity in the age of AI. Such discussions are invaluable for teacher trainees, as they prepare them to address similar questions in their future classrooms. They illustrate how integrating mathematical art and GenAI not only enriches subject matter learning but also provides a platform for critical digital literacy.

Conclusion

In conclusion, each of the three case studies presented in this chapter—Kitchen Chemistry, Modeling for Understanding, and Mathematical Art and Generative AI—offers a distinct, practice-based insight into how STEM and STEAM teacher education can evolve in response to emerging technologies and digital pedagogies.

In Kitchen Chemistry, educators explored how everyday materials and digital platforms can democratize access to experimental science. By integrating remote learning methods with hands-on inquiry, the course aligned with both the Finnish LUMA (STEM) Strategy and the EU's emphasis on equitable, inclusive STEM education. This approach enabled students to cultivate scientific thinking in non-traditional environments, fostering a sense of agency and curiosity that extended beyond the classroom.

The Modeling for Understanding course exemplified how advanced tools, such as molecular visualizations and computational chemistry, can deepen conceptual learning in STEM subjects. This aligns with DigCompEdu's emphasis on pedagogical innovation and the development of educators' digital competence. It also reflects Finnish education's commitment to inquiry-based learning and interdisciplinary integration. Here, teacher trainees were not just users of tools but also reflective practitioners learning to support higher-order thinking and knowledge transfer.

Finally, the World at Play mathematical art and GenAI case study illustrated how creativity and emerging technologies can intersect meaningfully to promote systems thinking and sustainability education. Through the co-creation of visual outputs with text-to-image AI platforms, students navigated complex concepts related to environmental futures, translating abstract ideas into expressive, multimodal representations. This practice not only activated mathematical reasoning but also supported digital and cultural literacies, aligning with the EU's Digital Education Action Plan (European Commission, 2020) and UNESCO's competencies for transformative education.

Taken together, these cases underscore a vital insight: effective STEAM education in the digital age demands multidimensional integration—of disciplines, modalities, spaces, and epistemologies. They show how national strategies (like Finland's STEM strategy and development plan) and supranational frameworks (such as DigCompEdu) can be enacted meaningfully at the ground level through thoughtful curriculum design, teacher training, and creative pedagogy.

For educators, these projects provide actionable models. They demonstrate that new technologies, whether generative AI, modeling software, or blended learning environments, are most impactful when anchored in strong pedagogical foundations and connected to real-world relevance. In doing so, they offer a roadmap for navigating the challenges of digital transformation, while reinforcing the role of teachers as agents of innovation and interdisciplinary learning.

As Finnish and European education systems move toward inclusive, future-oriented, and sustainable STEM and STEAM education, these cases offer grounded, research-informed pathways forward. They remind us that fostering scientific and technological literacy is not just about tools or content but about nurturing empowered learners and teachers capable of engaging with the ethical, societal, and creative dimensions of knowledge in the age of AI.

Disclaimer Generative AI tools (e.g., ChatGPT) were utilized to enhance language refinement and stylistic clarity during the preparation of this chapter. All AI-generated suggestions were critically reviewed and validated by the authors to ensure accuracy and academic integrity.

References

Abrahams, I. (2017). Minds-on practical work for effective science learning. In K. S. Taber & B. Akpan (Eds.), *Science education. New directions in mathematics and science education*. SensePublishers. https://doi.org/10.1007/978-94-6300-749-8_29

Ainsworth, S., Tytler, R., & Prain, V. (2020). Learning by construction of multiple representations. In P. van Meter, A. List, G. Lombardi, & P. Kendeou (Eds.), *Handbook of learning from multiple representations and perspectives* (pp. 92–101). Routledge.

Akkas, E., & Eker, C. (2021). The effect of phenomenon-based learning approach on students' metacognitive awareness. *Educational Research and Reviews, 16*(5), 181–188. https://eric.ed.gov/?id=EJ1296934

Aksela, M. (2019). Towards student-centred solutions and pedagogical innovations in science education through co-design approach within design-based research. *LUMAT: International Journal on Math, Science and Technology Education, 7*(3), 113–139. https://doi.org/10.31129/LUMAT.7.3.421

Aksela, M., & Lundell, J. (2008). Computer-based molecular modelling: Finnish school teachers' experiences and views. *Chemistry Education Research and Practice, 9*(4), 301–308. https://doi.org/10.1039/B818464J

Alrashidi, O., Phan, H. P., & Ngu, B. H. (2016). Academic engagement: An overview of its definitions, dimensions, and major conceptualisations. *International Education Studies, 9*(12), 41–52. https://doi.org/10.5539/ies.v9n12p41

Ang, J. W. J., Ng, Y. N., Lee, L. H.-W., & Yong, J. Y. (2024). Exploring students' learning experience and engagement in asynchronous learning using the community of inquiry framework through educational design research. *Education Sciences, 14*(3), 215. https://doi.org/10.3390/educsci14030215

Barlow, A., & Brown, S. (2020). Correlations between modes of student cognitive engagement and instructional practices in undergraduate STEM courses. *International Journal of STEM Education, 7*(1), 18. https://doi.org/10.1186/s40594-020-00214-7

Bekteshi, E., Gollopeni, B., & Avdiu, E. (2023). The challenges of conducting online inquiry-based learning among tertiary level education. *Journal of Technology and Science Education, 13*(1), 92–103. https://doi.org/10.3926/jotse.1700

Cavinato, A. G., Hunter, R. A., Ott, L. S., & Robinson, J. K. (2021). Promoting student interaction, engagement, and success in an online environment. *Analytical and Bioanalytical Chemistry, 413*(6), 1513–1520. https://doi.org/10.1007/s00216-021-03178-x

Citrin, R. A., & Kney, A. D. (2013, June). *Future Worlds: Development of assessment methods for an interactive cyberlearning platform for informal explorations in sustainability for students ages 9–12.* Paper presented at the 2013 ASEE Annual Conference & Exposition, Atlanta. https://peer.asee.org/future-worlds-development-of-assessment-methods-for-an-interactive-cyberlearning-platform-for-informal-explorations-in-sustainability-for-students-ages-9-12-research-to-practice.pdf

Clegg, T. L., Gardner, C., & Kolodner, J. L. (2010). Playing with food: Moving from interests and goals into scientifically meaningful experiences. In *Proceedings of the Ninth International Conference of the Learning Sciences (ICLS)* (Vol. 1, pp. 1135–1142). https://repository.isls.org/bitstream/1/2657/1/1143-1150.pdf

Delaney, P. (2023, December 12). *Limited access to hands-on learning experiences: Navigating the challenges of online lessons.* Online Lesson. https://www.onlinelesson.co.uk/challenges-of-distance-learning-limited-access-to-hands-on-learning-experiences

DeVille, K. (2024, May 31). *Student engagement strategies: Boosting participation in the classroom.* STEM Education Guide. https://stemeducationguide.com/student-engagement-strategies/

European Commission. (2020). D*igital Competence Framework for Educators (DigCompEdu)*. https://joint-research-centre.ec.europa.eu/digcompedu_en.

European Commission. (2024). *2030 digital decade: Report on the state of the digital decade 2024*. Publications Office of the European Union. https://digital-strategy.ec.europa.eu/en/factpages/state-digital-decade-2024-report

European Commission. (2025). *Living guidelines on the responsible use of generative AI in research*. https://research-and-innovation.ec.europa.eu/document/download/2b6cf7e5-36ac-41cb-aab5-0d32050143dc_en?filename=ec_rtd_ai-guidelines.pdf

Fredricks, J. A., Blumenfeld, P. C., & Paris, A. H. (2004). School engagement: Potential of the concept, state of the evidence. *Review of Educational Research, 74*(1), 59–109. https://www.jstor.org/stable/3516061

Gendjova, A. (2007). Enhancing students' interest in chemistry by home experiments. *Journal of Baltic Science Education, 6*, 5–15. https://www.scientiasocialis.lt/jbse/files/pdf/vol6/5-15.Gendjova_JBSE_Vol.6_No.3.pdf

Gilbert, J. K., & Treagust, D. (Eds.). (2009). *Multiple representations in chemical education*. Springer.

Gilbert, J. K., Reiner, M., & Nakhleh, M. (Eds.). (2008). *Visualization: Theory and practice in science education*. Springer.

Grosser, A. E. (1984). Cooking with chemistry. *Journal of Chemical Education, 61*(4), 362. https://doi.org/10.1021/ed061p362

Hilner, E. (2025, January 3). 13 active learning strategies for online classes. *Engageli*. https://www.engageli.com/blog/12-plus-active-learning-strategies-for-elearning-2025

Holmes, W., & Tuomi, I. (2022). State of the art and practice in AI in education. *European Journal of Education, 57*(3). https://doi.org/10.1111/ejed.12533

Hong, J.-C., Liu, Y., Liu, Y., & Zhao, L. (2021). High school students' online learning ineffectiveness in experimental courses during the COVID-19 pandemic. *Frontiers in Psychology, 12*. https://www.frontiersin.org/articles/10.3389/fpsyg.2021.738695

Jacobsen, E. K. (2011). Kitchen chemistry. *Journal of Chemical Education, 88*(8), 1018–1019. https://doi.org/10.1021/ed200359x

Jegstad, K. M. (2023). Inquiry-based chemistry education: A systematic view. *Studies in Science Education, 60*(2), 251–313. https://doi.org/10.1080/03057267.2023.2248436

Johnstone, A. H. (2000). Teaching of chemistry—Logical or psychological? *Chemistry Education Research and Practice, 1*, 9–15. https://doi.org/10.1039/A9RP90001B

Jones, L. L., Jordan, K. D., & Stillings, N. A. (2005). Molecular visualization in chemistry education: The role of multidisciplinary collaboration. *Chemistry Education Research and Practice, 6*(3), 136–149. https://doi.org/10.1039/B5RP90005K

Kelly, R., & Akaygun, S. (2019). Visualizations and representations in chemistry education. *Chemistry Education Research and Practice, 20*(4), 657–658. https://doi.org/10.1039/C9RP90009H

Khine, M. S., & Saleh, I. M. (Eds.). (2011). *Models and modelling: Cognitive tools for scientific inquiry*. Springer.

Lerman, Z. M., & Morton, D. (2009). Using the arts and computer animation to make chemistry accessible to all in the twenty-first century. In M. Gupta-Bwohon, S. Jhaumeer-Laulloo, H. L. Kam Wah, & P. Ramasami (Eds.), *Chemistry education in the ICT age* (pp. 31–39). Springer.

Lin, X., & Gao, L. (2020). Students' sense of community and perspectives of taking synchronous and asynchronous online courses. *Asian Journal of Distance Education, 15*(1), 169–179. https://doi.org/10.5281/zenodo.3881614

Little Scientist Foundation. (n.d.). *International dialogue on STEM education (IDoS)*. https://www.stiftung-kinder-forschen.de/en/international-dialogue-idos/

Ministry of Education and Culture, Finland (2021). *The National Roadmap for Research, Development and Innovation*. https://okm.fi/en/rdi-roadmap.

Ministry of Education and Culture. (2023a). *Finnish national STEM strategy and action plan: Experts in natural sciences, technology and mathematics in support of society's welfare and growth* (p. 22). Publications of the Ministry of Education and Culture, Finland.

Ministry of Education and Culture. (2023b). *National LUMA (STEM) strategy and action plan*. https://okm.fi/luma-strategia

Ness, I. J., & Glăveanu, V. (2019). Polyphonic orchestration: The dialogical nature of creativity. In R. A. Beghetto & G. E. Corazza (Eds.), *Dynamic perspectives on creativity: New directions for theory, research, and practice in education* (pp. 189–206). Springer.

Nguyen, J. G., & Keuseman, K. J. (2020). Chemistry in the kitchen laboratories at home. *Journal of Chemical Education, 97*(9), 3042–3047. https://doi.org/10.1021/acs.jchemed.0c0062k

Nuora, P., & Välisaari, J. (2020). Kitchen chemistry course for chemistry education students: Influences on chemistry teaching and teacher education: A multiple case study. *Chemistry Teacher International, 2*(1), 20180021. https://doi.org/10.1515/cti-2018-0021

OECD. (2023). *OECD digital education outlook 2023: Towards an effective education ecosystem* [White paper]. OECD Publishing. https://doi.org/10.1787/c74f03de-en

Oliveira, H., & Bonito, J. (2023). Practical work in science education: A systematic literature review. *Frontiers in Education, 8*, 1151641. https://doi.org/10.3389/feduc.2023.1151641

Pasquinelli, E., Borde, B., Dhar, J., Filtzinger, B., Franz, F., Henke, N., Lundell, J., Oberthür, J., O'Donnell, C., Pahnke, J. & Sadadou,

D. (2024). *Early STEM education in the digital age*. International Dialogue on STEM Education. White paper available online https://www.stiftung-kinder-forschen.de/en/international-dialogue-idos/idos-publications/white-paper-early-stem-education-in-the-digital-age/

Permatasari, M. B., Rahayu, S., & Dasna, I. W. (2022). Chemistry learning using multiple representations: A systematic literature review. *Journal of Science Learning, 5*(2), 334–341. https://doi.org/10.17509/jsl.v5i2.42656

Pohl, A. J. (2020). Strategies and interventions for promoting cognitive engagement. In A. L. Reschly, A. J. Pohl, & S. L. Christenson (Eds.), *Student engagement*. Springer. https://doi.org/10.1007/978-3-030-37285-9_14

Quayson, C., Kwarteng, T. A., Koranteng, E., & Hanson, R. (2022). Chemistry teacher trainees' use of molecular models in learning spiro and bicyclic compounds. *Science Education International, 33*(3), 291–295. https://doi.org/10.33828/sei.v33.i3.4

Redecker, C., & Punie, Y. (Eds.). (2017). *European framework for the digital competence of educators: DigCompEdu*. Publications Office of the European Union. https://doi.org/10.2760/159770

Roponen, A., Välisaari & Lundell, J. (2025). Engagement through experiments: Experiences from online kitchen chemistry courses. LUMAT*: International Journal on Math, Science and Technology Education, 13(1), 11.* https://doi.org/10.31129/LUMAT.13.1.2683.

Schaffar, B., & Wolff, L. A. (2024). Phenomenon-based learning in Finland: A critical overview of its historical and philosophical roots. *Cogent Education, 11*(1). https://doi.org/10.1080/2331186X.2024.2309733

Sjöström, J., Eilks, I., & Talanquer, V. (2020). Didaktik models in chemistry education. *Journal of Chemical Education, 97*(4), 910–915. https://doi.org/10.1021/acs.jchemed.9b01034

Taber, K. S. (2017). Models and modelling in science and science education. In K. S. Taber & B. Akpan (Eds.), *Science Education* (pp. 263–278). Sense Publishers.

Tewari, S. (2025, April 28). *Engaging minds: The power of active learning in the classroom*. UMass Amherst IDEAS. https://www.umass.edu/ideas/digest/engaging-minds-power-active-learning-classroom

The Princeton Review. (n.d.). AI and STEM: How artificial intelligence is transforming education. https://ws.princetonreview.com/ai-education/ai-and-stem

Thuneberg, H., Salmi, H., & Fenyvesi, K. (2017). Hands-on math and art exhibition promoting science attitudes and promoting science attitudes. *Education Research International*, 9132791. https://doi.org/10.1155/2017/9132791

Treagust, D. F. (2018). The importance of multiple representations for teaching and learning science. In M. Shelley & S. A. Yıldırım (Eds.), *Education research highlights in mathematics, science and technology 2018* (pp. 215–223). ISRES Publishing.

Tucci, M. (2025, March 9). *How artificial intelligence is transforming STEM education: Unlocking potential for all students.* Newark Academy STEM Blog. https://blogs.newarka.edu/stem/2025/03/09/how-artificial-intelligence-is-transforming-stem-education-unlocking-potential-for-all-students/

UNESCO. (2024). *AI Competency Framework for Teachers.* https://www.unesco.org/en/articles/ai-competency-framework-teachers

UNICEF. (2021). *Policy guidance on AI for children (version 2.0).* UNICEF Office of Global Insight. https://www.unicef.org/innocenti/media/1341/file/UNICEF-Global-Insight-policy-guidance-AI-children-2.0-2021.pdf

van Enckevort, C. (2014, August 18). *Students' learning of molecular modeling in science education: The case of computer-aided drug design against malaria disease* [Unpublished research project]. Utrecht University. https://studenttheses.uu.nl/bitstream/handle/20.500.12932/18107/Students%27%20learning%20of%20molecular%20modeling%20in%20science%20education_C.M.W.%20van%20Enckvort_3785688.pdf

van Heerden, S. (2024, March 26). *Disadvantages of e-learning: The challenge of hands-on disciplines.* E-Student. https://e-student.org/elearning-disadvantages-hands-on-disciplines/

Vuorikari, R., Kluzer, S., & Punie, Y. (2022). *DigComp 2.2: The Digital Competence Framework for Citizens—With new examples of knowledge, skills and attitudes (EUR 31006 EN).* Publications Office of the European Union. https://doi.org/10.2760/115376

United Nations. (2015). *The 17 Sustainable Development Goals.* https://sdgs.un.org/goals.

World Bank. (2024). *AI, the Next Generation.* https://www.worldbank.org/en/about/unit/unit-dec/impactevaluation/events/measuring-development/2024.

Wu, H.-K., Krajcik, J. S., & Soloway, E. (2001). Promoting understanding of chemical representations: Students' use of a visualization tool in the classroom. *Journal of Research in Science Teaching, 38*(7), 821–842. https://doi.org/10.1002/tea.1033

Yip, C., Clegg, T., Bonsignore, E., Lewittes, B., Guha, M. L., & Druin, A. (2012). *Kitchen chemistry: Supporting learners' decisions in science.* Paper presented at the ICLS International Conference of the Learning Sciences. Sydney. https://repository.isls.org//handle/1/2191

York, J. (2020, May 8). *8 ways to make lessons fun and engaging in an online classroom.* Teach Away. https://www.teachaway.com/blog/increase-engagement-in-online-classroom

Zhai, X., & Krajcik, J. (2024). *Uses of artificial intelligence in STEM education.* Oxford University Press. https://doi.org/10.1093/oso/9780198882077.001.0001

CHAPTER 11

STEM-Reforming Standard Curriculum Concepts Through AI Integration: A Case of Chemical Reactions

Martin Bilek and Monika Pelikanova

Introduction

Reforming the standard or traditional curriculum of both individual STEM disciplines (Science, Technology, Engineering, and Mathematics) and their integrated forms is essential to prepare students for the challenges of the modern world, where rapid technological advances continually change the job market and everyday life. STEM fields are now crucial to innovation and addressing global problems, including climate change, sustainability, and healthcare. Therefore, it is essential that the curriculum is relevant to these current needs (National Academy of Engineering & National Research Council, 2014). Technological advances, such as those in artificial intelligence (AI), robotics, and quantum computing, require that education in these fields be flexible and constantly evolving. This requires greater integration of disciplines, as it is no longer possible today to strictly separate mathematics, computer science, physics, chemistry,

M. Bilek (✉) • M. Pelikanova
Charles University, Prague, Czechia
e-mail: martin.bilek@pedf.cuni.cz

R. Elmas, M. Akarsu (eds.), *STEM Education*, Maarif Global Education Series,
https://doi.org/10.1007/978-981-95-9023-0_11

biology, and engineering, or other disciplines, not only in the STEM spectrum. Students should have the opportunity to learn how to integrate different areas of science and technology and apply these concepts to real-world problems. Overlap with the social sciences, humanities, or the arts is also essential (Beers, 2011).

Today's labour market demands specialists with knowledge and skills that go beyond traditional education. A reformed curriculum should emphasize practical applications of STEM knowledge, such as programming, data analysis, and project management (Duncan, 2010). Students should become more involved in real-world projects and practical tasks, which will allow them to understand better how their theoretical knowledge can be applied in practice. It is also essential to incorporate new challenges and trends into the curriculum, such as ethics in technology, digital security, and AI, which are increasingly integral to today's technological and work environments (Schneider, 2016; Aroch et al., 2024, Feldman-Maggor et al., 2025).

Another key area is promoting diversity and inclusion in STEM fields. Traditionally, some groups, particularly women and disadvantaged groups, have been underrepresented in these fields. Curriculum reform should then remove prejudices and barriers that may discourage students from pursuing careers in STEM fields, promoting equality of opportunity. An inclusive curriculum that considers diverse cultural and gender perspectives can contribute to greater diversity in these fields and, consequently, to a broader range of innovative solutions to global problems (National Research Council, 2012; Chroustová et al., 2021).

Reforming the STEM curriculum is therefore not just about modernising technical knowledge, but about creating an education system that prepares students for a future in which technology is increasingly intertwined with all aspects of our lives. It is essential that students not only understand technology but also develop the critical thinking, creativity, and collaboration skills that will enable them to meet the challenges of modern society and the labour market (National Research Council, 2012; Elmas et al., 2020; Koldová et al., 2024).

This chapter aims to explore how the traditional theme of teaching chemistry as one of the STEM disciplines can be modified to address the current challenges of education oriented towards its relationship to everyday life, the application of acquired knowledge, and the development of creativity and respect for the broader context. To do this, the implementation of AI is essential (Yuriev, Wink, & Holme, 2024). It can fundamentally change the way various topics are taught, including the topic of chemical reactions, which is the focus of this chapter. In the standard teaching of chemistry or science as an integrated subject, chemical reactions are often explained using static models and demonstration experiments that can be difficult for students to understand. Thus, we aim to focus on how various application examples, including the use of AI, can provide interactive and dynamic tools that allow students to better understand and visualize the complex processes related to the topic. The use of experimental cycles, effective experiments with an engineering emphasis, and personalized learning systems, along with AI, can not only help streamline learning but also foster creative and deeper student understanding of these fundamental processes. We will also present several practical applications and will seek to highlight the potential of the STEM approach in transforming the traditional educational model.

Integrating Artificial Intelligence into STEM Education

AI, as mentioned in the introduction, is having an increasing impact on the field of education, particularly in the natural sciences, where its applications are introducing new approaches to learning, teaching, and research. In science education, AI enables personalised learning that responds to the specific needs of students. Thanks to adaptive systems, students can progress at their own pace, improving learning efficiency and increasing motivation. For example, AI can analyse students' responses and recommend individualized study materials or assignments based on their performance, targeting their weaknesses (Woolf, 2010, Blonder, & Feldman-Maggor, 2024).

Another area where AI plays a significant role is in simulating complex natural processes. The use of AI in simulations allows students to interactively explore natural phenomena that would otherwise be difficult to observe or accessible only in theoretical terms. Examples include simulations of chemical reactions or ecological systems that help students better understand the complex relationships among the components of nature (Baker et al., 2018, Humphry, & Fuller, 2023).

AI also improves the efficiency of scientific research. In the life sciences, the advanced AI tools are increasingly being used to analyse big data, predict trends, or find patterns in complex sets of information (Jordan &

Mitchell, 2015). In this way, AI not only accelerates scientific progress but also serves as a valuable tool for preparing students for real-world research problems.

AI has the potential to significantly enrich science teaching, streamline learning processes, and contribute to the advancement of scientific research. However, it is essential that its use is properly implemented, and teachers are trained in how to use AI effectively in their teaching. These emerging capabilities of AI provide a strong foundation for transforming not only how students learn but also how future educators are trained—particularly in complex domains such as chemical reactions.

Teaching Chemical Reactions: Present and Challenges for Teaching and Teacher Training

Chemical reactions are a standard topic in chemistry education, and therefore, initial teachers need to be prepared for it. However, in common science classrooms, this topic is often presented by showing one example and then explained through chemical equations without a direct connection to experimental activities. Traditional subtopics include the classification of chemical reactions, the formal aspects of their description, and a strong connection with chemical nomenclature and mathematical skills (STEMkey, 2023).

The aim of this traditional approach transformation is to develop innovative teaching materials on the topic of chemical reactions for Initial Teacher Education (ITE) and to influence teaching practices at different levels of educational systems. It means in our case to prepare a conceptual and pedagogical improvement of the topic, identified as key and also as a critical point of early Chemistry/Science/STEM education. Starting from examples from real teachers' and learners' experiences and activities, as well as an analysis of textbooks and other instructional materials, a guide, worksheets, and a leading presentation were provided for the education of pre-service Chemistry, Science, and STEM teachers. Our main focus was on how to transform the teaching and learning of the topic' chemical reaction' into rich learning experiences for school students, which nurture the development of key competences through the sensible implementation of AI.

Future teachers need to gain and reflect on the necessity of promoting the development of key competences when teaching chemistry/science/

STEM subjects with a focus on chemical reactions. They are introducing our proposals into the key competence framework of the EU, including three elements of key competence: knowledge, skills, and attitudes, with concrete examples related to chemical reactions in everyday life and their support for sustainable development (industry, agriculture, transport, etc.).

Our focus is on introducing future teachers to educational approaches that support the transformation of standard "content" on the topic of chemical reactions and related tasks into tasks that give students an active role, utilize real-life contexts, connect different disciplines, and take into account students' diversity. This is also demonstrated through experimental cycles. There are fewer common alternatives to organizing and conducting school experiments based on the mutual transformations of chemicals, while the starting material of the cycle is also its final product. The cycles are associated with significant colour changes in substances, changes in the reactant state, and a variety of reaction types.

The following examples of teaching practice in the area of chemical reactions aim to enhance the connection with the acquisition of key competences in initial STEM education. It is innovative in that it addresses the acquisition of key competences and transversal competences. It not only is a showcase of "fancy" experimental activities without deeper explanations but also starts from the fundamental daily basics of Chemistry, Science, and STEM teaching, aiming to enable future teachers to teach these basics innovatively.

Chemical reactions are central themes in the chemistry curriculum, with a direct impact on the curricula of other science and engineering subjects. The topic can be considered both a critical and key point of both general and vocational science and technology education (Vondrová et al., 2015; Rychtera et al., 2018). This fact is related to the multifaceted nature of this issue, particularly the large number of reactions of different types, as well as their inconsistent classification, among other factors.

The fundamental question related to teaching the thematic unit on chemical reactions is why some chemical substances react with each other while others do not. Whether it is possible to predict which chemicals will react with each other, whether there are generally valid rules relating to reactivity, or whether it is necessary to rely on real experiments, etc. A deductive approach is often employed in teaching chemical reactions, which is a problematic issue for both pupils and teachers. Pupils can be informed about the course of a chemical reaction, which can be verified through a real experiment, and they can try to characterise the reactions more closely according to various criteria. This approach is reflected in traditional textbooks from primary to university level, in which the classification of chemical reactions is presented according to familiar criteria: according to external changes: composition reactions (synthesis, chemical fusion), decomposition reactions (analysis, chemical decomposition), chemical substitution (substitution) and double substitution (conversion); according to state: homogeneous reactions and heterogeneous reactions; by thermal colour: exothermic reactions and endothermic reactions; or by type of particles transferred: acid-base reactions, oxidation-reduction reactions and complex-forming reactions. Individual reactions can then be classified simultaneously according to different criteria.

In the innovative tendencies of chemistry teaching related to the topic of chemical reactions, the so-called experimental cycles approach is often proposed. These are closed reaction cycles involving different types of transformations of chemical substances in which the starting substance is also the final product of the cycle. This circumstance points to the interconnectedness of chemical transformations, their classification, and consequently the understanding of their nature, including their relevance to everyday life. Experimental cycles can be very diverse. They can focus on the transformations of inorganic substances, or the reactions of organic compounds related to everyday production practices. In other cases, they involve recycling waste or simulating cyclic processes that occur in nature.

Examples of STEM Approach in Teaching the Topic Chemical Reactions

STEM approach implemented in the topic chemical reaction, based on the support of the key competence formation. The key competences are defined as a combination of applied knowledge, skills, and attitudes (COM, 2019). In this topic, we employ student-centred learning approaches, real-life connections, trans-disciplinarity, and ICT-supported learning.

For science or STEM education, the knowledge component comprises an understanding of basic principles of the natural world, fundamental scientific concepts, theories, laws, and methods, technology and technological products and processes, as well as an understanding of the impact of science, technology, engineering, and human activity in general on the natural world. Altogether, these competences should enable individuals to better understand the advances, limitations, and risks of scientific theories and applications, as well as technological advances in relation to decision-making, values, moral questions, and culture (COM, 2019). In the module chemical reactions, students are introduced to the basic principles and consequences of chemical changes, which are the basis for understanding the real world around us. Through the inclusion of an experimental cycles approach, students are exposed to different types of transformations of chemical substances in which the starting substance is also the final product of the cycle. This circumstance highlights the interconnectedness of chemical transformations, their classification, and, consequently, the understanding of their nature, including their relevance in everyday life.

Skills in science, technology, and engineering education involve understanding science as a process for investigation through specific methodologies, including observations and controlled experiments. This includes the ability to use logical and rational thought to verify a hypothesis and the readiness to discard one's own convictions when they contradict new experimental findings. Skills are also related to an ability to use and handle technological tools and machines, scientific data to achieve a goal or to reach an evidence-based decision or conclusion, and the recognition of the essential features of scientific inquiry and the ability to communicate the conclusions of reasoning that led to them (COM, 2019).

Skills are developed in module chemical reactions through using experimental activities in all three approaches "in vivo" (real environment of everyday life), "In vitro" (laboratory setting of the environment) and "in silico" (models and simulations).

In addition to cognitive and practical skills, fostering appropriate attitudes is essential in STEM education. Such attitudes in chemistry and a broader sense in STEM education includes competences in critical thinking and curiosity, and a concern for ethical issues and support for environmental sustainability.

The proposals for improving the standard topic chemical reactions are developed in a staggered approach, from "warm-up" activities through "immersion" to practical consequences. The aims are focusing to develop competences in critical thinking on sustainability and a concern for ethical issues through the exploration and discussion of examples from different areas of everyday life.

Student-centred learning approaches, real-life connections, trans-disciplinarity, and ICT supported learning are offered through a context-based approach to introduce principles of chemical changes in everyday life. Our proposals in the chapter include numerous student activities, including working with information, group discussions, hands-on interaction in experimental activities, etc. In this way, chemical knowledge is connected with knowledge from other STEM disciplines in real-life contexts. Students also connect their understanding of chemical reactions with the sustainable and ethical use of resources.

Our proposal of an STEM approach to the topic of chemical reactions emphasizes the interconnection between all STEM subjects (STEMkey, 2023). Not only are strong links between chemistry and other science disciplines demonstrated here, but also important links to mathematics, technology, and engineering, as well as digital approaches, are also highlighted. It is implemented through current pedagogical trends:

- Inquiry-based learning, experimental learning, project-based learning,
- Real-life context,
- Interdisciplinary connections between STEM subjects,
- Considering sustainability and ethical approaches,
- Students' diversity,
- Digital tools for deepening insight.

Students will acquire during the solving of proposed activities:

- Understanding of the concept of chemical reaction and its consequences in everyday life,
- Examples of chemical reactions connected with different areas of everyday life,
- Reflection on specific characteristics of chemical reactions and their relation to mathematics and science education, beliefs on changes of substances in relation to market and their influence on human being,
- Examples of own connections with chemical reaction in cases from everyday life,
- Skills to compare different ways of dealing with changes of substances, that is, acting in the sense of active and critical citizenship versus ignoring evidence and following "traditions",
- Reasons for including changes of substances in STEM education,
- Experience in chemical reactions in classroom practice: examples of realisable classroom experimental activities,
- Reflections on what students learn when dealing with such a task,
- Readiness to include changes in substances in day-to-day teaching.

The proposed content is divided into three sections, each comprising distinct activities. It includes lecture parts, group discussions, student presentations, and lab work. The structure is as follows:

- Introduction into the topic ("warm-up" activities)
- Immersion tasks ("multi-context" activities)
- Application tasks ("everyday life connections" activities).

The emphasis on linking chemistry to real-life situations and other STEM domains enhances students' understanding of the subject matter while fostering the development of critical thinking and a responsible attitude. The following section outlines the specific structure of the proposed educational activities and their distribution across the various phases of instruction.

Introduction to the Topic: Chemical Reaction

For ambiance evocation, we propose a warm-up activity titled "Changes of Substances Around Us". The intention is to introduce some facts about chemical reactions in everyday life, which concern different consequences resulting from changes in the substances. In this case are changes of substances used as examples of positive and negative influences on everyday life and to explore students' previous knowledge, beliefs, experiences, and attitudes related to this topic. The teacher educator introduces the topic in connection with the "world of substances". He /she can offer different examples of physical bodies in the classroom (equipment, subjects, teaching aids, etc.) and discuss which substance(s) they include. The activity can also be situated in different places outside the classroom or school, such as in the garden, at a company (excursion), or in the food waste collection yard, among other locations (as a demonstration of everyday, self-evident things for us). At the start, he/she initiates a short discussion through the questions: What is the impact/effect of substances and bodies produced from them on our life/the lives of people? Or can substances and bodies produced from them change our lives? Or what can we contribute to sustainability, concerned with the consumption of the substances and bodies produced from them? And then present the interactive activity. The activity consists of working in pairs to prepare a title and short commentary on an elected picture related to connected substances and their changes in both positive and negative senses (construction industry, metal industry, plastic production, food production, corrosion, cleaning processes, etc.). After work, in pairs, students present their own results and discuss them with others. Substances and their changes are very complex phenomena. Their changes are based on chemical reactions, but their use is a multidisciplinary problem. For example, in the case of corrosion, students can discuss the negative influence of substances' changes (e.g., iron corrosion) and the positive influence (e.g., metallic passivation), as well as the social (e.g., allegory), cultural (e.g., patina), and environmental (e.g., waste) consequences. There is considerable scope for AI involvement in this activity in various ways that encourage students' creativity, critical thinking, and technical skills. For example, AI can be used to help identify substances in photographs. Students can use computer vision tools, such as Google Lens or mobile assistants (Siri, Google Assistant, etc.), to identify materials displayed in images. This will allow them to more accurately determine whether they are plastic, concrete, or metal, for example, which will then enrich the discussion on their properties and environmental impacts. The advantage is that most students already own a smartphone, so implementing this technology does not require the purchase of additional special equipment.

Another option is to use the implementation of large language models, where personalised AI assistants (such as CustomGPT within the ChatGPT platform) can serve well, or structured prompts can be provided to systematically question students, gain different perspectives, add facts, or pose new questions to consider. For example, for the intended question: "How would society be affected if we completely replaced plastics with bioplastics?", a suitable prompt for students might be: "You are a discussion partner for a student thinking about the question: How would society be affected if we completely replaced plastics with bioplastics? Your task is to guide an open-ended conversation, ask curious and expanding questions (e.g., regarding environmental, economic, social, technological, or global impacts), reflect on the student's answers, and encourage deeper thought. Never answer on the student's behalf — always wait for their response before continuing with the next question. Your tone should be friendly, natural, and focused on supporting critical thinking". Generative AI can also be used to create specific visualizations based on precise prompts. Instead of searching for photos, students can use generative AI (such as ChatGPT 4o, DALL-E, or Midjourney) to create custom images based on a description of the change in the substance. An example of a suitable prompt is: "Create a detailed scientific illustration of an iron structure submerged in seawater, visibly affected by electrochemical corrosion. The image should show a cut-through section of the iron surface, with areas covered by rust (iron oxides and hydroxides), including both early and advanced stages of corrosion. Highlight the areas where corrosion begins and progresses. Show the presence of seawater with green chloride ions, arrows indicating the movement of electrons and ions, and the formation of rust layers on the surface. Use a realistic scientific style with semi-shading, clear color-coded areas (blue for oxidation zones, red for reduction zones), and no artistic effects. Background should be clean and neutral, and the image should look like something from an educational textbook or scientific paper".

Students can then critically analyse the extent to which generative AI can represent complex chemical processes and discuss its limits. Another option is to use AI models or simulations to analyse the effects of different substances. For example, climate change applications or energy consumption calculations can show the practical implications of using a particular substance—for example, comparing concrete and wood in construction. In this way, cross-curricular links (physics, environmental education, civics) can be strengthened, and pupils can be encouraged to think about the implications of scientific decisions.

AI can also serve as a tool for collaboration and evaluation. It can summarise learner outcomes, offer alternative perspectives, or provide constructive feedback. This will encourage critical thinking and deeper reflection on the topic. The specific teaching scenario of this activity for a second grade lower secondary or upper secondary school, in which AI will be used as an interactive discussion partner as well as a tool for knowledge dissemination, aims to make pupils aware of how changes in substances (e.g., corrosion, fermentation, incineration, and use of plastics) affect our daily lives. They learn to distinguish the positive and negative consequences of these changes, develop critical thinking skills, work with modern technologies, and look for sustainable solutions. The activity can take place in several phases. The teacher opens the topic with questions such as, "What changes in substances do we know from everyday life?" or "Do you think these phenomena affect society, culture or the environment?" In pairs, students are given several photographs of different phenomena, such as rusting metal construction, plastic production, food fermentation, or solar panels. Their task is to describe what is happening in the picture, identify the change in the substance and determine its positive and negative impacts. Each pair will briefly present what they found out from the picture, what new information the AI offered them, and whether their view of the phenomenon has changed. Next, students may be asked to create a short text or presentation on the topic "How can we change the world for the better by changing substances?" using the help of AI, for example, to generate ideas which they then evaluate. For this activity, they need to prepare several thematic images on substance change, AI-enabled devices (laptops, tablets, phones), and simple instructions on how to ask AI (for example, how to formulate a specific question and verify the information obtained). In addition, it is useful to have a whiteboard or flipchart for a joint summary. This activity can be further extended to include, for example, using AI to generate new images as described by the students, or an ethical discussion on "Should decisions about ecology be left to AI alone?" In this way, learning can be linked to civic education and digital literacy. An alternative task could be to design an appropriately structured prompt to help them understand more about the phenomenon. A prompt for the question "What are the advantages and disadvantages of plastics in the food industry?" could look like this: "You are a discussion partner for a student reflecting on the question: What are the advantages and disadvantages of plastics in the food industry? Your task is to guide an open conversation by asking thoughtful questions that help the student explore aspects such as food safety, storage, environmental impact, economic efficiency, and alternative materials. Never answer on the student's behalf. Always wait for their response before continuing with a new question, reflection, or challenge. Your tone should be friendly and supportive, aimed at encouraging critical thinking". A prompt for the question "Can rust also be positive?" could be: "You are a discussion partner for a student thinking about the question: Can rust also be a good thing? Your task is to ask curious, open-ended questions that broaden the student's perspective on this topic—such as rust as a protective layer, its use in art, biology, or technology. Do not give direct answers. Wait for the student's reply before responding further. Maintain a natural and encouraging tone, and help the student discover unexpected angles to this often-overlooked phenomenon". For less experienced students, prompt creation tools can be used, such as SAM, The Prompt Creator (CustomGPT). It is important to guide students to not just require encyclopaedic knowledge from AI, but to use it more effectively as a tool for active learning—for example, through guided discussion, hypothesis formulation, or critical analysis of the information gathered, which promotes the development of their own scientific thinking.

The preceding activity underscores the pedagogical value of integrating modern technologies, everyday phenomena, and critical reflection in the initial exploration of chemical reactions. This foundation enables a more comprehensive understanding of the subject matter, which is further developed in the subsequent section through structured immersion tasks focused on experimental reaction cycles and their broader scientific and interdisciplinary relevance.

Immersion Tasks in the Topic Chemical Reaction

For a deeper understanding of the topic chemical reaction, we pay attention to so-called reaction cycles in connection with STEM approaches under the title "Different kinds of reaction cycles". Experimental cycles are closed cycles of chemical reactions involving different types of transformations of chemical substances in which the starting substance is also the final product of the cycle. This circumstance shows the interconnectedness of chemical transformations, their classification, and consequently, the understanding of their nature. Experimental cycles can be very diverse. They can be focused on the transformations of inorganic substances or on the reactions of organic compounds. In other cases, they are recycling of waste or simulations of cyclic processes occurring in nature (e.g., the water cycle in nature). In particular, cycles based on the interconversion of inorganic substances are common. Probably the most well-known is the copper cycle (Condike, 1975; Todd and Hobey, 1985; Walker et al., 2012), the most commonly used version of which in pedagogical practice is shown in Fig. 11.1. The starting substance is copper, a reddish-brown solid that reacts with nitric acid to form a dark blue solution of copper nitrate. This reacts with sodium hydroxide to form a light blue precipitate of copper hydroxide. Copper hydroxide, on heating, is converted to an insoluble black substance, copper oxide, which reacts with sulphuric acid to give a blue solution of copper sulphate. The action of iron on copper sulphate produces ferrous sulphate (green solution) and the reddish-brown solid copper, and the reaction series is completed. The cycle also represents a set of diverse chemical transformations, such as oxidation–reduction reactions, precipitation reactions, etc. In the literature, we can find different examples of experimental cycles and tasks for students that we can formulate based on the mentioned aspects by their cognitive levels.

The experimental cycle of copper is based on the changes in colour and state of individual chemicals. Working with the experimental cycle scheme

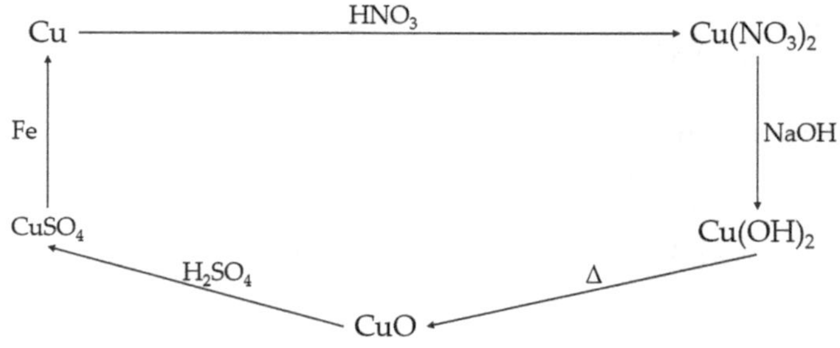

Fig. 11.1 Experimental copper cycle (Condike, 1975)

can be based on the description of the individual steps at the phenomenological level, that is, the description of the products of the individual steps, and on the description of the relevant chemical reactions in the individual steps, including the characterization of the reaction conditions and the development and quantification of chemical equations or their schemes (use of pictures, photographs, etc.). The tasks related to experimental cycles should be linked to sources of information such as professional literature and relevant internet resources (searching and comparing information on different types of cycles, chemicals involved in cyclic changes, etc.). Students read the experimental cycle scheme and describe each of the individual steps (Kolář et al. 2018):

1. At the phenomenological level, that is, the description of the products,
2. On the description of the relevant chemical reactions, including the characterization of the reaction conditions,
3. At the development and quantification of chemical equations,
4. At the illustration of the scheme related to everyday life (use of pictures, photographs, etc.). See Fig. 11.2.

To further enhance students' engagement and deepen their understanding of chemical reaction cycles, AI tools can be integrated into the learning process. AI can assist students in multiple ways:

- Image Recognition and Analysis: Students can use AI applications to analyse photographs or videos of chemical experiments to identify substances based on colour changes, state of matter, or even text recognition on labels and instruments.

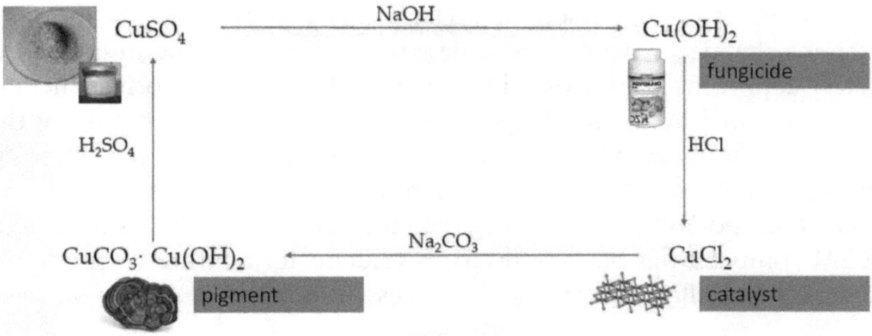

Fig. 11.2 Modified copper cycle with relation to everyday life

- Reaction Prediction Tools: By using AI-based chemistry platforms (e.g., MoleculeNet, IBM RXN), students can input chemical formulas or reactants and explore potential reactions, including predictions of products and balanced equations.
- AI Chatbots and Virtual Assistants: Learners can interact with AI tutors to ask questions about chemical cycles, reaction types, or stoichiometry. These virtual assistants can guide them step-by-step through reaction mechanisms or help verify their chemical equations. Integration of such assistants into the educational process enables students to receive immediate feedback even outside the classroom or during self-study, when they don't have direct access to a teacher. This allows students to work at their own pace and dedicate more time to more difficult concepts, which supports an individualized approach to learning.
- Simulation and Modelling: AI-powered simulations can model the copper cycle and other experimental cycles in virtual environments, allowing students to visualize molecular-level changes that occur during each step of the process.
- Data Collection and Interpretation: Using AI tools for processing experimental data (e.g., temperature change, reaction time, and pH) can help students analyse trends and draw conclusions more efficiently. For example, specialized AI assistants such as Data Analyst ChatGPT allow students to upload their experimental data, which the system then automatically analyses, identifies key trends and correlations, and creates visualizations. This enables students to better understand the relationships between reaction conditions and experimental results without having to spend time learning advanced statistical methods and programming languages.

Experimental cycles based on different changes of substances, meaning different chemical reactions, provide a prominent place for an interdisciplinary approach. To stay only in chemistry and mainly in formal chemistry, represented by chemical equations and their stoichiometry, is not the motivation for students. A broader sense of various STEM connections is the right way to increase students' interest and effort in understanding the topic. Incorporating AI into experimental cycles not only strengthens STEM competencies but also prepares students for a future where digital literacy and intelligent technologies are essential.

Application Tasks in the Topic Chemical Reaction

Application tasks on the topic of chemical reactions offer various examples from different areas of everyday life. One example deals with thermochemistry by the title "Are you cold? Warm up!". The topic of this activity is connecting with the heat balances of chemical reactions. It contains information on products, for example, "hand warmers", which make some everyday activities easier, help in breaking through health problems, in rehabilitation, sports, etc.

The materials named "chemical warmers" which are becoming popular, but their structure and content are not known to the users, are divided into two groups: reversible and irreversible warmers. The former group contains so-called super-cool gels, the latter one works mainly on the principle of exothermic iron oxidation or exothermic hydration of some salts, for example, strong exothermic hydration of burnt $CaCl_2$ which is used, for example, in "drink heaters" called "Caldo Caldo" (Fig. 11.3). It is an aluminium bag containing a drink, such as coffee, placed in a plastic cup. The space between the bag and cup is filled with $CaCl_2$ and water. After activating the heating mechanism, that is, enabling $CaCl_2$ and water to mix by making holes in the plastic foil, the drink gets hot. The same principle is used with "Freddo Freddo" products which use the strong endothermic reaction of $Na_2S_2O_3$ and water as a cooling mechanism.

The proposal of the task/small project enables pupils to:

- See how to use "chemical warmers" in common life,
- Gain information on "chemical warmers" by their self-activity, that is, the principles they operate on, about the production, availability on the market, their advantages and disadvantages,
- Discover principles of "chemical warmers" in experimental activities,

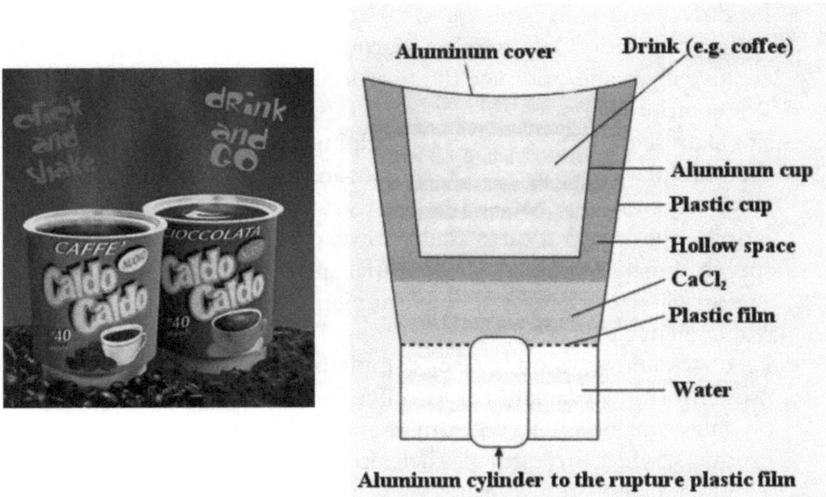

Fig. 11.3 Principle of Caldo Caldo "chemical warmer." (Source: https://enviroexperiment.zcu.cz/chemie-2-stupen-zs/horky-led)

- Cooperate actively in solving problems,
- Prepare your product based on "chemical warmer".

To enhance the project and reflect current technological trends, AI tools are incorporated into several phases of the activity:

- Research Assistance: Pupils can use AI-based research assistants (e.g., ChatGPT or similar tools) to gather relevant information on chemical reactions, product designs, and environmental impact. However, it is essential that they also learn to critically evaluate the obtained information, verify it from reliable sources such as scientific publications, textbooks, or scientific databases, and recognize potential inaccuracies or simplifications that AI may provide. In this way, AI not only helps in accessing information but also serves as a tool for developing critical thinking and information literacy.
- Data Analysis and Simulation: Pupils can simulate chemical reactions using AI-supported educational software to visualize heat production, reaction rates, and safety parameters.

- Product Design: AI tools can be used for brainstorming and prototyping the design of chemical warmers, helping students generate creative ideas and optimize the functionality of their product.
- Presentation Support: AI can assist students in preparing professional presentations or reports by helping with structure, design, and even language translation or correction. Additionally, by using the voice functions in ChatGPT, students can practice the actual presentation—they can rehearse their delivery, receive feedback on their speech, practice answers to potential questions, and improve their rhetorical skills in a safe environment without the stress of a real audience.
- Feedback and reflection: After completing the project, students can use AI tools to evaluate the effectiveness of their design and reflect on the environmental, economic, and ethical aspects of using such products, which strengthens their prompting abilities to create both a language and professional reviewer. In this way, they learn to formulate precise and effective questions, which is a skill transferable to many other areas of study and professional life.

The school project with topic of "chemical warmers" brings directly the relation between the science and technical education and everyday life connections. This kind of project contributes to the implementation of so-called dynamic points of curriculum, which means new technologies and new processes used in everyday life. By including AI in this project, pupils are introduced to modern digital tools that foster creativity, critical thinking, and technical skills. This integration strengthens the connection between science education, technology, and real-life applications, which is in line with the dynamic aspects of the curriculum.

Benefits and Challenges of the STEM Approach and the Integration of AI into the Teaching of Standard Science Education Topics

The STEM approach implemented in contemporary science education allows students to develop knowledge and skills across disciplines and to better understand the connections between theoretical knowledge and its practical application. In the context of teaching chemical reactions, the integration of STEM approaches and AI tools offers new opportunities for fostering students' creativity, critical thinking, and technical literacy. At the same time, however, it also brings some challenges.

One of the advantages of this connection is the ability to use real-life phenomena to teach basic chemical principles. For example, the activity "Changes in substances around us" allows students to analyse changes in substances (e.g., corrosion, fermentation, combustion, plastic production) from different perspectives—chemical, environmental, social or cultural. Using AI, such as image recognition tools (e.g., Google Lens), pupils can easily identify materials and discuss their properties and impacts. Generative AI can also be engaged to create visualisations of chemical phenomena (e.g., illustrations of electrochemical corrosion), promoting understanding of complex processes. However, it should be emphasized that the visualizations produced in this way may not always be chemically completely accurate and should primarily serve as a starting point for discussion and critical analysis. Students should be encouraged to compare AI-generated visualisations with scientifically validated models.

AI can also play an important role as a discussion partner. Using thoughtfully structured prompts, students can be led to think more deeply about questions such as "How would society change if we replaced plastics with bioplastics?" or "Can corrosion be a positive thing?". In this way, AI supports the development of critical thinking without directly providing ready-made answers.

The STEM approach is also effectively applied to so-called experimental chemical reaction cycles, such as the well-known copper cycle. This allows pupils to observe different chemical transformations and the connections between them. Using AI, students can analyse photographs of experiments, model reactions, predict products or check the accuracy of chemical equations. In addition, the use of AI tools for data collection and analysis (e.g., temperature, pH, and reaction time) allows for more efficient interpretation of results without the need for advanced statistical knowledge.

Another application could be a project focused on "chemical heaters"—products that use exothermic reactions to generate heat. Here, pupils combine chemical knowledge with the technical design of their own product. AI can help not only with information research and design, but also with making presentations or reflecting on the project. This promotes pupils' independence and the development of their digital competencies.

Despite all these advantages, the challenges that AI integration brings must be seen. One of these is the need to guide students to approach information critically—AI can provide inaccuracies or simplifications. Teachers, therefore, need to show students how to verify sources and evaluate the reliability of data. An effective approach is to create practical tasks where pupils analyse AI-generated content that contains minor inaccuracies—for example, looking for errors in chemical equations. An important part of this is also systematically comparing the information obtained from AI with validated expert sources and then discussing the differences found. Another challenge is developing the ability to work effectively with prompting—that is, the ability to pose meaningful and well-formulated instructions to the AI. It is also important to remember that every school or other educational institution should set rules for the use of AI. This means neither banning nor rejecting AI but guiding students to use these tools correctly. It is also important to be realistic about the differences between paid and unpaid versions of AI tools. Unpaid versions often provide less reliable information and may be more likely to "hallucinate" (convincingly present inaccuracies or untruths). The advantage of paid versions, for example, when creating specialised CustomGPT/assistants, is the possibility to upload your own tutorials and expert resources, significantly reducing the risk of inaccuracies. Thus, a teacher can create a high-quality teaching assistant in the paid version and then share it with students who use the free version of the service.

Overall, it can be said that the combination of the STEM approach and AI can significantly enrich the teaching of traditional science education topics, as in the case of the chemical reactions presented here. This enables the linking of theory with practice, the promotion of interdisciplinary thinking, and the preparation of students for the digital world of the future. However, the key to success remains pedagogical guidance that helps pupils to use technology in a meaningful and active way.

Conclusion

In an era of rapid technological advancement and increasing demands for digital literacy, combining a STEM approach with AI represents one of the most significant ways to transform traditional educational practices. Thus, in this chapter, we wanted to show that even a standardized and traditionally conceived curriculum, such as the topic of chemical reactions, can be enriched with new dimensions of meaningful learning when set in the context of everyday life, connected to other disciplines, and complemented by modern technologies, including AI.

The examples of activities presented across the three phases—warm-up, immersion, and application—show how to offer students learning experiences that are not only conceptually rich but also practically and ethically anchored. The use of AI is not an end in itself, but serves as a tool to improve learning, promote differentiation, develop key competences, and, last but not least, foster skills that are essential for civic and professional life in the twenty-first century.

The emphasis on experimental cycles, work with visualizations, discussions using structured prompts, and project activities linking science with technology and ethical contexts are concrete steps towards teaching that reflects the complexity of the contemporary world. In this chapter, we have also sought to emphasize the need for focused preparation of future science and engineering teachers who should be able to use new technologies not only effectively but also in pedagogically meaningful and critical ways.

In the context of the broader framework of this monograph—that is, the search for and shaping of future educational practices in the age of AI—this chapter confirms that innovation in teaching does not necessarily mean radical changes in content. Often, a change in approach, emphasis, and connections is sufficient. This is where the potential of STEM education for the twenty-first century lies: to offer students the opportunity to understand the world in its interconnectedness, to develop creative thinking, and to prepare them for a world in which coexistence with technology is an everyday reality.

However, the expansion of these approaches into mainstream school practice also faces certain limitations, including in particular insufficient methodological guidance, technical background and uneven digital competences of teachers. It is therefore essential that curriculum innovation is accompanied by systematic professional development, accessible teaching tools, and support from school authorities. AI and digital technologies alone will not guarantee quality teaching—the key remains the teacher as a facilitator of learning who can create meaningful learning situations, foster curiosity, and lead students to reflect thoughtfully on information and their own knowledge.

If STEM education is truly to be a tool for shaping the future society, it must strike a balance between technological prowess and humanity—between the effectiveness of the tools and the value-laden nature of their use. The very topic of chemical reactions in the context of AI-integration can be an example of how to seek this balance already at the level of primary and secondary education.

References

Aroch, I., Katchevich, D., & Blonder, R. (2024). Modes of technology integration in chemistry teaching: Theory and practice. *Chemical Education Research and Practice, 25*, 843–861. https://doi.org/10.1039/D3RP00307H

Baker, R. S., Corbett, A. T., Koedinger, K. R., & Roll, I. (2018). Improving learning outcomes through AI-driven interactive simulations in science education. *Journal of Educational Computing Research, 56*(2), 143–161. https://doi.org/10.1177/0735633117708290

Beers, S. Z. (2011). *21st century skills: Preparing students for their future*. ASCD.

Blonder, R., & Feldman-Maggor, Y. (2024). AI for chemistry teaching: Responsible AI and ethical considerations. *Chemistry Teacher International, 6*(4), 385–395. https://doi.org/10.1515/cti-2024-0014

Chroustová, K., Bílek, M., Machková, V., Kmeťová, J., Kučková, Š., & Smirnova, T. A. (2021). The Identification of Girls' Attitudes to STEM via Science Camp "Lab Technology and Everyday Life Consumption". In *PBE 2021 - Project-based education and other student-activation strategies and issues in science education XIX* (pp. 114–125). Charles University, Faculty of Education.

COM. (2019). *Key competences for life-long learning*. Publications Office for the European Union.

Condike, G. F. (1975). Near 100% student yields with the "cycle of copper reactions" experiment. *Journal of Chemical Education, 52*(9), 615.

Duncan, A. (2010). *Science, technology, engineering, and mathematics (STEM) education: A national perspective*. U.S. Department of Education.

Elmas, R., Rusek, M., Lindell, A., Nieminen, P., Kasapoğlu, K., & Bílek, M. (2020). The intellectual demands of the intended chemistry curriculum in Czechia, Finland, and Turkey: A comparative analysis based on the revised bloom's taxonomy. *Chemical Education Research and Practice, 21*, 839–851. https://doi.org/10.1039/D0RP00058B

Feldman-Maggor, Y., Blonder, R., & Alexandron, G. (2025). Perspectives of generative AI in chemistry education within the TPACK framework. *Journal of Science Education and Technology, 34*, 1–12. https://doi.org/10.1007/s10956-024-10147-3

Humphry, T., & Fuller, A. L. (2023). Potential ChatGPT use in undergraduate chemistry laboratories. *Journal of Chemical Education, 10*(4), 1434–1436. https://doi.org/10.1021/acs.jchemed.3c00006

Jordan, M. I., & Mitchell, T. M. (2015). Machine learning: Trends, perspectives, and prospects. *Science, 349*(6245), 255–260. https://doi.org/10.1126/science.aaa8415

Kolář, K., Bílek, M., Machková, V., Rychtera, J., & Chroustová, K. (2018). Experimental cycles as alternative approach to teaching the topic chemical reactions. In K. Miecznikowski (Ed.), *14th European Conference on Research in Chemical Education – Book of Abstracts* (pp. 7–8). Warszaw University.

Koldová, H., Janík, T., Bílek, M., Dostál, J., Mádlová, M., Mentlík, P., Rokos, L., Šimik, O., & Trnová, E. (2024). *Přístupy k integraci vzdělávacího obsahu z pohledu kurikula, výuky a učitelského vzdělávání* [Approaches to the integration of educational content in terms of curriculum, teaching and teacher education]. Jihočeská univerzita v Českých Budějovicích, Pedagogická fakulta.

National Academy of Engineering & National Research Council. (2014). *STEM integration in K–12 education: Status, prospects, and an agenda for research.* National Academies Press.

National Research Council. (2012). *A framework for K–12 science education: Practices, crosscutting concepts, and core ideas.* National Academies Press.

Rychtera, J., Bílek, M., et al. (2018). Která jsou klíčová, kritická a dynamická místa počáteční výuky chemie v České republice? [What are the key, critical and dynamic points of initial chemistry education in The Czech Republic?]. *Arnica—Acta Rerum Naturalium Didactica, 8*(1), 35–44.

Schneider, M. (2016). *The role of higher education in STEM education reform.* National Science Foundation.

STEMkey. (2023). *STEMkey - teaching standard STEM topics with a key competence approach.* ICSE Freiburg. Retrieved from https://icse.eu/international-projects/stemkey/

Todd, D., & Hobey, W. D. (1985). An improvement in the classical copper cycle experiment. *Journal of Chemical Education, 62*(2), 177.

Vondrová, N., Rendl, M., et al. (2015). *Kritická místa matematiky základní školy v řešení žáků* [Critical points of primary school mathematics in students' solutions]. Karolinum.

Walker, M., Noll, M., Panders, J., Winnebeck, K., & Courtney, M. E. (2012). *Types of reactions: The copper cycle.* New York State Pollution Prevention Institute. Retrieved from https://hhw.uvlsrpc.org/files/2613/7218/4212/NYSP2I_Green_Chem_Module_Types_of_Reactions_Copper_Cycle.pdf

Woolf, B. P. (2010). *Building intelligent interactive tutors: Student-centered strategies for revolutionizing e-learning.* Morgan Kaufmann.

Yuriev, E., Wink, D. J., & Holme, T. A. (2024). The Dawn of generative artificial intelligence in chemistry education. *Journal of Chemical Education, 101*(8), 2957–2959. https://doi.org/10.1021/acs.jchemed.4c00836

CHAPTER 12

Opportunities and Challenges in STEM Education in the Era of AI

Muhammet Arıcan and Pierre Sutherland

Introduction

STEM education is highly effective in developing students' critical thinking, creativity, and problem-solving, and helps them address complex societal issues through dynamic and integrative methods (Stehle & Peters-Burton, 2019). Educators can effectively integrate STEM disciplines by adopting different approaches to tailor classroom contexts and instruct students on addressing real-life challenges. There are some essential features of classroom STEM activities. For instance, STEM activities should allow students to learn from their errors and facilitate continuous improvement (Moore et al., 2014a). They should be well-aligned with school STEM curriculum and educational standards (Fortus et al., 2004; Harris & Felix, 2010). Furthermore, STEM activities should incorporate research and exploratory tasks to enhance students' conceptual understanding (Smith et al., 2005). In addition, student-centered approaches

M. Arıcan (✉)
Boğaziçi University, Istanbul, Türkiye
e-mail: muhammet.arican@bogazici.edu.tr

P. Sutherland
Sparx Learning Ltd., Exeter, UK

R. Elmas, M. Akarsu (eds.), *STEM Education*, Maarif Global Education Series,
https://doi.org/10.1007/978-981-95-9023-0_12

should be followed when implementing these activities. Student-centered approaches not only encourage collaboration among students (Carlson & Sullivan, 2004; Selingo, 2007) but also improve their social and communication skills (Katehi et al., 2009).

While STEM education is crucial for helping students develop essential skills, its effectiveness hinges on instructors' ability to design and implement high-quality STEM materials. Hence, the challenge of designing these high-quality STEM materials must be addressed to ensure successful STEM education. High-quality STEM materials possess several key characteristics: First, they should incorporate engaging and motivating contexts to attract students' interest (Moore et al., 2014b). Additionally, they should facilitate evidence-based reasoning processes (Smyrnaiou et al., 2015) and present challenging real-world tasks that promote critical thinking, creativity, and problem-solving skills (Morrison, 2006). Emerging technologies, specifically artificial intelligence (AI), can be used by educators to design and implement these high-quality STEM materials. While AI presents many opportunities in STEM education, it also poses several challenges. Crompton et al. (2024) conducted a systematic review focusing on the benefits and challenges of AI within K–12 environments. The review highlights three main advantages of AI: integration into teaching methods (such as gaming and personalized learning), support for administrative tasks (like diagnostic tools), and enhancement of subject content. They also reported several challenges, including negative perceptions toward using AI technology in education, the need to develop technological and AI skills among teachers and students, privacy and ethical concerns, and usability issues with AI tools.

In this chapter, we discuss the opportunities and challenges of integrating AI into STEM education. The chapter focuses on how advancements in AI technology can enhance STEM learning experiences of students through personalized learning opportunities and increasing student engagement and motivation. The chapter also discusses how teachers can use AI in classrooms to facilitate interdisciplinary connections, expand access to STEM education in schools, and assist instructors in assessing students' work throughout the teaching-learning process. Furthermore, we also present challenges faced in effectively integrating AI in STEM education. These challenges include limited resources, gaps in access to AI technology, the need for teacher preparedness and professional development, ethical and privacy concerns, and research and implementation gaps.

Opportunities in the Integration of AI into STEM Education

STEM education empowers students to effectively integrate science, technology, engineering, and mathematics disciplines to solve real-life challenges. Although technology has great potential in this integration, its implementation often fails to adequately contribute to developing essential skills in students required for a quality STEM education (Komalawardhana & Panjaburee, 2024). Moreover, commonly used digital tools, such as presentation software, websites, and learning management systems, tend to be static and may not adequately support educators in designing and implementing high-quality STEM materials. On the other hand, adaptive and interactive features of AI have the potential to develop necessary STEM skills in students and address design and implementation shortcomings by offering several key benefits.

The benefits of AI integration include the adaptability to individual differences in learning (Nikhil, 2025; Sun et al., 2025), enhancing student engagement and motivation (Zhang et al., 2024), facilitating interdisciplinary connections (Gibson et al., 2023), enabling automatic assessment of learning (Divasón et al., 2023; Ouyang et al., 2023), and expanding access to STEM education (Nikhil, 2025; Sun et al., 2025). Integrating AI into STEM classes can benefit educators by enhancing student learning and reducing workload. The opportunities of AI in STEM education are briefly discussed in the following pages.

AI-Enhanced Personalized Learning in STEM Education STEM education equips students for the future by emphasizing the development of critical thinking, problem-solving, and analytical skills. However, due to large class sizes, fixed curricula, limited classroom time, and resource constraints, teachers often face challenges in delivering personalized attention to every student (Al Darayseh & Mersin, 2025). Recently, the use of AI in educational settings has led to revolutionary changes in personalized learning (Nikhil, 2025). Sun et al. (2025) stated that personalized learning involves customizing educational experiences to align with the specific needs, preferences, and capabilities of each student. STEM education presents unique challenges due to content complexity and the diverse backgrounds of learners, and these challenges necessitate integrating AI technologies in instruction (Nikhil, 2025). AI has the potential to transform traditional STEM education by adapting to the individual needs, preferences, and experiences of students by analyzing their learning styles, strengths, and weaknesses (Nikhil, 2025; Sun et al., 2025). By analyzing students' strengths and weaknesses, educators can modify their teaching approaches to enhance the overall success and create personalized learning paths that allow students to understand complex STEM concepts at their own pace (Huang et al., 2023). According to Zhang et al. (2024), personalized learning paths enable students to face grade-level STEM challenges. For them, these learning paths enhance student motivation and reduce their frustration. Through personalized learning paths, educators can provide students with tasks at varying difficulty levels and use data to inform instruction, so students can achieve a sense of accomplishment while improving their skills. Additionally, AI technologies can assist educators in presenting abstract ideas and concepts in a more concrete and accessible manner. This approach enhances students' conceptual understanding and increases knowledge retention (Chen et al., 2020).

AI-driven personalized learning can enhance the effectiveness of STEM classes through adaptive learning platforms, intelligent feedback mechanisms, and data-informed instructional methods (Zhang et al., 2024). Sun et al. (2025) reviewed the applications of AI tools in STEM education. They reported the following applications of AI: learning prediction, intelligent tutoring systems, personalized feedback, educational robotics, behavior detection, automation, and conversational AI. Sun et al. (2025) noted that these applications can enhance students' learning experiences and personalize education. Hence, they demonstrate AI's transformative potential across various aspects of teaching and learning. Two key applications of AI in STEM education are integrating intelligent tutors and tutees in instruction and incorporating AI as a learning partner (Sun et al., 2025). While students can receive immediate feedback from intelligent tutors, they can also take on tutoring roles. These two features help students deepen their understanding of STEM concepts while honing their critical thinking and problem-solving skills through active learning. Moreover, AI can serve as an intelligent learning partner by assisting students with data collection, analysis, and presentation, which fosters engagement and comprehension. For example, intelligent concept mapping tools developed for this purpose assist learners in the mapping process by offering immediate feedback and evaluations, which improve their ability to organize and understand information (Nugumanova et al., 2021).

AI-Enhanced Student Engagement and Motivation in STEM Education Fredricks et al. (2004) stated that student engagement consists of three dimensions: cognitive, affective (emotional), and behavioral. According to Bond and Bedenlier (2019), indicators of cognitive engagement involve learning from others, engaging in critical thinking, exercising self-regulation, fostering positive self-perceptions, and gaining understanding. On the other hand, affective engagement includes excitement, curiosity, interest/fun, enjoyment, satisfaction, positive social interaction, and a positive attitude toward learning. Finally, indicators of behavioral engagement include activities that reflect students' involvement, effort, attention, confidence, development of multidisciplinary skills, study routines, completion of tasks, and the amount of time dedicated to those tasks. Lo et al. (2024) conducted a systematic review of the effects of ChatGPT on student engagement and found strong evidence supporting its impact on behavioral engagement. The results regarding affective engagement were mixed; students reported experiencing both satisfaction and anxiety. Additionally, while there were indications of improved cognitive engagement of students, there were also concerns about reduced critical thinking skills and an overreliance on technology.

Zhang et al. (2024) noted that integrating AI into STEM education provides students with immediate feedback and helps teachers to adapt their teaching methods based on students' progress. AI integration also improves the effectiveness of instruction by enhancing student engagement and motivation. Similarly, Yang et al. (2024) noted that cutting-edge technologies such as educational robots can be used in STEM classes to facilitate student motivation and performance. The use of AI in STEM classes fosters computational thinking, enhances academic success, and promotes scientific exploration (Yang et al., 2024). Educators can enhance student engagement and motivation in STEM education by utilizing interactive simulations, virtual labs, game-based learning methods, real-time feedback mechanisms, and chatbots. For example, Huang et al. (2023) reported that they developed an AI-enabled personalized video recommendation system and found that it improved students' motivation, engagement, and learning outcomes.

Facilitating Interdisciplinary Connections through AI in STEM Education The interdisciplinary nature of STEM education requires students to integrate knowledge from various disciplines, as well as the development of critical thinking and computational skills (Ortiz-Revilla et al., 2022; Yang et al., 2024). STEM education promotes an integrated approach that emphasizes joining STEM disciplines to address real-world challenges. The STEM disciplines can be integrated using a disciplinary, multidisciplinary, interdisciplinary, or transdisciplinary approach (English, 2016). English (2016) summarized these approaches as follows: In the disciplinary approach, concepts and skills are learned in isolation within each discipline. The multidisciplinary approach involves learning concepts and skills separately in each discipline, but centers around a common theme. The interdisciplinary approach focuses on interconnected concepts and skills from multiple disciplines to enhance knowledge and skills. Finally, the transdisciplinary approach involves applying knowledge and skills learned from various STEM disciplines to tackle real-world issues and projects. STEM Task Force Report (2014) emphasizes that STEM education goes beyond just mixing the four disciplines; it involves "real-world, problem-based learning" that connects these fields through engaging teaching methods (p. 9). The report states that these disciplines should not be taught separately since that is not how they function in real life or the job market. The Common Core State Standards for Mathematics and the Next Generation Science Standards also encourage deeper connections among STEM disciplines (English, 2016).

Determining a real-life problem with an interesting and engaging context during STEM education can be challenging for instructors. AI can effectively identify these real-life contexts, collect and analyze data from various disciplines, and assist students in integrating their knowledge of these disciplines to address real-world challenges. As Yang et al. (2024) noted, incorporating AI into STEM education offers a dynamic teaching approach spanning multiple disciplines. This approach enables students to explore interdisciplinary concepts while actively engaging with relevant AI tools. While AI technology in STEM education is rising, the focus often remains on individual subjects rather than integrating knowledge across disciplines, neglecting the interdisciplinary approach essential for a comprehensive understanding (Sun et al., 2025). According to Sun et al. (2025), this narrow approach misses the vital interdisciplinary aspect of STEM education, which is essential for helping students understand concepts comprehensively and develop various skills for real-world situations. Currently, AI in education mainly revolves around general adaptive learning and tutoring systems, which often do not cater to the specific needs of STEM learners. STEM education prioritizes hands-on, experiential learning and needs tools that can help visualize and manipulate complex ideas within an interdisciplinary context. Sun et al. (2025) noted that existing AI educational technologies do not sufficiently address these requirements.

Instructors can benefit from AI by facilitating students' understanding of interdisciplinary connections in STEM education. For example, AI technology can assist instructors in identifying relevant interdisciplinary resources, including research papers, videos, student assignments, and worksheets on the STEM topics they teach in the classroom. Similarly, virtual workshops and simulations can be easily created using AI. These workshops and simulations allow students to experience interdisciplinary concepts interactively. Furthermore, AI can be used to design interdisciplinary projects that focus on real-world challenges. Additionally, AI can provide personalized and adaptive learning experiences by analyzing individual student performance and suggesting more tailored learning pathways. These tailored learning pathways enable dynamic adjustments to learning content and help connect STEM disciplines. Various AI tools have been developed with the advancement of this technology. One notable example that can be used in STEM settings is Khanmigo, an AI teaching assistant developed to assist students with mathematics problem-solving and to support teachers in their instructional process (Khan Academy, n.d.).

Expanding Access to STEM Education Through AI Technologies STEM education and degree programs are often more expensive than other fields (Chirikov et al., 2020). Costs are often associated with the necessity of various materials and resources, such as laboratories, hands-on materials, and the technological software and tools necessary for effective STEM education (Stohlmann et al., 2012). Consequently, establishing a school culture and environment supporting an integrated STEM education approach can be costly and time-consuming (Thibaut et al., 2018). Integrating AI-powered technology can reduce reliance on hands-on materials and laboratory equipment and lower the costs in STEM education. However, the successful implementation of AI technology mainly depends on the quality of school infrastructure, particularly on the technology and students' access to internet service. Additionally, certain advanced features of AI technology can only be accessed through paid subscription methods. Thus, while AI technology can enhance students' access to quality STEM education and provide more affordable learning opportunities by reducing the need for physical materials, a lack of essential resources may hinder their ability to access this education.

Research (e.g., Mruthyunjaya et al., 2024; Saw & Charlotte, 2021) indicates that students from rural and small towns are significantly underrepresented in STEM fields compared to their urban counterparts. It is crucial to focus on rural areas due to the significance of STEM education in fostering scientifically literate individuals and addressing workforce demands (Mruthyunjaya et al., 2024). Integrating AI technology in STEM education allows students to learn from anywhere and enhances their learning experience by providing online access to STEM materials from home. For instance, MATHia is a one-to-one online intelligent mathematics coaching system designed for grades 6 to 12. Developed by researchers at Carnegie Mellon University, MATHia employs a unique machine learning method to analyze students' interactions with mathematics and their thought processes (Sloan, 2024). In terms of higher education, *Knewton Alta* serves as an excellent example. This adaptive learning platform offers comprehensive instruction, targeted remediation, and detailed explanations of answers tailored to students' responses and identified knowledge gaps (Wiley, n.d.). Therefore, as in these examples, students with limited access to quality STEM education can participate in online STEM opportunities that help educators and policymakers work toward closing the gap in equitable access to high-quality education. However, as stated above, this feature of AI technology significantly depends on access to the internet and computers, which requires infrastructure quality in schools and countries. AI technology allows instructors and students to convert text-based materials into audio or video formats. Moreover, it offers real-time language translation, which enhances access to STEM education, especially for non-native speakers and individuals with disabilities.

As an example of research that aimed at making education more accessible, affordable, and effective, Goel (2020) developed an AI-based system to enhance online learning in higher education. The results showed that this system supported inquiry-based learning, answered student questions, promoted interactions, and quickly generated educational content. Moreover, Goel (2020) reported significant improvements in student engagement and substantial time savings for instructors.

AI-Driven Educational Assessment in STEM Education Formative and summative assessments are used in STEM education to monitor and evaluate student learning and the effectiveness of instructional methods. Formative assessment encompasses ongoing teacher, peer, and self-feedback on student learning and evaluation of individual and group projects and presentations. Formative assessment is crucial for identifying areas where students may need additional support during instruction. On the other hand, summative assessment is used at the end of a learning unit or module and primarily serves grading purposes. While these assessments are important for evaluating overall student performance, they can pose significant challenges, such as tracking individual student progress and effectively assessing individual and group performance (Jiao et al., 2022). Providing timely feedback during formative assessments and the heavy workload from extensive summative assessments limit instructors' time for personalized, student-centered instruction (Ouyang et al., 2023). As stated above, AI technologies offer promising solutions to these challenges by facilitating a more personalized and adaptive learning environment in STEM education. AI can help alleviate instructors' workloads while enhancing student learning experiences.

Ouyang et al. (2023) conducted a systematic review of 17 empirical studies published between January 2011 and April 2023 to understand how AI-driven educational assessment is applied in STEM education. They identified that AI-driven assessments have been mainly used for three purposes in STEM education: determining academic performance, tracking learning status, and determining instructional quality. The review also noted the use of various algorithms for assessment purposes, ranging from traditional methods like natural language processing and machine learning to advanced techniques such as deep learning and neural fuzzy systems. As stated above, there are also challenges in integrating AI into STEM education. The following section briefly discusses these challenges.

Challenges in Integrating AI in STEM Education

Integrating AI into STEM learning environments can significantly enhance students' educational experiences. However, this integration comes with significant challenges that require thoughtful consideration from educators and institutions. Some of these challenges stem from the inherent nature of STEM education, such as difficulties in integrating disciplines, gaps in resources and infrastructure, and the availability of real-life contexts and data. Integrating AI-driven technology into STEM education can help educators address some of these challenges. On the other hand, further challenges may arise during this integration, including limited technical and infrastructure resources, disparities in access to AI technology, teacher preparedness and professional development, and ethical and privacy concerns with AI integration. While the integration of AI into STEM education offers many opportunities for enhancing student learning, it is essential to address the associated challenges to ensure the effectiveness of this integration. Therefore, this section examines the challenges and critical barriers to effective AI integration into STEM education, intending to offer considerations for researchers and practitioners in this field.

Resource and Infrastructure Constraints in Integrating AI into STEM Education A solid STEM education infrastructure is essential for effective AI implementation. Many educational institutions struggle with inadequate infrastructure and limited access to necessary AI technologies. Establishing a robust AI foundation often necessitates significant investment in expensive hardware like graphics processing units (GPUs), resulting in higher electricity usage and the need for specialized technical assistance. Even subscription-based AI services present financial challenges, with licenses currently costing $20 per user monthly, and require high-speed internet, especially for image and video generation. The demands of AI infrastructure further complicate the already slow and expensive progress of STEM education infrastructure development (Stohlmann et al., 2012). Although AI could potentially reduce the reliance on physical laboratories, the initial infrastructure expenses make it inaccessible for numerous schools (Chirikov et al., 2020). Furthermore, Joseph and Uzondu (2024) noted substantial investment requirements for AI and machine learning (ML) technology, encompassing infrastructure and necessary training, posing a considerable obstacle for many schools.

AI and the Digital Divide: Equity Challenges in STEM Education Integrating AI into STEM education presents a significant challenge: the potential to deepen existing educational inequalities. While not inevitable, mitigating this risk requires substantial and intentional effort. Kohnke and Zaugg (2025) highlighted the potential for AI model training methods to worsen social inequalities by under-representing marginalized groups. They also caution that addressing algorithmic bias requires significant and careful attention. Rural communities face challenges, such as limited resources, funding, poor internet connectivity, and a lack of understanding of STEM disciplines, which have led to the underrepresentation of STEM education among these communities (Mruthyunjaya et al., 2024; Saw & Charlotte, 2021). The digital divide will also likely impede AI integration in under-resourced schools lacking the necessary technology for online educational opportunities. Nuangchalerm and Prachagool (2023) noted an unequal distribution of AI-powered STEM education resources based on socioeconomic status and geographic location. This disparity between well-resourced and under-resourced schools adds additional strain on AI infrastructure, affecting instruction, curriculum development, connectivity, and subscription costs. Limited access to AI technology, information technology (IT) support, and reliable internet connectivity can slow down the integration of AI, significantly impacting opportunities for rural students.

Challenges in Teacher Preparedness and Professional Development for Integrating AI into STEM Education Beyond infrastructure, another major challenge in integrating AI into STEM settings is ensuring teachers are trained to use AI and teach students to do the same effectively. STEM education is inherently a long process that demands significant time and effort from teachers and students. Although AI integration can ease the work of teachers and students, a lack of knowledge and experience with this technology can cause confusion and eventually may lead to its avoidance. Hrastinski et al. (2019) reported that teachers had a limited understanding of AI, negatively affecting their knowledge of utilizing this technology in educational settings. Similarly, Thibaut et al. (2018) reported that many educators feel unprepared to use STEM applications, a problem potentially worsened by teachers' existing beliefs about teaching and learning, which could make them resistant to AI's possibilities.

Besides a limited understanding of AI, teachers' negative attitudes toward using this technology in classrooms also affect AI integration (Beege et al., 2024). For instance, Tissenbaum and Slotta (2019) noted a teacher's negative attitude toward using AI technology in teaching, who described AI as distracting, which affected his willingness to integrate AI into classroom settings. These negative attitudes toward AI integration and the lack of technological knowledge highlight the importance of the professional development of STEM educators. Providing teachers with guidance can eliminate these negative attitudes and enhance their technological skills. However, Beege et al. (2024) stated that using AI in teaching does not automatically equate to good teaching, emphasizing the crucial role of instructor readiness for successful implementation. Similarly, students may lack the requisite skills to effectively engage with AI technology, further complicating its integration into STEM education. Hence, this mutual unpreparedness of teachers and students highlights the necessity for implementing comprehensive training programs focusing on AI technology. For this purpose, various initiatives and organizations provide professional guidance for teachers on understanding AI technology and methods for using it in classroom teaching. For instance, the International Society for Technology in Education (ISTE, n.d.) provides professional development opportunities for K–12 teachers to promote safe and responsible use of AI in education. In addition, Google Labs is another platform where students can explore and test out the newest AI products from Google and influence the direction of AI technology (Google Labs, n.d.).

Ethical and Privacy Concerns in AI-Enhanced STEM Education Although AI offers significant benefits in STEM education, researchers have had some ethical and privacy concerns about its potential risks to individuals. Borenstein and Howard (2021) stated that a common justification for integrating AI in education is that it can perform better than human decision-makers. However, Borenstein and Howard (2021) also emphasized that AI tools are still developed by humans, who can unintentionally introduce biases and errors. Hence, they noted that defining "fairness" in AI is a complex and subjective issue since perceptions of fairness can differ and evolve over time. Furthermore, integrating AI into educational settings presents ethical challenges related to data privacy, algorithmic bias, and transparency (Nuangchalerm & Prachagool, 2023).

A major ethical concern is the extensive collection and use of student data during the training of AI applications. The processing of data, especially from children, presents additional risks. Moreover, individuals may unintentionally upload sensitive information, which can lead to AI tools processing this data. For example, Duffy (2020) reported on technologies that collect images of people from the internet. Such technologies have raised concerns about privacy violations and the challenges in monitoring who uses these technologies and for what purposes (Borenstein & Howard, 2021). Furthermore, training AI models can perpetuate societal inequalities because it often relies on data from demographics with greater access to AI (Dwivedi et al., 2021). This skew can influence personalized recommendations and potentially affect how algorithms interact with or filter individuals from underrepresented demographics (Huang et al., 2023; Vargas-Murillo et al., 2023). As a result, it may lead to systematic and unfair treatment of individuals, influenced by social biases related to gender, race, ethnicity, and sexuality. For this purpose, the United Nations (UN) Secretary-General (2025) emphasized the importance of prioritizing human agency and rights in the development of AI technology. The UN urged for guarantees that all users possess the necessary tools and knowledge to utilize this technology "wisely, safely, and ethically."

The "black box" problem surrounding the transparency and accountability of paid AI models is a significant issue (Joseph & Uzondu, 2024). Some AI models operate with limited accountability. Schorcht et al. (2024) note that the reasoning processes within large language models (LLMs) can be obscured, amplified, or distorted by hallucinations, further complicating their operation. Vargas-Murillo et al. (2023) emphasize the risk of this "hallucination effect" in educational AI, where invented information could mislead students. Joseph and Uzondu (2024) identify this as a primary challenge in AI and education.

Integrating AI into STEM education can offer educators access to real-life data and allow them to upload student information for grading and other purposes. However, it is essential to understand privacy concerns and take appropriate measures to protect student information. There is a need to focus on people first to address ethical concerns with AI. Hence, Borenstein and Howard (2021) suggest educating individuals at the early stages of their interaction with AI technology, and that ethics should be included in every stage of the AI design process.

Research and Implementation Gaps in AI-Integrated STEM Education The review of STEM literature indicates several issues with the research and implementation of STEM activities. First, while STEM integration is increasingly advocated, multidisciplinary studies are limited and lack clear guidance on common STEM goals (English, 2016). Second, research findings in STEM implementation are often context-specific, highlighting a disconnect and lack of coherence in STEM education discussions (English, 2016). Third, the effectiveness of integrated STEM initiatives in improving students' understanding of core content remains understudied (English, 2016). Finally, existing studies often lack sufficient detail about the curriculum or program and provide inadequate documentation regarding the nature and support of STEM integration (Thibaut et al., 2018). These foundational issues also create challenges for integrating AI into STEM classes.

With the rapid development of AI technology, many scholarly papers have been published across various disciplines that highlight the importance of integrating AI technology into different fields. However, our literature review indicates that many of the existing studies are primarily systematic reviews (e.g., Chen et al., 2020; Crompton et al., 2024; Komalawardhana & Panjaburee, 2024) or opinion papers (e.g., Borenstein & Howard, 2021; Dwivedi, 2021). Studies on AI integration into STEM education usually focus on several key areas: They often report on student engagement and motivation (e.g., Huang et al., 2023; Lo et al., 2024) and teachers' perspectives (e.g., Al Darayseh & Mersin, 2025), explore various insights, trends, and opportunities in the field (e.g., Yang et al., 2024), and focus on specific features of AI such as personalized learning (e.g., Nikhil, 2025; Sun et al., 2025; Zhang et al., 2024). Implementing STEM education necessitates a significant amount of time and effort from educators. The existing studies mentioned here offer only a limited view of AI integration. Therefore, to fully evaluate the potential of AI integration, there is a clear need for rigorously designed research that incorporates AI throughout the entire STEM education process rather than focusing on specific components. Such research can help us identify key benefits and challenges associated with AI integration.

Conclusion and Suggestions

Implementing STEM education in classrooms presents several challenges. For example, integrating different disciplines requires significant time and effort. STEM education necessitates understanding the engineering perspective and effective use of technology. Additionally, it demands the identification of context-relevant real-life issues. Although the integration of AI into STEM education offers opportunities for educators to address these challenges, the rapid advancement of this technology stands out as a significant obstacle for teachers in understanding and integrating it into their lessons. However, despite the rapid advancement of AI technology, limited reliable resources are available for gathering adequate information about this technology and effective integration methods. Therefore, it is essential for policy-makers and school administrations to organize professional development seminars for teachers. These seminars should help teachers become familiar with AI technology and learn how to integrate it into STEM contexts, eventually assisting students in understanding this technology. For this purpose, the United Nations Educational, Scientific, and Cultural Organization (UNESCO, n.d.) published "Artificial intelligence and education: Guidance for policy-makers" and "AI competency frameworks for students and teachers" guidelines to generate a common understanding among policymakers, teachers, and students in understanding the opportunities and challenges that AI offers for education.

Besides a general lack of familiarity with AI technology, there is a significant gap regarding the integration of AI into learning objectives and classroom activities in STEM-related curricula. Furthermore, there is also insufficient research on the efficacy of existing STEM programs and how they can be updated in light of recent developments in AI technology. While institutions and publishers have been updating their rules for AI use in academic work and publications, there is an urgent need to update school curricula to keep pace with the rapid developments in this field. Therefore, it is critical to develop a comprehensive framework for AI-integrated STEM education. This framework should include clear curricular guidelines and be developed through collaboration among STEM disciplines. Furthermore, it is important to invest in longitudinal research to address the challenges associated with AI integration effectively. This research should gather data on the most effective strategies for AI integration. Additionally, stakeholders from academia, industry, and policy-making should be encouraged to take a collaborative approach to address the challenges associated with AI integration in STEM settings.

Disclaimer We acknowledge that we used generative AI tools, ChatGPT and Grammarly, for language editing, idea generation, and resource identification while writing this chapter.

References

Al Darayseh, A., & Mersin, N. (2025). Integrating generative AI into STEM education: Insights from science and mathematics teachers. *International Electronic Journal of Mathematics Education, 20*(3), em0832. https://doi.org/10.29333/iejme/16232

Beege, M., Hug, C., & Nerb, J. (2024). AI in STEM education: The relationship between teacher perceptions and ChatGPT use. *Computers in Human Behavior Reports, 16*, 100494. Article 10.1016/j.chbr.2024.100494.

Bond, M., & Bedenlier, S. (2019). Facilitating student engagement through educational technology: Towards a conceptual framework. *Journal of Interactive Media in Education, 2019*(1), 1–14. https://doi.org/10.5334/jime.528

Borenstein, J., & Howard, A. (2021). Emerging challenges in AI and the need for AI ethics education. *AI and Ethics, 1*(1), 61–65.

Carlson, L., & Sullivan, J. (2004). Exploiting design to inspire interest in engineering across the K-16 engineering curriculum. *International Journal of Engineering Education, 20*(3), 372–378.

Chen, L., Chen, P., & Lin, Z. (2020). Artificial intelligence in education: A review. *IEEE Access, 8*, 75264–75278. https://doi.org/10.1109/ACCESS.2020.2988510

Chirikov, I., Semenova, T., Maloshonok, N., Bettinger, E., & Kizilcec, R. F. (2020). Online education platforms scale college STEM instruction with equivalent learning outcomes at a lower cost. *Science Advances, 6*(15), eaay5324. https://doi.org/10.1126/sciadv.aay5324.

Crompton, H., Jones, M. V., & Burke, D. (2024). Affordances and challenges of artificial intelligence in K-12 education: A systematic review. *Journal of Research on Technology in Education, 56*(3), 248–268.

Divasón, J., Martínez-de-Pisón, F. J., Romero, A., & Sáenz-de-Cabezón, E. (2023). Artificial intelligence models for assessing the evaluation process of complex student projects. *IEEE Transactions on Learning Technologies, 16*(5), 694–707. https://doi.org/10.1109/TLT.2023.3246589

Duffy, C. (2020, May 29). The ACLU sues Clearview AI, calling the tool an 'unprecedented violation' of privacy rights. *CNN Business*. https://edition.cnn.com/2020/05/28/tech/clearview-ai-aclu-lawsuit

Dwivedi, Y. K., Hughes, L., Ismagilova, E., Aarts, G., Coombs, C., Crick, T., et al. (2021). Artificial Intelligence (AI): Multidisciplinary perspectives on emerging challenges, opportunities, and agenda for research, practice and policy. *International Journal of Information Management, 57*, 101994. https://doi.org/10.1016/j.ijinfomgt.2019.08.002

English, L. D. (2016). STEM education K-12: Perspectives on integration. *International Journal of STEM Education, 3*(3), 1–8.

Fortus, D., Dershimer, C., Krajcik, J., Marx, R., & Mamlok-Naaman, R. (2004). Design-based science and student learning. *Journal of Research in Science Teaching, 41*(10), 1081–1110. https://doi.org/10.1002/tea.20040

Fredricks, J. A., Blumenfeld, P. C., & Paris, A. H. (2004). School engagement: Potential of the concept, state of the evidence. *Review of Educational Research, 74*(1), 59–109. https://doi.org/10.3102/00346543074001059

Gibson, D., Kovanovic, V., Ifenthaler, D., Dexter, S., & Feng, S. (2023). Learning theories for artificial intelligence promoting learning processes. *British Journal of Educational Technology: Journal of the Council for Educational Technology, 54*(5), 1125–1146. https://doi.org/10.1111/bjet.13341

Goel, A. (2020). AI-powered learning: Making education accessible, affordable, and achievable. *arXiv preprint* https://arxiv.org/abs/2006.01908. https://doi.org/10.48550/arXiv.2006.01908.

Google Labs. (n.d.). *Google Labs*. Retrieved April 8, 2025, from https://labs.google/about

Harris, J., & Felix, A. (2010, March 29). *A project–based, STEM–integrated team challenge for elementary and middle school teachers in alternative energy* [Conference presentation]. The Society for Information Technology & Teacher Education International Conference, pp. 3566–3573.

Hrastinski, S., Olofsson, A. D., Arkenback, C., Ekström, S., Ericsson, E., Fransson, G., et al. (2019). Critical imaginaries and reflections on artificial intelligence and robots in post digital K-12 education. *Postdigital Science and Education, 1*(2), 427–445. https://doi.org/10.1007/s42438-019-00046-x

Huang, A. Y., Lu, O. H., & Yang, S. J. (2023). Effects of artificial intelligence–enabled personalized recommendations on learners' learning engagement, motivation, and outcomes in a flipped classroom. *Computers & Education, 194*, 104684. https://doi.org/10.1016/j.compedu.2022.104684

International Society for Technology in Education. (n.d.). *AI professional development*. Retrieved April 9, 2025, from https://iste.org/ai

Jiao, P., Ouyang, F., Zhang, Q., & Alavi, A. H. (2022). Artificial intelligence-enabled prediction model of student academic performance in online engineering education. *Artificial Intelligence Review, 55*, 6321–6344. https://doi.org/10.1007/s10462-022-10155-y

Joseph, O. B., & Uzondu, N. C. (2024). Integrating AI and machine learning in STEM education: Challenges and opportunities. *Computer Science & IT Research Journal, 5*(8), 1732–1750.

Katehi, L., Pearson, G., & Feder, M. (Eds.) (2009). *Engineering in K-12 education: Understanding the status and improving the prospects*. Committee on K-12 engineering education, National Academy of Engineering, and National Research Council. National Academies Press.

Khan Academy. (n.d.). *Khanmigo*. Retrieved April 7, 2025, from https://www.khanmigo.ai/

Kohnke, S., & Zaugg, T. (2025). Artificial intelligence: An untapped opportunity for equity and access in STEM education. *Education Sciences, 15*(1), 68. https://doi.org/10.3390/educsci15010068

Komalawardhana, N., & Panjaburee, P. (2024). Trends and development of technology-enhanced personalized learning in science education: A systematic review of publications from 2010 to 2022. *Journal of Computers in Education, 11*(3), 721–742. https://doi.org/10.1007/s40692-023-00276-w

Lo, C. K., Hew, K. F., & Jong, M. S. Y. (2024). The influence of ChatGPT on student engagement: A systematic review and future research agenda. *Computers & Education*, Article 105100.

Moore, T. J., Guzey, S. S., & Brown, A. (2014a). Greenhouse design to increase habitable land: An engineering unit. *Science Scope, 37*(7), 51–57.

Moore, T. J., Stohlmann, M. S., Wang, H. H., Tank, K. M., Glancy, A. W., & Roehrig, G. H. (2014b). Implementation and integration of engineering in K-12 STEM education. In S. Purzer, J. Strobel, & M. Cardella (Eds.), *Engineering in pre-college settings: Synthesizing research, policy, and practices* (pp. 35–60). Purdue Press.

Morrison, J. (2006). *Attributes of STEM education: The student, the school, the classroom* [Monograph]. Teaching Institute for Excellence in STEM.

Mruthyunjaya, V., Burcin, R., Evans, C., & Dolan, J. M. (2024, February). *Mind the gap: Enhancing AI and STEM accessibility in rural Pennsylvania* [Conference presentation]. IAFOR International Conference on Education, Hawaii.

Nikhil, V. (2025). A comprehensive study on AI-enhanced personalized learning in STEM courses. In T. Murugan, K. Periasamy, & A. M. Abirami (Eds.), *Adopting artificial intelligence tools in higher education* (pp. 149–170). CRC Press. https://doi.org/10.1201/9781003469315

Nuangchalerm, P., & Prachagool, V. (2023). AI-driven learning analytics in STEM education. *International Journal of Research in STEM Education, 5*(2), 77–84.

Nugumanova, A. B., Tlebaldinova, A. S., Baiburin, Y. M., & Ponkina, Y. V. (2021). Natural language processing methods for concept map mining: The case for English, Kazakh, and Russian. *Journal of Mathematics, Mechanics & Computer Science, 112*(4), 93–108. https://doi.org/10.26577/JMMCS.2021.v112.i4.08

Ortiz-Revilla, J., Greca, I. M., & Arriassecq, I. (2022). A theoretical framework for integrated STEM education. *Science & Education, 31*(2), 383–404. https://doi.org/10.1007/s11191-021-00242-x

Ouyang, F., Dinh, T. A., & Xu, W. (2023). A systematic review of AI-driven educational assessment in STEM education. *Journal for STEM Education Research, 6*(3), 408–426.

Saw, G. K., & Charlotte, A. A. (2021). STEM pathways of rural and small-town students: Opportunities to learn, aspirations, preparation, and college enrollment. *Educational Researcher, 50*(9), 595–606. https://doi.org/10.3102/0013189X211027528

Schorcht, S., Buchholtz, N., & Baumanns, L. (2024). Prompt the problem–investigating the mathematics educational quality of AI-supported problem solving by comparing prompt techniques. *Frontiers in Education, 9*, 1386075. https://doi.org/10.3389/feduc.2024.1386075

Selingo, J. (2007). Powering up the pipeline. *ASEE Prism, 16*(8), 24–29.

Sloan, K. (2024, July 23). *AI in MATHia: It's more than right and wrong answers.* Carnegie Learning. https://www.carnegielearning.com/blog/mathia-ai

Smith, K. A., Sheppard, S. D., Johnson, D. W., & Johnson, R. T. (2005). Pedagogies of engagement: Classroom-based practices. *Journal of Engineering Education, 94*(1), 87–101. https://doi.org/10.1002/j.2168-9830.2005.tb00831.x

Smyrnaiou, Z., Petropoulou, E., & Sotiriou, M. (2015). Applying argumentation approach in STEM education: A case study of the European student parliaments project in Greece. *American Journal of Educational Research, 3*(12), 1618–1628.

Stehle, S. M., & Peters-Burton, E. E. (2019). Developing student 21st century skills in selected exemplary inclusive STEM high schools. *International Journal of STEM Education, 6*(39), 1–15.

STEM Task Force Report. (2014). *Innovate: A blueprint for science, technology, engineering, and mathematics in California public education.* Californians Dedicated to Education Foundation.

Stohlmann, M., Moore, T., & Roehrig, G. (2012). Considerations for teaching integrated STEM education. *Journal of Pre-College Engineering Education Research, 2*(1), 28–34. https://doi.org/10.5703/1288284314653

Sun, D., Cheng, G., Yu, P. L. H., Jia, J., Zheng, Z., & Chen, A. (2025). Personalized STEM education empowered by artificial intelligence: A comprehensive review and content analysis. *Interactive Learning Environments*, 1–23. https://doi.org/10.1080/10494820.2025.2462156

The United Nations Secretary-General. (2025, January 24). Global education must integrate AI, centred on humanity. *UN News.* https://news.un.org/en/story/2025/01/1159381

Thibaut, L., Ceuppens, S., De Loof, H., De Meester, J., Goovaerts, L., Struyf, A., et al. (2018). Integrated STEM education: A systematic review of instructional practices in secondary education. *European Journal of STEM Education, 3*(1), 1–12.

Tissenbaum, M., & Slotta, J. D. (2019). Developing a smart classroom infrastructure to support real-time student collaboration and inquiry: A 4-year design study. *Instructional Science, 47*(4), 423–462. https://doi.org/10.1007/s11251-019-09486-1

UNESCO. (n.d.). *Artificial intelligence in education.* Retrieved April 8, 2025, from https://www.unesco.org/en/digital-education/artificial-intelligence

Vargas-Murillo, A. R., de la Asuncion, I. N. M., & de Jesús Guevara-Soto, F. (2023). Challenges and opportunities of AI-assisted learning: A systematic literature review on the impact of ChatGPT usage in higher education. *International Journal of Learning, Teaching and Educational Research, 22*(7), 122–135.

Wiley. (n.d.). *Knewton Alta.* Retrieved April 7, 2025, from https://www.wiley.com/en-gb/education/alta/features

Yang, Y., Sun, W., Sun, D., & Salas-Pilco, S. Z. (2024). Navigating the AI-enhanced STEM education landscape: A decade of insights, trends, and opportunities. *Research in Science & Technological Education*, 1–25. https://doi.org/10.1080/02635143.2024.2370764

Zhang, L., Qu, J., & Zhang, Z. (2024). The impact of personalized learning driven by artificial intelligence on STEM education in primary and secondary schools. *International Journal of New Developments in Education, 6*(10), 29–34.

CHAPTER 13

Generative AI in Initial STEM Teacher Education: A Transformative and Situated AI Literacy

Antonio Quesada and Marta Romero-Ariza

Introduction

The idea that a new technology, such as artificial intelligence (AI), can generally produce profound and unpredictable transformations in various sectors, including education, is not new (Schwab, 2017; Salmon, 2019). AI is rapidly transforming various sectors, including educational settings, leaving no place for doubt. Generative AI (in this text, genAI), a branch of AI that can create (we will not enter into the debate about the authenticity of creation) new content such as text, images, and code, has emerged and expanded as a powerful tool with the potential and expectation to revolutionize teaching and learning practices unpredictably (Bae et al., 2024; Al-Abdullatif, 2024; Pratschke, 2024, or Lee & Zhai, 2024). In this regard, however, voices are raised suggesting that this situation is not new and that what is "expected from an emerging technology" may not be as predicted or that, as has happened before, the

A. Quesada (✉) • M. Romero-Ariza
University of Jaén, Jaén, Spain
e-mail: antquesa@ujaen.es; mromero@ujaen.es

R. Elmas, M. Akarsu (eds.), *STEM Education*, Maarif Global Education Series,
https://doi.org/10.1007/978-981-95-9023-0_13

"hype" generated in the initial phases is overblown after some time. Cuban (1999) described an analysis of this effect and the different phases through which the adoption of a technology occurs. However, the integration of genAI in education in general and its application in specific domains (herein subjects or domains such as STEM) requires a deep understanding of its characteristics, functionalities, and knowledge about its limitations. For example, the generation of content (through lesson plans) for use in early-age education and aimed at STEM education must be carefully reviewed to avoid, for instance, biases, stereotypes, or inadequate or naïve views of science (Powell & Courchesne, 2024; Sun et al., 2024). The plethora of literature on genAI applied to various knowledge fields is growing exponentially, and any individual can access this material—from open science books to research articles (Ogunleye et al., 2024; Zhang et al., 2024). Initiatives like Google's free "Generative AI for Educators" course offer hands-on training on using genAI tools to save time, personalize instruction, and improve lessons. Additionally, OpenAI, in conjunction with Common Sense Media, has launched a free training course for teachers aimed at demystifying AI and prompt engineering, providing guidance on using ChatGPT to create lesson content and manage classroom activities. Other studies highlight how guided use of genAI can structure lesson sequences, support reflective refinement, and align outputs with curriculum goals (Gurl et al., 2025; Kalenda et al., 2025). Therefore, as educators, we are still eager for—and often miss—specific programs or interventions focused on teacher education (Tunjera & Chigona, 2023; Leahy et al., 2025; Zaugg, 2024). In addition, some initiatives are sometimes criticized for lacking systematic interventions or robust AI frameworks focused on educational purposes. In this scenario and throughout the chapter, we will provide some key insights and relevant information about genAI and its role in STEM education, focusing on initial teacher education and recognizing teachers as key actors in innovation and educational transformation. In the second part, we will present the design principles that guided the design of the intervention for teachers, focusing on the teachers' roles acquired during the different tasks. In the final part of this chapter, we will show some selected preliminary results from the intervention analysis.

Teacher Education and AI Literacy: Does AI Undermine Teacher Competencies?

As genAI becomes more ubiquitous in educational and social settings as part of our daily lives, it seems essential that teachers, especially those at the initial stages of their careers, must develop critical and situated AI literacy. AI literacy is part of technological literacy and can be defined as the set of competencies that enable individuals to understand, evaluate, and use (AI) technologies in a critical and responsible manner (Al-Abdullatif, 2024; Velander et al., 2024).

Research on education highlights the significance of fostering AI literacy and the appropriate use of intelligent TPACK frameworks (Tram, 2024; Ning et al., 2024) as a means to more effectively prepare educators for the age of AI (Al-Abdullatif, 2024). From these frameworks, teachers should have a fundamental understanding of the basic concepts of AI, including the basic aspects of how machine learning, deep learning, and neural networks work (Al-Zahrani, 2024). This understanding should emerge from an informed perspective, though not necessarily as an expert with deep knowledge and understanding of AI. Teachers must effectively use genAI tools to support teaching and learning, including understanding not only the features and functionalities but also the limitations of the tools as well as the didactic strategies for their proper integration into the classroom and their importance in the teachers' own professional development (knowledge about AI applications) (Bae et al., 2024; Lee & Zhai, 2024). Teachers must be able to critically evaluate the didactic usefulness, effectiveness, and suitability of AI applications and consider their possible limitations. We must take into account the great responsibility as teachers of young children (ages 6–12) regarding the negative impact that the use of AI-generated resources could have if they had not been tested or critically analyzed.

However, the increasing use of AI raises serious concerns about data privacy, bias, and the potential loss of human agency in decision-making processes. Therefore, excessive AI use can weaken thinking by providing users with instant answers that could dissuade deeper thinking and further analysis. Studies have found that over-reliance on AI has a detrimental effect on students' cognitive skills, compromising their critical thinking, creativity, and ability to resolve complex problems (Zhai et al., 2024). Therefore, at all times, the user (in our case, STEM teachers) must be aware of the appropriate use of genAI in the preparation of their classes or materials for their students or of the adoption of this technology for their professional growth.

STEM Teachers and Critical Thinking

Developing twenty-first-century competences through education requires a re-examination of instructional processes and transitioning from teaching knowledge to encouraging learning that fosters critical thinking. Preparing teachers to incorporate critical thinking into STEM education is crucial, but a critical aspect of this effort is initial teacher education, a stage where teachers must be adequately oriented, inspired, and motivated to address this issue effectively (Romero-Ariza et al., 2021), avoiding in the context of the use of genAI what we have previously mentioned as possible undermining of the development of their competencies.

Critical thinking is a fundamental competency for STEM teachers because it allows them to rigorously analyze, question, and evaluate both the content they teach and the methodologies they use, which has a positive impact on the teacher profession. This capacity is key to promoting quality teaching in the classroom, based on inquiry, problem-solving, and informed decision-making, which are essential aspects in scientific, technological, engineering, and mathematical disciplines, that is, in STEM approaches. It is a key element in integrated STEM (iSTEM). In addition, a teacher with critical thinking skills is better equipped to guide students in constructing significant knowledge, promoting autonomy and their ability to face complex challenges, especially at an early age, by cultivating inquiring minds. The educational proposal has been considered among other aspects, including the development of critical thinking as an implicit and explicit integration into various activities and pedagogical strategies that enhance its application both in teaching practice and student learning.

Design Principles

We follow a DBR (Design-Based Research) approach for the design and optimization of the intervention intended at the first stages, in which decisions and processes are research-based (The Design-Based Research Collective, 2003; Tinoca et al., 2022; Haagen-Schützenhöfer et al., 2024; Obczovsky et al., 2025). The intervention proposal described here is in its first design cycle.

Three Roles for Pre-service Teacher Educational Proposals

The design principles related to the professional dimension are deeply rooted in the idea of the triple role of the teacher; that is, "the teacher" as student (or apprentice), the teacher as a reflective practitioner, and the teacher as a designer. These three roles do not act separately but intersect within pre-service teacher education. An example of the application of these principles is provided by Quesada et al. (2019). The relevance of the initial teacher education was stressed to ensure that the teacher participants had opportunities to perceive (and improve) the use of genAI in their daily educational settings. This section focuses on the various roles that pre-service teachers assume at different stages throughout the learning episodes.

When we account for the pedagogical and technological component of AI (Celik, 2023), this triple role expands into a more complex set of roles within the context of STEM education. The guiding question for the design is: How can we apply genAI in the development of STEM practices for primary educational settings? And how to engage pre-service teachers in such a development process? We will use a practical and immersive experience to partially answer those design questions.

Immersion experiences were defined as those that positioned teachers as learners within a topic or "subject". This subject can be a science investigation, a technological problem, a mathematical context, or any other interdisciplinary approach in a learning scenario. Through these investigations, participants (within their double role of STEM teacher-STEM student) are provided, explicitly or implicitly, with curriculum orientations, pedagogies, and assessment practices that align with the syllabus they will have to face in their teaching practices, at the same time they experience what their future students will do as STEM learners. Thus, for example, in one of the tasks we propose below, the teacher in training implicitly introduces the concept of "big idea in science" (Harlen, 2015) as an element that structures the Science curriculum. These immersive settings require knowledgeable and experienced facilitators to guide the tasks, a crucial aspect of these pedagogies (Urbani et al., 2017). Therefore, the role of the educator as a guide in the implementation of this intervention is crucial for its success.

STEM Pre-service Teachers as Students

Allowing teachers to act as future students enhances their education. Teachers can experience their students' "first-person" learning processes and episodes by situating themselves in learners' shoes. These episodes and processes relate not only to STEM content knowledge but also to scientific processes that embrace approaches based on inquiry and practical learning, both of which are essential to STEM education. Experiencing these same "challenges" equips them to meet the needs of their pupils. This immersive experience also helped the students learn more about their pedagogical component. We can cite a few successful examples in which teachers take the role of students if we shift our attention from the classroom to actual STEM settings. A study conducted in STEM academies (Lamberg & Trzynadlowski, 2015) for primary education teachers who were exposed to STEM proposals as students demonstrated a better understanding of how to incorporate interdisciplinary activities into their lessons. These educators also gained more self-confidence in undertaking STEM-related projects. Immersion practices (teacher as student) are widely used in the initial stages of teacher education and have a positive impact not only on the aspects of mastery of content and processes but also on increasing teachers' self-confidence in implementing pedagogies that promote iSTEM (Wu et al., 2024).

This experience as learners directly informs their subsequent roles as instructional designers, equipping them with deeper insights into classroom dynamics. By facing learning challenges directly, pre-service teachers acquire content knowledge and cultivate an understanding of the pedagogical complexities essential for designing effective learning experiences. This immersion sets the stage for a change in perspective, from being students to being designers of meaningful, interdisciplinary STEM learning environments. As a result, the change from learners to designers could be regarded as a natural step in the professional development of STEM teachers.

STEM Pre-service Teachers as Designers

When future researchers experience the iSTEM task in their role as "students" they can identify, assimilate, and recognize processes, as well as the challenges and difficulties that will be of utmost relevance in their role as teachers and designers. The emphasis on the role of teacher as designer in an iSTEM context highlights teaching skills (Pedagogical Content

Knowledge) to craft interdisciplinary learning experiences that suit national curricular STEM standards and foundations of twenty-first--century skills. This role approach emphasizes the significance of future teachers going beyond "the content" or scientific or engineering processes, equipping them with a range of resources that ensure and foster scientific processes, inquiry skills, and problem-solving (Krajcik & Delen, 2017). Effective integration of subjects or "Big Ideas" in STEM must be a truly cohesive experience aligned with students' learning goals. Thus, the role of the teacher as designer intersects with that of the teacher as a reflective practitioner.

Assuming the role of pre-service teachers as designers enables them to move beyond STEM content. It requires them to thoughtfully integrate interdisciplinary concepts, pedagogical strategies, and twenty-first-century skills into coherent and engaging STEM learning experiences. However, the ability to design effective instruction does not emerge in isolation because it is deeply intertwined with the capacity for reflective practice. As future teachers engage in designing educational resources, they simultaneously should confront issues such as effectiveness, relevance, and ethical implications of their choices. This need for constant questioning and critical thinking is what will help them become reflective practitioners.

STEM Pre-service Teachers as Reflective Practitioners

The literature reports that, despite the established trajectory of the STEM movement, many educators, especially those in the initial phase of their career or final course of their education courses, find the process of designing educational resources for iSTEM relatively new (Shernoff et al., 2017). Within the context of our pre-service teacher's course, for example, they reported that they did not recognize the main characteristics of the iSTEM, and more worryingly, almost all of them reported not knowing how to make proposals for their future students; thus, the role of teachers as designers and teachers as reflective practitioners are linked and interconnected. This situation underscores the importance of recognizing teachers as not only lesson plan designers but also reflective practitioners who engage in ongoing inquiry about their own practices. When genAI is introduced into this process, new possibilities and challenges emerge for resource creation, requiring teachers to make informed, ethical, and pedagogically sound decisions. Therefore, integrating genAI into educational design explicitly strengthens the interconnectedness among the creative, reflective, and professional dimensions of teaching (Lee & Zhai, 2024).

Learning Objects: Granularity

Another design principle is related to the structure of sessions and tasks. The granularity of tasks, which are conceived and understood as learning objects (Ariza & Quesada, 2011), forms another design principle for this proposal. In this way, designing learning objects with the right level of detail helps to break down knowledge into smaller, meaningful parts, making it easier to use in thoughtful and relevant and contextualized practices.

Tasks can be performed as a learning progression or broken down into "learning objects", which could be used standalone, depending on, among other factors, the educational context, pre-service teacher experiences or mastery, time, educator objectives, etc. In the context of initial teacher education, we believe that aspects related to AI require a diversification of pedagogical strategies. This diversification triggers the promotion of a progressive, reflective, and critical understanding of more complex concepts. Often these concepts seem to be enough detailed or unfamiliar to teachers' and with a perceived insufficient time dedicated in previous courses for teachers' competency development. For example, in our case, task about the "features for an effective prompt" has enough granularity by itself to be considered and standalone learning object.

From a situated learning perspective—such as the one proposed by the apprentice model—these learning objects can be designed to present more authentic situations. For example, these situations may include ethical dilemmas related to the use of genAI in education. Those learning objects can also involve the analysis of products generated through iterative human–genAI cycles. In addition, this granular approach promotes the integration of genAI with classical teaching practices. For example, genAI can be used to enrich or personalize traditional lesson sequences by incorporating AI-generated products or responses using the appropriate "prompt".

Granularity, therefore, allows these experiences to be adapted to the level of the teacher-learner, promoting a gradual appropriation of the content without losing sight of its educational relevance. A balance between specificity and flexibility is essential for learning objects to be meaningful and transferable considering the right level of granularity (Willey, 2000 cited in Dahl & Morch, 2025). Once we have drawn the framework and the design principles, we present in the following sections the result of the educational pre-service teacher proposal, expanding and commenting on some aspects that support the design and clarify the intervention.

Proposal Contextualization

This proposal was designed to be implemented in the fourth year of a teacher education program for primary school teachers, as part of a specific course titled "Science Education for Primary School". In previous courses, pre-service teachers gained experience using iSTEM educational resources, practices, and pedagogies (Ariza et al., 2018). Throughout the semester, student-teachers had also experienced various activities related to iSTEM (Quesada et al., 2019). The proposal includes activities distributed in three sessions of approximately two hours each (each of which could be considered as one learning object). Figure 13.1 presents a flowchart illustrating the various tasks involved in this intervention.

The granularity of these allows them to be carried out, depending on teachers' needs, the level of competence of the teachers, the time of training available, the objectives of the program, and the educator. The formative proposal was implemented in the first cycle of the DBR approach with a sample of 228 fourth-year primary education teachers (137 female and 90 male students). As mentioned above, the proposal is sufficiently adaptive to be conducted working alone (not recommended) or in small groups. In our case, groups of four students were proposed for the phase of sharing products and outcomes. However, exploring previous ideas about genAI and generating "suitable prompts for a STEM sequence" was strongly recommended to do it alone and then co-construct and share in

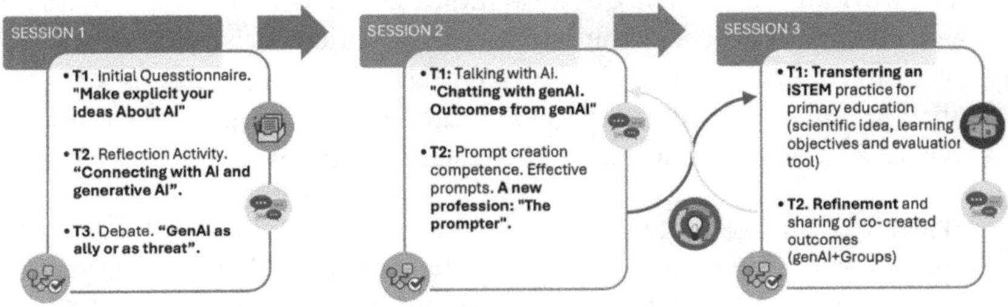

Fig. 13.1 Flowchart of the sessions and tasks. (Designed by the authors)

student groups. The proposal was reviewed by three expert researchers in science education before implementation, and their suggestions were included in the first version, which was part of the first implementation cycle.

Proposal Aims

This educational proposal was developed to achieve different interconnected objectives to promote the development of genAI and STEM-- competencies for pre-service teachers as follows: (a) Understand the characteristics, functionalities, implications, and limitations of genAI in educational settings in general and STEM scenarios in particular. (b) Recognize the need to be an AI literate teacher and carry out professional development for teachers that allows for its appropriate use in teaching and learning processes. (c.) Reflect on the use of genAI to enhance teaching (and instead learning), providing creativity and critical thinking. (d) Involve future teachers in the critical, "good", and "suitable" ways to use genAI in different situations, including specific lesson plans or rubrics for primary education in STEM subjects. (e) Reflect on the social, educational, and ethical implications of genAI. The sequence of session tasks and activities is described below.

Make Explicit Your Ideas About AI

In this initial task, the pre-service teachers were required to answer some questions in relation to genAI. To this end, an open-ended questionnaire was designed to guide the initial reflection. One question related to a dilemma was also included to determine to what extent a pre-service teacher critically self-positioned in his or her role in the educational field, with or without genAI. The question indicates whether the teacher will be replaced by AI. This part of the activity was designed based on the "theory of necessity", motivation, and the potential of use dilemmas as key elements for engaging and triggering contextual learning. Questionnaire (Table 13.1) was administered to the teacher as follows:

> "There is a talk and debate of scientific, environmental, or digital literacy, among others." The exponential growth in the use of AI in practically all areas of our daily lives makes it a main component of digital literacy itself.

Table 13.1 Questionnaire for pre-service teachers in task 1 (in brackets intended categorization)

Questions	*Categories*
As a future teacher, what will I know about AI and generative AI? What should I know as a teacher about how to make appropriate, ethical, and professional use of this technology?	**Previous Knowledge about genAI**
Have you used AI at any point in your career? What was the purpose of your use of AI?	**Current uses of gen AI**
What do you think the impact of AI integration on initial teacher training programs?	**Perceived educational impact of AI**
How do you think generative AI can be an "ally" regarding developing our profession as teachers?	**AI as a pedagogical tool**
How can we effectively integrate AI into initial educational programs using a technological, didactic, and pedagogical framework?	**Perceived educational impact of AI**
What type of training would you like to receive related to generative AI? How can we effectively integrate these concepts?	**Perceived educational impact of AI.**
What do you think artificial intelligence cannot do? Could I improve you as a teacher?	**Human aspects and limits of teaching**
Can AI replace you in the classroom? What" features" do you think you have as a teacher that make you irreplaceable for genAI?	**Dilemma or controversy (Deductive Emerging Categories)**

Recently, the term "AI literacy" has been coined, but from our point of view, it is one component of digital literacy, but "Have you thought about some implications that genAI have (or will have) in your training or future profession as a teacher?"

From a granularity point of view, the task can be expanded for a much more in-depth discussion with pre-service (or in-service) teachers using Gouvea (2024) article as a reference. The author highlights how the use of machine learning (ML) to evaluate student work can amplify racial and linguistic biases and calls on the educational community to critically question not only the technical utility of ML but also its social and ethical implications. These aspects can also be connected to the analysis of AI-generated images and possible biases in task 3.

Connecting with a "New" Technology for Education

In the context of a possible significant transformation of pedagogical practice through the use of AI or genAI, a series of questions arise to investigate teachers' beliefs. These questions encompass the possibility that educators may choose to modify and enrich their practices to incorporate such technologies. However, perceived barriers as well as their degree of fear regarding, in some cases, the imposition to adopt such technologies (Sun et al., 2024; Tunjera & Chigona, 2023; Díaz & Nussbaum, 2024, Hazzan-Bishara et al., 2025). Posing questions encourages reflection and might anticipate difficulties encountered when using genAI, for example, when creating lesson plans. Finally, pre-service teachers reflect on the specific functions or contributions that are attributed to AI and genAI regarding the optimization of their pedagogical work and the support of their professional responsibilities (Fig. 13.2.).

The activity from the chapter "Technology, change and You" (de la Higuera & Iyer, 2024) has been subjected to minor adaptation and served as the basis for this task. In that activity, "a new" technology was presented as "the motion picture" in 1922, which would revolutionize education, with the belief that it would replace textbooks. Herein, pre-service teachers are required to leverage their previous experience with other "new technologies" in general and in education to make comparisons and address the subsequent reflection, initiated by the overarching question:

> If a new technology emerges soon that could aid in your teaching duties, what attributes would you look for?

This task emphasizes the teacher's role as a designer and reflective practitioner aiming to identify characteristics that facilitate the creation of AI-mediated products. We also posed the following questions:

> "Would you change your teaching practice to use it?" Are you afraid of facing pressure to alter your teaching methods?" What challenges do you think you should face? Regarding AI and genAI, what can AI and genAI do to help you and improve your teaching?"

The closing activity involved sharing responses through dialogue in the classroom. This sharing can reveal relevant aspects and themes that highlight agreed-upon topics and those with discrepancies. To that end, we

I. Session 1

Task 2. Reflection activity (Connecting with AI and generative AI)

Time: 20 minutes

Read this excerpt carefully and answer questions at the end. Then we share it.

"In 1922, Thomas Edison declared that the moving image would revolutionize education. He believed that it would replace all textbooks.
However, the teachers' use of the films was limited. The photographic slide projector was adopted by many masters, from the 50s to the late 90s."

Unlike when film was used, teachers could:

1. Prepare your own slides at a low cost.
2. Using the projector as a whiteboard, a tool that did not change the way he taught.
3. Reuse slides, reorder them, and refine them."

If there was a new technology that could help you in your teaching tasks:

- What features would you (as teacher) look for?
- If the use of it meant a significant change in your teaching practice, would you change your teaching practice to use it?
- Would you be afraid of being "forced" to change? What challenges do you think you should face?
- **In relation to AI, what things do you think it can do for you to "help" in your teaching task?**

Fig. 13.2 Example of Handouts for pre-service teachers′ task 2 in session 1. (Adapted from de la Higuera & Iyer, 2024)

could introduce to pre-service teachers' task three or continue discussions with them linking some responses with specialized literature. At this point, the task can be enriched by providing articles or critical essays from the specialized literature in which, for example, it is emphasized that "AI for teaching and learning in schools" also needs for "pedagogical intelligence", highlighting the need for a conscious and reflective pedagogical integration (Díaz & Nussbaum, 2024). These authors argue that inadequate pedagogy limits the effectiveness and relevance of AI in school contexts.

Debate "genAI as Ally or as Threat"

In initial teacher education programs, we should offer opportunities to develop critical thinking, reasoned argumentation, and the ability to understand and adopt reflective positions regarding contemporary challenges to promote key competencies. In this sense, structured debates on controversial topics constitute a highly valuable pedagogical strategy (Romero-Ariza et al., 2021; Savitz, 2021; Hendratmoko et al., 2024). This activity proposes a dialogic debate among pre-service teachers regarding genAI in education, comparing its benefits and risks and considering it as either an "ally" or a "threat". Controversies and dilemmas allow ethical reasoning to be activated, promote informed decision-making, and assist the social construction of knowledge. Embedded in the sequence of tasks, this debate is centered on the use of genAI in education, a topic that "directly challenges future teachers as professionals responsible for the ethical and contextualized didactic use of their knowledge". Debate requires previous preparation, prior research in search of data and evidence, analysis of evidence, empathy toward diverse positions, and the reasoned expression of opinions. These elements are fundamental to consolidating a reflective and critical professional identity. Pre-service teachers are grouped into two opposing groups of three to four members. Each group of students defends a different perspective on the use of genAI for primary education. The open debate will last 45 minutes, with each group having 10 minutes to initially present their arguments and 25 minutes for counterargument or refute. One prerequisite of the task is that all arguments have to refer to "good-quality" evidence. In this sense, it is compulsory to share with others the source of that evidence. The educator will act as a moderator to ensure fair turn-taking and time management, but also as an observer to provide "on-the-spot" and formative feedback during the debate and afterward, the closure activity.

To avoid initial biases or preferences, topics are randomly assigned to groups, and each group must seek information to position itself against the other teams. To increase the aspect of controversy, "negative" themes are offered that must also be "defended". This generates complex and rich full situations, but at the same time, it is relevant to promote critical thinking and the search for evidence because, occasionally, many students do not agree with this position. The main dimensions established for the debate in relation to genAI as "ally" are: "personalization", "access", "innovation", "promotion of twenty-first-century skills", "efficiency",

Table 13.2 "Triggering" questions for debate

Questions	*Dimension*
How can AI adapt educational content to each student's individual needs and learning pace to maximize their potential?	**Personalization**
In what ways can AI democratize education by providing access to high-quality resources and tools to students regardless of their location or circumstances?	**Access**
How can AI optimize administrative and repetitive tasks, freeing up teachers' time to focus on more creative and meaningful activities?	**Efficiency**
How can AI optimize pedagogical tasks and promote more creative and meaningful activities?	**Efficiency**
What new forms of learning and teaching could emerge thanks to AI, and how might these transform educational experiences?	**Innovation**
How can AI provide instant and personalized feedback to students and teachers to improve their understanding and motivation?	**Feedback**
How can AI support the development of students' critical thinking skills?	**(21st C. Skills)**

and "feedback". The main dimensions established for the debate concerning "genAI" as a "threat" are: "de-humanization", "cost" (not in terms of economics but personal and professional investment), and "dependency".

At the end of the debate, each group must submit a summary of the main points of agreement and disagreement related to the topics previously exposed and a report in which they provide evidence of their arguments. These pieces of evidence refer to expert and specific literature or bibliography on the subject. Pre-service teachers can start the debate using some of the guiding questions presented in Table 13.2.

Chatting with genAI: Outcomes from genAI—Are They What I Need?

This task is divided into two parts. First, pre-service teachers reflected on their prior beliefs about "prompts". Then, they applied these ideas by interacting with genAI tools within a chosen teaching scenario focused on primary science education. Educators left space for the student-teachers to determine their own "scenario" of teaching and learning, emphasizing science for primary education. Table 13.3 presents the guiding questions, actions, and dimensions that we intend to promote as educators. This task aims to familiarize pre-service teachers with the creation of meaningful

Table 13.3 Guiding questions and strategies for task 4

Question or action	*Role*	*Domain*
What do you think are the main features that make a prompt effective?	TRP, TS	Teacher believes
What effective questions (prompts) would you ask genAI to optimize the T-L process in science?	TRP, TD	Teacher believes
Consider: (a) inquiring with AI about the characteristics that define a good prompt and (b) inquiring with AI about the characteristics that define a good prompt to optimize T-L processes in science.	TS	Human responses (individual o social) and AI responses
Analyze the image shown below and try to "guess" what prompt the AI was given to generate it. Do you think it meets the requirements of a good prompt? Create your own image of a Primary Education classroom (in a context of teaching and learning sciences)	TS, TRP, TD	Human responses (individual o social) and AI responses
Consider developing "prompts" that compare your "prompt with these made by genAI" and start a new conversation looking for provide you with examples on specific topics of science teaching in primary education	TRP, TD	Human interactions with genAI

TRP Teacher as reflective practitioner, *TS* Teacher as student, *TD* Teacher as Designer

*Pre-service teachers are invited to make at least a three-iteration cycle in the AI conversation.

interactions with genAI tools like ChatGPT from a critical, effective, and pedagogically sound perspective. This focus is specifically on the design of optimized prompts, which could be considered as well-formulated questions, statements, or instructions that enable the full potential of these technologies to be realized in educational settings, particularly in the teaching of science subjects at the primary school level. The following text contains quotes from how the task was presented to pre-service teachers.

> But we must not forget that to obtain good results, we must communicate effectively and efficiently with AI. […] The expansion of the concept of prompting skills, especially within educational contexts, has gained relevance with the emergence and popularization of advanced artificial intelligence tools, such as ChatGPT. This phenomenon highlights the importance of developing and refining the ability to formulate questions and prompts effectively to interact with AI systems, with the aim of optimizing teaching and learning processes. […] for example, the way prompts are presented in

> science can greatly influence the type of answers offered by ChatGPT. (López-Simó & Rezende, 2024)

A set of guiding questions (Table 13.3) was used to define this task, highlighting the technological and pedagogical dimensions of knowledge.

Individual and collaborative responses aimed to foster critical thinking, not only in terms of analyzing and evaluating their answers but also comparing them with those provided by genAI. The comparison between the individual and group's initial ideas and those responses generated by genAI promotes a metacognitive process and attempts to engage pre-service teachers to analyze their competence in formulating questions as well as on the need for precision, clarity, and purpose in the dialog with these systems. We should note that in science education, "questioning" is a key competence (Promoting Research and Innovation in Mathematics and Science Education [PRIMAS] Consortium, 2025, adapted from Swan & Pead, 2008), and here, it is extrapolated to the use of genAI. Knowledge of how to ask "good questions" in the context of science education is linked to those researchable questions" (Chin & Osborne, 2010; Tena & Couso, 2023). In this sense, we can also establish that performing and knowing how to make good prompts allows teachers to optimize their questioning skills. Moreover, by incorporating the analysis of genAI images, the task broadens the understanding of the prompt concept toward visuals, allowing teachers to experiment with instructional design strategies that creatively, critically, and contextually integrate emerging concepts. Figure 13.3 shows an example of a genAI image that pre-service teachers have to analyze and inquire about "an effective prompt", and through different iterations, they have to improve.

A New Profession: "Prompter"

Knowing how to develop good prompts could make the design of STEM materials more effective in terms of not only quality but also the time spent preparing them (both for students and support materials: rubrics, learning objectives, narratives, identification of alternative ideas of students, conceptual clarification, etc.). In this part of the task, the pre-service teacher must formulate an initial prompt that allows a conversation to be held with genAI (chat) to obtain the following three products (not necessarily in the same conversation):

Fig. 13.3 What do you think about the prompt to generate this image? (AI generated image using a "personalized" prompt)

(a) Conceptual clarification or extension of a "great scientific idea" adapted to primary education.
(b) Writing Learning Objectives for a practical work for primary education related to some topics of that "great scientific idea".
(c) In co-creation with genAI, design a rubric related to the evaluation of a product that your students would give you, related to the topic defined in section.

Each of these parts has to be reviewed in at least a cycle of three interactions in order to improve each of the products; that is, after the first genAI response, pre-service teachers have to continue asking or including more information in the original prompt, looking for the improvement of the genAI outcomes. Pre-service teachers are invited to review the quality and coherence of those products, considering, among others, the appropriateness of the primary STEM curriculum.

In this activity, emphasis is placed on the double role of the pre-service teacher as designer and expert in the Primary Education curriculum, and as a reflective practitioner. Roles are highlighted since in the different genAI iterations processes the teacher must critically evaluate the outcomes. Connections with STEM education, knowledge of the "great scientific idea" (Harlen, 2015), practical work at an early age (Oliveira & Bonito, 2023); and the evaluation of learning in STEM (Gao et al., 2020) are explicitly introduced.

Transferring an iSTEM Practice for Primary Education

This final task can be considered a transfer and application task. It is intended that the pre-service teacher applies what he has learned in the previous tasks in relation to the characteristics of a good and effective prompt. They have to integrate and improve some of the genAI outcomes (scientific idea, learning objectives and rubrics; or other evaluation instruments) in one final product. This is a guide for teachers related to a practical iSTEM work for students aged 10–12 years in primary education. The following task is presented to pre-service teachers as:

> You must co-design a document of practices to work in the primary classroom (third cycle) under an iSTEM or iSTEAM approach.

This collaborative design (human–human; human–genAI) of a product not only promotes teamwork skills but also places genAI as a co-creative agent, generating a space for "cognitive" co-construction between humans and technology (Webster et al., 2024). In this sense, it is intended that the activity favors a dialogic relationship with genAI, in which its products are not passively received, but are "negotiated" through different cycles of iteration, validated and adjusted by the participants through the critical and pedagogical use of the tool. This dynamic situated learning simulates an emerging professional practice in which the teacher not only consumes technology but also "actively interacts" with genAI to generate meaningful educational materials. This last task has been proposed as an open activity for teachers to continue emphasizing dialogic strategies with genAI to obtain the desired product.

Preliminary Results

Preliminary findings of the implementation study are presented below using a thematic structure. This organization allows for a clearer understanding of the diversity of data collected during the intervention and highlights key dimensions of interest, such as prior perceptions, gender-related patterns, and outcomes of iSTEM task design.

The overall analysis of all pre-service teachers and genAI outcomes from implementation is beyond the scope and extent of this chapter. We collect some "spots" that have emerged from the preliminary analyses and that will be invaluable to improve the design in forthcoming cycles of refinement. This has helped us to understand how this cohort of pre-service teachers conceptualized the use of genAI (ChatGPT) and how it has integrated it into the design of specific iSTEM tasks.

Grouping results will facilitate the reader's navigation through the complex and multifaceted set of data we have, making the logical progression of findings more accessible and aligned with the objectives of the intervention. This organization allows for a clearer understanding of the data and highlights key dimensions of interest, such prior perceptions and pre-service teacher's needs, gender-related patterns, and outcomes of iSTEM task design.

Because each activity has its own outcomes, this section will show the outstanding findings or those that we believe have been the most significant and representative to understand how the intervention has worked. Methodologically, a content analysis was applied through the identification of keywords in the answers and through the support of a language model (GPT-4) and the expert vision of two researchers, which allowed the identification of interesting nodes of information. This assisted approach combines elements of human analysis with automated tools to strengthen the validity and traceability of the analysis of this topic.

Below are some foreseen analyses through certain quotations and how we, as designers of the proposal, value these results.

- *Pre-activity teachers' believes and training needs*

The detailed exploration of the answers will offer us valuable information to modify and adapt the tasks of the intervention, being able to personalize them if we consider the different "profiles" or levels of competence in AI reported by these teachers (future line of research). Therefore, we must say that the proposal has been contextualized so that the students themselves are aware of its degree of acceptance, their motivations, and their perception of self-efficacy (to generate dialogs with AI and develop materials in this case related to integrative iSTEM approaches that can be used with their future primary school students).

In relation to the pre-activity of exploring previous ideas collected through an open questionnaire, questions related to knowledge, use, perceptions, and training needs-based AI and generative genAI in the educational context were selected.

- *General attitudes towards genAI*

The preliminary analysis revealed a broad and nuanced perception of genAI among future STEM teachers. The responses revealed a growing interest in integrating these technologies into their educational practice, valuing their possibilities for the personalization of learning, the creation of resources, and support for teaching tasks.

This finding aligns with previous research (Yang & Appleget, 2024; Hava & Babayiğit, 2025) concluding that ChatGPT has a positive and significant impact on teachers' TDC (Teacher Digital Competency), acting as a tool that fosters adoption and familiarity with emerging technologies. Other previous research emphasizes the role of attitudes toward new technologies in the formation of digital skills. The use of ChatGPT can help teachers develop self-efficacy and familiarity with emerging technologies, thus contributing to the development of their digital competence.

- *Relationship, ethical and gender-based insights*

Ethical and human concerns were identified (perhaps biased by the formulation of some statements in the questionnaire). For this perspective, aspects related to personal human–human relationships are especially mentioned, in relation to empathy, professional judgment and the irreplaceable role of the teacher (an aspect that is taken up in another of the tasks of the intervention).

Results also suggest differences in relation to gender. In this way, women's responses tend to emphasize the need for useful, personalized, and applicable AI education in the classroom more, showing interest in a realistic and empathetic integration of AI. The need to understand specific aspects of AI that can help in day-to-day life as teachers is also frequently expressed.

On the other hand, the men's responses included more proposals on the use of AI in planning, automation of teaching tasks, and efficiency improvement, reflecting a somewhat more instrumental or technical approach in several cases.

- *Emerging educational demands and pedagogical awareness*
  Other aspects that have emerged from the content analysis are:

  - *High demand for more specific training*: Pre-service teachers state the need to receive genAI training in their careers. Curiously, this expression of interest does not hinder them from reporting that they have already used the genAI for different purposes throughout their academic career (c.a., 98% indicate that they used genAI mainly to "ask" questions). After the intervention, this "ask" action seems to evolve aligned to "making efficient prompts".
  - *Emerging awareness of the pedagogical dimension*: Even though many responses are focused on utility, it is recognized that the integration of AI should not be limited to the technological level but should be framed within solid pedagogical and psycho-pedagogical principles.
  - *Situated and adaptive training*: Pre-service teachers value educational models that connect with real classroom practices, allowing them to apply contextualized AI tools. This may be an interesting line for future research since it has not been discussed in-depth or we do not have more information to determine what they mean by "contextualized".

- *iSTEM design outcomes and genAI*

In relation to the specific part of iSTEM design, we can indicate that they generally value positively, from the point of view of curriculum design, that AI offers them the learning goals and assessment instruments for scientific practice in primary education.

In some cases, they also establish different dialogic iterations to modify some of the aspects of the assessment instruments.

> During the creation of this material, AI provided valuable insights and structures to organize the content of the science topic in a clear and coherent way." For example, AI helped us devise learning objectives in a [...] way that was tailored to the level of the learners, ensuring that they were understandable and achievable. In addition, the AI provided us with a detailed assessment rubric, which made it easier to organize the assessment of the tasks efficiently and objectively. (Student A)

However, it should be noted that only in some cases do students make an approximation or really make a critical judgment about the STEM proposal co-created with genAI and its peers (even when this was an explicit part of the task).

Thus, in most cases, they do not question that the AI offers them "experiments" and not authentic iSTEM proposals. This is due to weakly defined or incomplete initial prompts. This is not related to the use of the genAI itself but to previous shortcomings in relation to key topics for science education, such as: What are the great scientific ideas? What are the attributes for a proposal to be truly considered as iSTEM?

Conclusion and Further Lines of Research

The integration of generative AI (genAI) into initial STEM teacher educational programs is an unavoidable necessity and a transformative challenge to equip future teachers with twenty-first-century skills. The designed experience is grounded on TPACK theory and the Unified Theory of Technology Acceptance and Use, and emphasizes the triple role of the teacher as a learner, designer, and reflective practitioner. This approach seeks to be a good practice useful for initial education programs. Preliminary findings from this intervention in the first DBR cycle revealed a nuanced perception of genAI among future STEM teachers, which includes, among other aspects, an interest in the personalization of learning and in the creation of STEM classroom resources and materials. These results, which show how teachers conceptualize and integrate genAI into STEM task design, will be crucial for refining the design in future improvement cycles and for better understanding critical and situated AI literacy in teacher education.

References

Al-Abdullatif, A. M. (2024). Modeling teachers' acceptance of generative artificial intelligence use in higher education: The role of AI literacy, intelligent TPACK, and perceived trust. *Education Sciences, 14*(11), 1209. https://doi.org/10.3390/educsci14111209

Al-Zahrani, A. M. (2024). Unveiling the shadows: Beyond the hype of AI in education. *Heliyon, 10*, e30696. https://doi.org/10.1016/j.heliyon.2024.e30696

Ariza, M. R., & Quesada, A. (2011). Interoperability: Standards for learning objects in science education. In F. Lazarinis, S. Green, & E. Pearson (Eds.), *Handbook of research on e-learning standards and interoperability: Frameworks and issues* (pp. 300–320). IGI Global, 2018. https://doi.org/10.4018/978-1-61692-789-9.ch015

Ariza, M. R., Gallego, A. M. A., & Quesada, A. (2018). Empowering teachers to bring authenticity and responsive action into the science classroom. *School Science Review, 100*(371), 40–45.

Bae, H., Hur, J., Park, J., Choi, G. W., & Moon, J. (2024). Pre-service teachers' dual perspectives on generative AI: Benefits, challenges, and integration into their teaching and learning. *Online Learning, 28*(3), 131–156. https://doi.org/10.24059/olj.v28i3.4543

Celik, I. (2023). Towards Intelligent-TPACK: An empirical study on teachers' professional knowledge to ethically integrate artificial intelligence (AI)-based tools into education. *Computers in Human Behavior, 138*, 107468. https://doi.org/10.1016/j.chb.2022.107468

Chin, C., & Osborne, J. (2010). Students' questions and discursive interaction: Their impact on argumentation during collaborative group discussions in Science. *Journal of Research in Science Teaching, 47*(7), 883–908. https://doi.org/10.1002/tea.20385

Cuban, L. (1999). The technology puzzle: Why is greater access not translating into better classroom use? *Education Week, 18*(43), 47–68.

Dahl, J. E., & Morch, A. A. (2025). A theoretical and empirical analysis of tensions between learning object principles and current educational practices. *Education and Information Technologies, 30*(5), 1234–1254. https://doi.org/10.1007/s10639-025-13636-z

de la Higuera, C., & Iyer, J. (Eds.). (2024). *AI for teachers: An open textbook.* Pressbooks. https://www.ai4t.eu/textbook/

Díaz, B., & Nussbaum, M. (2024). Artificial intelligence for teaching and learning in schools: The need for pedagogical intelligence. *Computers and Education, 217*, 105071. https://doi.org/10.1016/j.compedu.2024.105071

Gao, X., Li, P., Shen, J., & Sun, H. (2020). Reviewing assessment of student learning in interdisciplinary STEM education. *International Journal of STEM Education, 7*(24), 1–14. https://doi.org/10.1186/s40594-020-00225-4

Gouvea, J. S. (2024). Ethical dilemmas in current uses of AI in science education. *CBE-Life Sciences Education, 23*(1). https://doi.org/10.1187/cbe.23-12-0239

Gurl, T. J., Markinson, M. P., & Artzt, A. F. (2025). Using ChatGPT as a lesson planning assistant with preservice secondary mathematics teachers. *Digital Experiences in Mathematics Education, 11*, 114–139. https://doi.org/10.1007/s40751-024-00162-9

Haagen-Schützenhöfer, C., Obczovsky, M., & Kislinger, P. (2024). Design-based research–Tension between practical relevance and knowledge generation–What can we learn from projects? *Eurasia Journal of Mathematics, Science and Technology Education, 20*(1), em2378. https://doi.org/10.29333/ejmste/13928

Harlen, W. (Ed.). (2015). *Working with big ideas of science education.* Science Education Programme of the InterAcademy Partnership (IAP). https://www.interacademies.org/publication/working-big-ideas-science-education

Hava, K., & Babayiğit, Ö. (2025). Exploring the relationship between teachers' competencies in AI-TPACK and digital proficiency. *Education and Information Technologies, 30*, 3491–3508. https://doi.org/10.1007/s10639-024-12939-x

Hazzan-Bishara, A., Kol, O., & Levy, S. (2025). The factors affecting teachers' adoption of AI technologies: A unified model of external and internal determinants. *Education and Information Technologies, 30*, 15043–15069. https://doi.org/10.1007/s10639-025-13393-z

Hendratmoko, A. F., Madlazim, M., Widodo, W., Suyono, S., & Supardi, Z. A. I. (2024). Inquiry and debate in science learning: Potential strategy for improving students' scientific argumentation skills. *International Journal of Education in Mathematics, Science, and Technology, 12*(1), 114–138. https://doi.org/10.46328/ijemst.3152

Kalenda, P. J., Rath, L., Abugasea Heidt, M., & Wright, A. (2025). Pre-service teacher perceptions of ChatGPT for lesson plan generation. *Journal of Educational Technology Systems, 53*(3), 219–241. https://doi.org/10.1177/00472395241301388

Krajcik, J., & Delen, I. (2017). How to support students in developing usable and lasting knowledge of STEM. *International Journal of Education in Mathematics, Science and Technology, 5*(1), 21–28.

Lamberg, T., & Trzynadlowski, N. (2015). How STEM academy teachers conceptualize and implement STEM education. *Journal of Research in STEM Education, 1*(1), 45–58. https://doi.org/10.51355/jstem.2015.8

Leahy, K., Ozer, E., & Cummins, E. P. (2025). AI-ENGAGE: A multicentre intervention to support student engagement, learning efficiency, and critical thinking through AI-focused, teacher-led interventions in higher education, including Initial Teacher Education. *Education Sciences, 15*(7), 807. https://doi.org/10.3390/educsci15070807

Lee, G.-G., & Zhai, X. (2024). *Using ChatGPT for science learning: A study on pre-service teachers' lesson planning* (Preprint). https://doi.org/10.48550/arXiv.2402.01674

López-Simó, V., & Rezende, R. T., Jr. (2024). Challenging ChatGPT with different types of physics education questions. *The Physics Teacher, 62*(4), 290–294. https://doi.org/10.1119/5.0160160

Ning, Y., Zhang, C., Xu, B., Zhou, Y., & Wijaya, T. T. (2024). Teachers' AI-TPACK: Exploring the Relationship between Knowledge Elements. *Sustainability, 16*(3), 978. https://doi.org/10.3390/su16030978

Obczovsky, M., Bernsteiner, A., Haagen-Schützenhöfer, C., & Schubatzky, T. (2025). Systematizing decisions in design-based research: From theory to design. *Science Education, 109*, 523–536. https://doi.org/10.1002/sce.21915

Ogunleye, B., Zakariyyah, K. I., Ajao, O., Olayinka, O., & Sharma, H. (2024). A systematic review of generative AI for teaching and learning practice. *Education Sciences, 14*(6), 636. https://doi.org/10.3390/educsci14060636

Oliveira, H., & Bonito, J. (2023). Practical work in science education: A systematic literature review. *Frontiers in Education, 8*, 1151641. https://doi.org/10.3389/feduc.2023.1151641

Powell, W., & Courchesne, S. (2024). Opportunities and risks involved in using ChatGPT to create first-grade science lesson plans. *PLoS One, 19*(6), e0305337. https://doi.org/10.1371/journal.pone.0305337

Pratschke, B. M. (2024). *Generative AI and education: Digital pedagogies, teaching innovation and learning design*. Springer. https://doi.org/10.1007/978-3-031-67991-9

Promoting Research and Innovation in Mathematics and Science Education (PRIMAS) Consortium. (2025). *PRIMAS teaching and learning materials: English modules*. https://primas-project.eu/modules/modules-english/

Quesada, A., Ariza, M. R., Abril, A. M., Cobo, C., & Ruiz, S. (2019). Pre-service teachers and STEM education. Immersion tasks and preliminary result. In *13th International Technology, Education and Development Conference* (pp. 3581–3590). IATED. https://doi.org/10.21125/inted.2019.0924

Romero-Ariza, M., Quesada Armenteros, A., & Estepa Castro, A. (2021). Promoting critical thinking through mathematics and science teacher education: the case of argumentation and graph interpretation about climate change. *European Journal of Teacher Education, 47*(1), 41–59. https://doi.org/10.1080/02619768.2021.1961736

Salmon, G. (2019). May the fourth be with you: Creating Education 4.0. *Journal of Learning for Development, 6*(2), 95–115. https://doi.org/10.56059/jl4d.v6i2.352

Savitz, R. S. (2021). Debate as a tool to develop disciplinary practices and disciplinary literacy in teacher education. *Teaching and Teacher Education, 101*, 103329. https://doi.org/10.1016/j.tate.2021.103341

Schwab, K. (2017). *The fourth industrial revolution*. Crown Publishing Group.

Shernoff, D. J., Sinha, S., Bressler, D. M., & Ginsburg, L. (2017). Assessing teacher education and professional development needs for the implementation of integrated approaches to STEM education. *International Journal of STEM Education, 4*(1), 13. https://doi.org/10.1186/s40594-017-0068-1

Sun, F., Tian, P., Sun, D., Fan, Y., & Yang, Y. (2024). Pre-service teachers' inclination to integrate AI into STEM education: Analysis of influencing factors. *British Journal of Educational Technology, 55*, 2574–2596. https://doi.org/10.1111/bjet.13469

Swan, M., & Pead, D. (2008). *Professional development resources*. Bowland Math Key Stage three, Bowland Trust/Department for Children, Schools and Families. http://www.bowlandmaths.org.uk

Tena, E., & Couso, D. (2023). Designing research questions with 10–12-year-old students. *Enseñanza de las Ciencias, 41*(1), 101–123. https://doi.org/10.5565/rev/ensciencias.5573

The Design-Based Research Collective. (2003). Design-based research: An emerging paradigm for educational inquiry. *Educational Researcher, 32*(1), 5–8. https://doi.org/10.3102/0013189x032001005

Tinoca, L., Piedade, J., Santos, S., Pedro, A., & Gomes, S. (2022). Design-based research in the educational field: A systematic literature review. *Education Sciences, 12*(6), 410. https://doi.org/10.3390/educsci12060410

Tram, N. H. M. (2024). Unveiling the drivers of AI integration among language Teachers: Integrating UTAUT and AI-TPACK. *Computers in the Schools, 42*(2), 100–120. https://doi.org/10.1080/07380569.2024.2441155

Tunjera, N., & Chigona, A. (2023). Investigating effective ways to use artificial intelligence in teacher education. In *Proceedings of the 22nd European Conference on e-Learning* (pp. 331–340). Academic Conferences International. https://doi.org/10.34190/ecel.22.1.1625

Urbani, J. M., Roshandel, S., Michaels, R., & Truesdell, E. (2017). Developing and modeling 21st-century skills with preservice teachers. *Teacher Education Quarterly, 44*(4), 27–50.

Velander, J., Otero, N., & Milrad, M. (2024). What is critical (about) AI literacy? Exploring conceptualizations present in AI literacy discourse. In A. Buch, Y. Lindberg, & T. Cerratto Pargman (Eds.), *Framing futures in postdigital education*. Springer. https://doi.org/10.1007/978-3-031-58622-4_8

Webster, R., Griffin, S. M., Chatterjee, I., Kuriakose, A., & Nguyen, H. (2024). A co-creation approach to using generative AI for enhanced learning gain. *Journal of Learning Development in Higher Education, 32*, 1434. https://doi.org/10.47408/jldhe.vi32.1434

Wu, X., Yang, Y., Zhou, X., Xia, Y., & Liao, H. (2024). A meta-analysis of interdisciplinary teaching abilities among elementary and secondary school STEM teachers. *International Journal of STEM Education, 11*(38). https://doi.org/10.1186/s40594-024-00500-8

Yang, S., & Appleget, C. (2024). An exploration of pre-service teachers' perceptions of generative AI: Applying the technological acceptance model. *Journal of Digital Learning in Teacher Education, 40*(3), 145–158. https://doi.org/10.1080/21532974.2024.2367573

Zaugg, T. (2024). *Embracing innovation in education: Investigating the use and impact of artificial intelligent assistants among pre-service teachers* (Preprint). https://doi.org/10.21203/rs.3.rs-5032655/v1

Zhai, C., Wibowo, S., & Li, L. D. (2024). The effects of over-reliance on AI dialog systems on students' cognitive abilities: a systematic review. *Smart Learning Environments, 11*, 28. https://doi.org/10.1186/s40561-024-00316-7

Zhang, X., Zhang, P., Shen, Y., Liu, M., Wang, Q., Gašević, D., & Fan, Y. (2024). A systematic literature review of empirical research on applying generative artificial intelligence in education. *Frontiers of Digital Education, 1*(3), 223–245. https://doi.org/10.1007/s44366-024-0028-5

CHAPTER 14

The Future of STEM: Socioecological STEM Education

Elizabeth Suazo-Flores, Christine McGrail, and Amber Adgerson

INTRODUCTION

The Science, Technology, Engineering, and Mathematics (STEM) Education movement started in the United States to enhance quantitative and computational literacy, motivate students to pursue STEM careers, and increase the United States' ability to compete globally (NRC & NAE, 2009). The STEM education movement has been embraced globally, with different emphases across countries (e.g., Costello et al., 2020; Ritz & Fan, 2015), influencing the teaching and learning of STEM disciplines (Sneider & Purzer, 2014). New standards, such as *A Framework for K–12 Science Education*, highlighted the need for science and engineering to be viewed as complementary, a combination of content and practice, and essential to addressing challenges faced by humans around the globe (Cunningham et al., 2020; NRC, 2012). Despite the focus on relevant STEM content to increase learners' participation in STEM, the context of

E. Suazo-Flores (✉) • C. McGrail • A. Adgerson
University of North Dakota, Grand Forks, ND, USA
e-mail: Elizabeth.suazo@und.edu; christine.mcgrail@und.edu; amber.adgerson@und.edu

R. Elmas, M. Akarsu (eds.), *STEM Education*, Maarif Global Education Series,
https://doi.org/10.1007/978-981-95-9023-0_14

STEM activities remains detached from learners' cultural backgrounds and family practices (i.e., funds of knowledge [FoK], Moll et al., 1992). Therefore, we propose considering learners' backgrounds and family practices, including their community contexts, when designing STEM activities and encouraging collaboration among learners rather than competition to increase participation in STEM (Cunningham et al., 2020). Building from our work with two different groups of children and existing literature (Calabrese Barton & Tan, 2018; Brown et al., 2018; Civil, 2016; Lee & Grapin, 2022), we argue that the future of STEM education is in shifting the focus from seeing the STEM movement as a technical issue (Stanic & Kilpatrick, 1992), that is, the lack of STEM practitioners, to focusing on curriculum reform that positions students' FoK as the starting point to work collaboratively on local or global challenges affecting their communities.

STEM education would then become a space for learners to understand and sustain themselves, others, and their local and global contexts (Davis et al., 2019). This would also require STEM education researchers to operate and interact in ways that acknowledge and build from learners' and their communities' FoK. In the chapter, we share an example of this type of praxis through our reflections and actions working with preschool and elementary-aged students in a STEM activity. The socioecological framework (Louie & Zhan, 2020; Vossoughi et al., 2023) informed our interactions with learners and educators in planning and teaching the activity. Socioecological approaches have the potential to broaden participation in STEM (Bosio & Waghid, 2023; Brown et al., 2018) by decentering the individual to one that is eco-centric and focuses on relationships among people and between people and nature. We describe how we interpreted the socioecological framework from Louie and Zhan (2020) in our planning and teaching of a composting activity in a Midwestern state of the United States. We hope to offer readers one example of how the socioecological framework could be used as a tool for STEM education researchers to create STEM activities inspired by learners' FoK.

Our Conceptualization of STEM

STEM education highlights skills such as creativity, curiosity, self-directed learning, reasoning, critical thinking, and problem-solving (Jang, 2016; Li, 2018; NRC, 2012) through students' active engagement and self-direction, which support their conceptual understandings of STEM disciplines (Eddy et al., 2015; Freeman et al., 2014; Grohs et al., 2018; Menon et al., 2024). STEM is an interdisciplinary construct integrating major concepts and practices from STEM disciplines (Li, 2018; Navy & Kaya, 2020). However, the conceptualization and operationalization of STEM have been varied and have yet to reach a consensus (Costello et al., 2020; Moore et al., 2020). One aspect of STEM where there is consensus is that engineering and real-world contexts are central to most definitions of integrated STEM (Bryan & Guzey, 2020; Roehrig et al., 2021). We build on Bryan and Guzey (2020) to characterize integrated STEM activities by having learning goals defined by content and practices from one or more STEM disciplines, authentic problems to be solved through collaboration, communication, and teamwork, and using engineering design practices. We define authentic problems as those that are informed by learners' backgrounds or lived experiences. We consider the practices of science and engineering to work in tandem as young learners engage in meaningful STEM learning experiences that require students to simultaneously ask questions, plan and carry out investigations, interpret data, develop and use models, use mathematics and computational thinking, obtain information, construct explanations, engage in argument from evidence, and define problems.

Efforts to Increase Participation

In 2014, policies worldwide began to focus on integrated STEM rather than separate approaches to the individual STEM disciplines. The accompanying policy shift was in response to a burgeoning sense of urgency to supply workers in the global STEM workforce (English, 2016). With the release of the National Academies Press *Toward Integrated STEM: Developing a Research Agenda* (*Integrated STEM,* NRC, 2014), efforts were underway to characterize existing practices in STEM and their impact on student achievement and motivation with the intent to increase the number of students interested in pursuing STEM careers.

As science, engineering, and STEM are intertwined in classrooms, research emerged on design as a pedagogical method. STEM learning experiences that typically employ design-based approaches are enjoyable for students and help increase student engagement (Moore et al., 2020; Struyf et al., 2019; Sulaeman et al., 2021). Further, design-based approaches are successful in promoting learning gains in science content areas such as physics (Fortus et al., 2004). While STEM integration has taken many different forms, a shared theme has been to focus on real-world problems or authentic activities. However, while there have been gains in discipline content and engagement, it has been found that the focus of design-based science and STEM is more on construction and considerations of surface-level constraints, such as time and materials, rather than on science reasoning or societal and environmental implications (NRC, 2012; Roehrig et al., 2021; Silk et al., 2009). Additional criticism remains as the design-based approach is typically implemented in classrooms through technical problems, with stereotypical and gender-specific projects such as race cars and egg drops that have little connection to students' lives (Purzer et al., 2014). For example, some female (e.g., Murphy et al., 2019), immigrant, or socioeconomically disadvantaged students in a Eurocentric country (Costello et al., 2020) may not feel engaged assembling model race cars or working with food (e.g., eggs) that they might lack access to. The scarcity of opportunities to engage with STEM in ways that feel meaningful and connected to learners' lived experiences, especially for learners from underrepresented backgrounds, remains an obstacle to STEM participation (Calabrese Barton & Tan, 2018).

What STEM Education Has Done and Needs to Do

Science and engineering education have already made great strides to move teaching and learning away from content transfer to active learning through the interplay of content and application. Concurrent with science education reform, engineering education reform efforts highlighted the need to shift from the paradigm of engineering as purely technical toward engineering curricula that would emphasize the roles of authentic tasks, critical thinking, and community involvement (Trbusic, 2014). We embrace this paradigm shift while adding crucial considerations for STEM curricula that support all learners, including marginalized and underrepresented students in STEM who might not be engaged or feel seen in traditional school curricula.

When we use the words marginalized and underrepresented students, we refer to students from populations with enduring "educational debt" (Ladson-Billings, 2006) and "opportunity gaps" (Darling-Hammond, 2007) whose assets have been overlooked because of "deficit-oriented" lenses (Harper, 2010). We join Civil (2016) in positioning STEM education curricula as opportunities STEM educators can create to enhance authentic and eco-centric learning for all students, including marginalized and underrepresented learners. One way STEM educators can identify or create curricula is by building from their learners' ([FoK], Moll et al., 2005).

FoK are assets and bodies of knowledge acquired from the lived experiences of families and students that have allowed them to function in society (Moll et al., 2005). FoK also serves as a filter for learners to identify what is valuable, interesting, or appealing in an activity. Presenting learners with a STEM activity that builds from their FoK has the potential to connect and engage with them and elicit their thinking (e.g., Civil, 2016). Using this approach, STEM education researchers spend time learning about learners' and their communities' FoK (bottom-up approach) to then design STEM activities informed by those FoK. This would be different from offering a STEM activity that might be far from learners' and their communities' FoK (top-down approach). Building from learners' and their communities' FoK is way to start STEM explorations that could later lead to enhancing learners' STEM knowledge and practices. Therefore, we join others in proposing learners' FoK as a legitimate way of engaging in STEM activities (Civil, 2016; Calabrese Barton & Tan, 2018).

Engaging in the work of addressing and designing solutions for societal challenges is not neutral; it is fraught with disparity, as designed solutions create disproportionate impacts on people and places (Lee & Grapin, 2022; Trbusic, 2014). Consequently, there is more work to be done to make STEM relevant to students' lives and "centered on the interests, strengths, and needs of diverse student groups, units and activities need to reflect the experiences and realities of diverse student groups with a focus on minoritized students groups (e.g., race and ethnicity, socioeconomic status, multilingual learners" (Lee & Grapin, 2022, p.7). We join Civil (2016) to propose that the future of STEM education could focus on engaging learners in developing an understanding of phenomena in the natural world around them, designing solutions for problems affecting their communities, and using their critical thinking and social responsibility. This work begins at the classroom (local level), for elementary children, wherein integrating STEM with their FoK will augment children's problem-solving skills and the resilience they need to face both local and global ecological challenges.

The Future of STEM Education: Using a Socioecological Framework

In the United States and globally, children's connectedness to nature is diminishing due to the increase in connectivity to technological devices and less availability to engage in embodied activities with others (Haidt, 2024). Children are less used to playing outside and wandering in nature. This has prompted them to have views of nature that are problematic for human survival. For instance, children might think that soil or earthworms are not useful or are disgusting (e.g., Gülay Ogelman, 2012; Pramling, 2010).

On the other hand, STEM education researchers are turning their attention to the role of the living world (Barwell et al., 2022; Vossoughi et al., 2023). The living world comprises trees, plants, soil, insects, birds, and earthworms (Barwell et al., 2022). For example, researchers in mathematics education are calling to replace mathematics in STEM with an "M" referring to "multispecies flourishing" (Khan, 2020; Khan & Armstrong, 2019) and to start educating learners so they become a generation that does not consider the living world as a fixed background (Coles, 2023) but instead essential for its existence and the existence of human beings (e.g., Khan, 2020). Thus, STEM educators are challenged to prepare future STEM practitioners to be stewards of their environments and the socioecological framework (Bronfenbrenner, 1977; Louie & Zhan, 2020; Vossoughi et al., 2023) could be used as a mechanism to do so.

The socioecological framework invites STEM educators to de-center from individuality and learning to center on relationships: relationships among people and between people and the living world (Vossoughi et al., 2023). When following the socioecological framework, researchers and educators focus on building relationships with other human beings and between human beings and the living world. We are inspired by Louie's and Zhan's (2020) layers of socioecological framework to inform our ways of being, interacting, and operating (i.e., Suazo-Flores et al., 2024) with learners and educators. Below, we describe how we have interpreted Louie's and Zhan's (2020) layers of socioecological framework and the work of Vossoughi et al. (2023) with two groups of school learners. Louie and Zhan (2020) built on Bronfenbrenner (1977) to describe five fluid and permeable layers that are bidirectional and inform one another (see Fig. 14.1). Below, we describe the layers of the socioecological framework

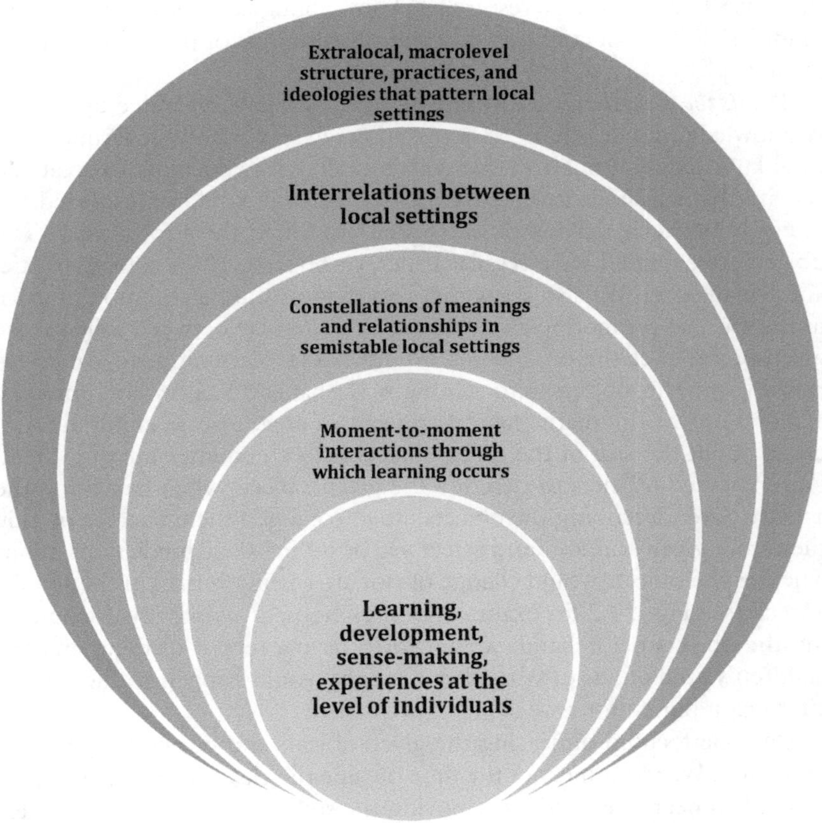

Fig. 14.1 Layers of socioecological framework

and provide evidence of how we interpreted them in our work planning and teaching a STEM activity for preschoolers and elementary learners.

Layer 1: Learning, Development, Sense-Making, Experiences at the Level of Individuals

STEM education researchers plan activities that are informed by learners' and educators' FoK and have the potential to elicit STEM practices and knowledge. This work is done before meeting the learners and is informed by what STEM education researchers know about learners' and their communities' FoK from conversations and observations of the places learners inhabit.

The *Is the Soil Alive?* STEM activity took two days, separated by a week. Acknowledging our privileged access to green spaces and the farming cultural knowledge in our area, we collaborated with educators to create an activity that turned learners' attention to the soil. Children explored the interplay between soil and objects when we buried them for a week. The objects we explored were regular paper, cardboard, 100% cotton, plastic, and banana peel. We also knew that recycling was not a common practice in the area, so promoting awareness of the interplay between soil and the given objects was the original STEM educators' learning goal. We intentionally cut the objects into geometrical shapes to allow for potential explorations of the mathematical concept of area, in case children wondered about the size of the given objects before and after burying them. We first asked learners to describe the given objects when launching the activity. After describing the objects, we encouraged them to trace or glue them to a given handout and select yes or no for their predictions about whether the objects would change or not after being buried in the soil for a week (see Fig. 14.2 for examples of preschoolers' work). By documenting the objects beforehand, we aimed to have a record of them and the children's predictions of what they thought would happen to the objects after being buried for a week.

Day one focused on tracing the given objects and burying the objects in the soil. We spent most of the time outside the classrooms, where a plot of soil had been prepared by removing the sod. We provided the learners with cameras they could use to photograph and video the things that interested them. Day two focused on exploring a STEM topic that learners showed interest in on day one, digging up the objects, and having a whole class discussion about what happened to the objects. We let the learners' interests guide the following days with us. The nine-year-old learners were very interested in earthworms. We planned two extra days where we explored earthworms by bringing containers with earthworms and reading books about them. Learners wanted to repeat the activity using the containers and the earthworms to explore the interaction between the earthworms and the objects.

We built from Bryan's and Guzey's (2020) definition of integrated STEM to position *Is the Soil Alive?* as a STEM activity. There are at least two STEM subjects involved in the activity: science and mathematics. We

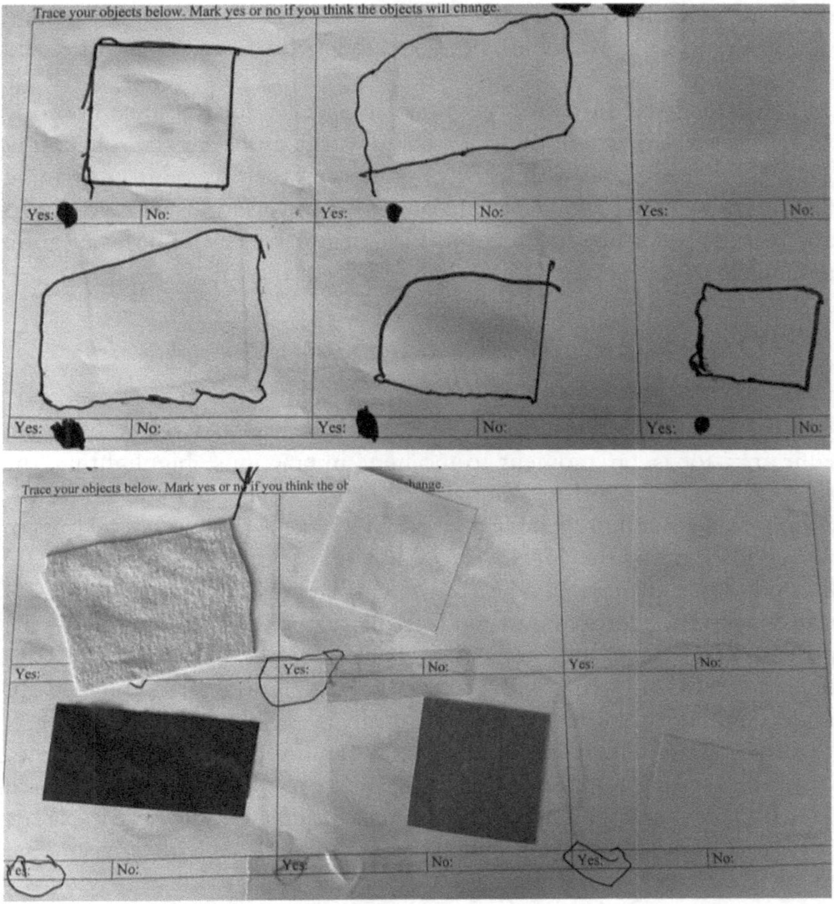

Fig. 14.2 Objects traced or glued by the preschoolers

anticipated having conversations with the learners about soil, composting, and insects. Regarding mathematics, we expected learners to wonder about shapes, arrays, counting, and using their early multiplication knowledge. We planned the tasks to be solved through collaboration, communication, and teamwork, and engaging learners in strength-based, holistic STEM learning. The activity was authentic to the learners as we lived it together. Our design process for the STEM activity was iterative as we listened to learners' conversations, planned the follow-up activities from our understanding of their interests, and allowed learners' understanding of the engineering problem to be solved to emerge from the activities and conversations.

The *Is the Soil Alive?* activity is socioecological as it is about composting and turns learners' attention to the living world, especially the role of soil. The activity resonated with learners as most of them came from farming families or were used to of playing outside. Conversations with educators and learners informed its creation. We saw our role as STEM education researchers as developing learners' awareness of the soil.

Layer 2: Moment-to-Moment Interactions Through Which Learning Occurs

When interacting with learners, socioecological STEM researchers and educators focus on moment-to-moment interactions, highlighting the role of belonging and emotional well-being in learners and educators (Louie & Zhan, 2020). The first group of learners was a preschool class of ten children from the University Children's Learning Center (UCLC) located at our institution. Most of the enrolled children were relatives of university staff, professors, or graduate students. As we were new to the preschoolers, we visited them before our first STEM activity and conducted a drawing activity with them. This visit allowed us to enhance our understanding of the contexts of the students and their FoK in ways that "evolved organically, always relating to authentic contexts, guided by student interest and community" (Hogg, 2011, p. 672). This also allowed preschoolers opportunities to play and talk with us. On the first day of the *Is the Soil Alive?* activity, we invited them to describe the objects, emphasizing their color and shapes. We then asked them to trace or glue the objects on paper and write down their names. Preschoolers enjoyed talking about the materials. They recognized objects such as paper, construction paper, and fabric. They did not know a word to describe the given red piece of plastic; they did recognize they could not rip it. We then moved outside to dig in the soil to make holes where we buried the objects. All the preschoolers wanted a turn with the shovels. Human-environment connections were created as the learners' discourse became more "eco-centric" and focused on their heightened awareness of commonalities between themselves and non-human living beings. The preschoolers were thrilled to discover earthworms and insect larvae when digging holes to bury the objects. They picked them up, showed them to each other,

Fig. 14.3 To the left, preschoolers' popsicle sticks, and to the right, a nine-year-old's stick

exclaiming, "It's alive!", and had impromptu discussions with the educators about what it means to be alive. These connections further emphasized "how all living things, including humans, soil, woodlands, wildlife, and vegetation are interlinked" (Bosio & Waghid, 2023, p. 7). Learners also took pictures of us, themselves, and the insects and earthworms they found. As stated, the objects were secured to the children's prediction charts, so we supplied an additional set of two of each object type for the whole class to bury in the soil. They placed large craft sticks as markers to show where the objects were buried (see Fig. 14.3). With the help of educators, preschoolers spelled the words representing each object type on the large craft stick.

The second group of learners was a class of 15 nine-year-old children from a rural school located 45 minutes from the authors' university. Most of the children's families had farms they cultivated and profited from. As we were new to the nine-year-old learners, we introduced ourselves and led learners to ask us questions to get to know us better. We then launched the activity by asking them to share their hypothesis about what would happen to objects such as regular paper, cardboard, plastic, 100% cotton, and banana peels after being buried for a week. The kids shared their ideas, and we then went outside to plant the objects. While walking with the learners, they had little chats with us about their friends and daily

Fig. 14.4 Nine-year old students working with the soil

activities. We brought two large shovels and used a space that one of the staff members from the school had cleared for us by the school parking lot. Every learner had five objects to bury. Before planting the objects, we asked learners how we should proceed so we could find the objects after a week. They suggested placing the objects in rows and columns. Some of them wanted to know how many objects in total were planted; we suggested exploring the answer to that question in our following meeting. Then, learners started making rows in the soil (see Fig. 14.4). As with the preschoolers, most of the learners wanted to take a turn with the shovels. We were there to listen to their conversations, make sure nobody was being hurt, and whoever wanted to have a turn could have it.

In the following paragraphs, we describe the next two layers of the socioecological framework and provide examples of how we interpreted them in our work. As documented elsewhere (Louie & Zhan, 2020), the layers of the socioecological framework are interconnected and bidirectional. It was difficult for us to provide evidence of the next two layers in a linear form and independent ways.

Layer 3: Constellations of Meanings and Relationships in Semistable Local Settings

In their interactions with learners or when planning STEM activities, STEM education researchers acknowledge that learners' meanings and ways of interacting around the STEM activity result from the accumulation of moment-to-moment interactions in semistable local settings (Louie

& Zhan, 2020). Preschoolers and nine-year-old learners had experiences gardening, farming, recognizing colors, and shapes outside of school settings, and they engaged in the *Is the Soil Alive?* activity using those experiences as lenses. STEM researchers account for the influence of local settings surrounding learners and teachers. How are those local settings informing the learners' and educators' meanings?

Layer 4: Interrelations Between Local Settings

STEM education researchers incorporate opportunities for learners and educators to identify interrelations of local settings such as classrooms, homes, and communal environments (i.e., parks, playgrounds, libraries) that inform learners' and educators' interactions (Louie & Zhan, 2020). Most of the preschoolers referred to the soil as "yucky" or "dirty." This form of referring to the soil is well documented (e.g., Gülay Ogelman, 2012; Pramling, 2010). Some families do not like their children to play with soil, so children probably hear expressions such as "you are dirty" or "the soil is yucky." Some of the nine-year-old learners, despite living on farms, described how their families did not let them play with the soil.

Evidence of how learners' conversations inform meanings and relationships built over time is evidenced below. We wrote more details about preschoolers' experiences with the activity elsewhere (Suazo-Flores et al., 2025b). Below, we share a piece of conversation between preschoolers and one of the authors in the *Is the Soil Alive?* activity.

Researcher: How is this piece of paper compared to this other one? Is it the same or is it different?
Kids: It is different.
Researcher: How is it different?
Child 1: It has orange!
Child 2: It is dirty!
Researcher: How about the fabric?
Child 1: It has a kind of pink.
Child 2: It is dirty.
Child 1: It has pink or purple. Purple. My grandma would know because she likes purple.

One of the main learning goals in preschool is to teach the names of colors and shapes and how objects are similar or different. In the transcripts, we can see how preschoolers are using their knowledge of colors and distinguishing how objects are similar or different. We can also see evidence of how their interactions with family members are brought to bear in the *Is the Soil Alive?* activity. Child 1 relates her knowledge of purple to her grandmother by saying that her grandmother likes purple.

Nine-year-old learners wondered about how many objects they buried given that every one of them had five pieces of different objects. We see this as evidence of learners using their mathematics knowledge to understand their world. We wrote about this exploration elsewhere (Suazo-Flores et al., 2025). These learners also shared how they saw themselves as farmers while digging in the soil; a learner said: "I feel like a real farmer," which is evidence of how they were interrelating their STEM knowledge and practices of local settings.

Layer 5: Extralocal, Macrolevel Structure, Practices, and Ideologies That Pattern Local Settings

STEM education researchers focus on turning learners' attention to macro-level structures, practices, and ideologies connected to the STEM activities they are engaged in (Louie & Zhan, 2020). We see this layer of the socioecological framework aiming at developing learners' critical consciousness. Critical consciousness, as first posited by Paulo Freire (1970), allows students to read and understand the world using "intellectual tools" (Ladson-Billings, 2021, p. 72) that allow them to recognize and respond to socio-political concerns within their everyday lives. In STEM classrooms, students' critical consciousness may be enhanced through the introduction of real-world issues such as heat islands, deforestation, decreasing biodiversity, climate displacement, or any other local or global concepts relevant to the experiences and futures of students (Duncan et al., 2023; Jones & Taylor, 2022).

Our preschoolers and elementary students started wondering about insects and earthworms they found as they buried and unburied the objects in the two days we worked together (See Figs. 14.5 and 14.6). These moment-to-moment interactions allowed us to engage in reflexive dialogue, which encourages learners to critically "examine their goals and values as seen through the lens of the topics they have discussed with their peers" such as "relationships among people, and between people and nature" (Bosio & Waghid, 2023, p. 4). As a result of this dialogue, we started Day two of the activity by reading a book about soil to the learners.

Fig. 14.5 Preschoolers noticing insects and earthworms during the activity

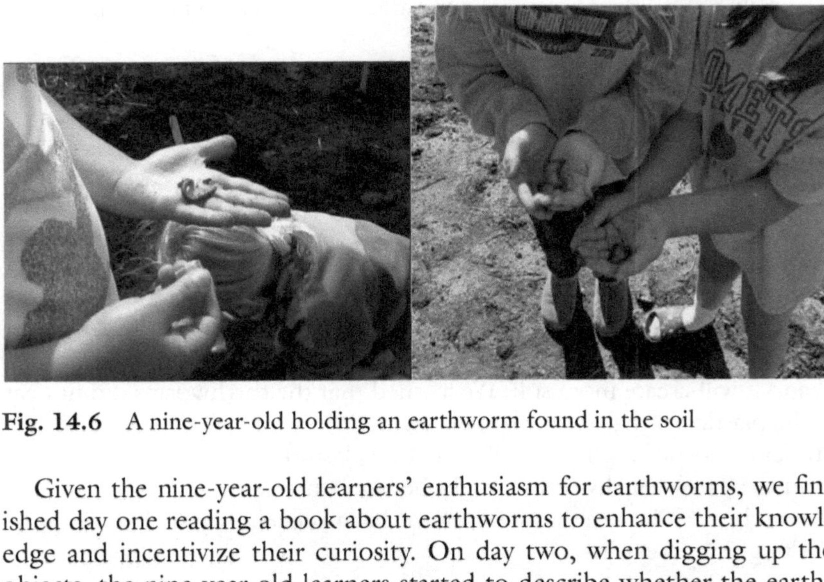

Fig. 14.6 A nine-year-old holding an earthworm found in the soil

Given the nine-year-old learners' enthusiasm for earthworms, we finished day one reading a book about earthworms to enhance their knowledge and incentivize their curiosity. On day two, when digging up the objects, the nine-year-old learners started to describe whether the earthworms they found were young or old using what they had learned from the book we read the week before. After day two, we asked the nine-year-old learners whether they would like to see more earthworms. They liked the idea and wanted to replicate the exploration to study the role of the earthworms and the objects in the soil. We planned for two extra days. On the first day, we brought two large containers: one with earthworms and another without earthworms, and we planted the objects in both. The second day happened a week later, and we dug up the objects to have a conversation about the role of earthworms, the objects, and the soil.

Below is a transcript of our conversations with the nine-year-old learners evidencing their emergent conscious awareness.

Researcher: We were able to find two pieces of plastic. Do you notice any changes?
Students: No, they are exactly the same thing.
Researcher: Should we put plastic in the soil? Is that a good thing to do?
Students: No.
Student 1: It is pollution.
Student 2: It will not dissolve.
Researcher: Exactly, so that is something we proved. What happened to the banana peel?
Student 1: We didn't find it.
Researchers: And we searched really hard.
Student 2: They [earthworms] ate it.
Student 3: They composted it.
Researcher: There is probably more compost in the bin [the one with the earthworms] than in the other one [the one without the earthworms].
Student 4: In the other bin, we found it [banana peel] and it is all black.
Researcher: So, is that a good thing [banana peel] to put in the soil?
Students: Yes.
Researcher: Any peel you have, put it on the soil. It is food for the worms, and it will create more soil. We learned that the earthworms did not eat the plastic, so that means something.
Student: Worms might take nibbles at it [plastic].
Student: They [earthworms] might be like, what is it?
Researcher: And when they ran into the banana peel, they are like, yummy! I'm going to eat it.

Sociological STEM Education and AI

At the time of writing this chapter, there were a wide range of access to non-regulated AI resources, and research studying their use is producing strong evidence that teachers play a key role in making decisions to adapt information given by AI (Díaz & Nussbaum, 2024; Gary, 2019; Muhisn et al., 2019). This is especially true when STEM educators plan and teach socioecological STEM activities with their students' FoK in mind. Educators and researchers more than ever need to understand well the contexts surrounding them and their learners' or participants' backgrounds. We end this chapter with a discussion sharing suggestions for how STEM educators and researchers could use AI in their planning and teaching of sociological STEM activities.

As STEM researchers or educators might not be familiar with local or social issues in their contexts or ideas for socioecological STEM activities, we consider AI useful as a tool to explore possibilities. In writing this chapter, we explored two AI tools: Perplexity and ChatGPT to search for information that would help to plan socioecological STEM activities. We found the AI suggestions useful in giving us some orientation for topics of activities to try with appealing contexts for learners. Using AI we could get ideas and activities plans that were generic to standardized learners (or one-size fit-all learners). We still consider it necessary to consult with teachers or local community members to confirm whether the AI-generated ideas resonate with them and their students. STEM educators need to contextualize the information given by AI resources to their students' or participants' FoK. As we have referred in this chapter, FoK are practices families and local communities have used to operate and navigate society (Moll et al., 2005) and AI, as a database continuously changing, does not have access to students' and communities' FoK. Thus, STEM educators and researchers need to adapt the information generated with AI tools.

Discussion

In this chapter, we shared how we integrated STEM knowledge and practices with the socioecological. The *Is the Soil Alive?* STEM activity resulted from being in relationships with preschool and elementary educators, building from our shared interest in imagining a new future for STEM education, and acknowledging the learners' and their communities' FoK. The socioecological framework (Louie & Zhan, 2020) allowed us to embrace learners' and educators' FoK, and carefully attend to their actions and expressed interests to plan the following emerging STEM activities. This way of operating is different from traditional STEM activities where challenges are created before meeting the learners. With a few examples of how socioecological STEM could look in classroom settings, we planned the activity, embracing uncertainty but being true to our relationships. We did not know what could happen, what the children would say, and whether the objects were going to stay in the soil after a week. Thus, conducting STEM activities within the socioecological requires flexibility, time, and awareness of learners' FoK.

We highlight the *Is the Soil Alive?* activity as an example of a socioecological STEM activity because we saw firsthand how it started turning learners' attention to the living world (Barwell et al., 2022). Students engaged with soil and the living beings living in it by observing and describing the influence of the soil on the given objects. We did not anticipate children's interest in the color changes associated with the objects or in touching and looking closely at the insects and earthworms we encountered. While traditional STEM education training does not prepare educators to know how to follow learners' interests or build from their FoK first, we see student discovery as fertile ground for development. In our example, we used our mathematics and science education backgrounds to plan for potential discussions around describing geometric shapes and sizes, living vs. non-living, and insect larvae vs. earthworms. We allowed students to have a discourse that surfaced their awareness of the interdependent nature of living soil, earthworms, and buried objects. This furthered students' knowledge of STEM knowledge and practices (e.g., Suazo-Flores et al., 2025) and critical consciousness (Ladson-Billings, 2021) as they grappled with their role as humans in the interconnectedness of other living beings and the environment. When they determined that plastic in the soil must be unhealthy for earthworms, they derived their understanding of the societal problem that could be addressed through engineering design. Our socioecological STEM activity yielded multiple directions for student-driven learning in both science and mathematics and sparked what we identified as early steps in critical consciousness. When students discussed the implications for the well-being of earthworms when plastic was put in the soil, we knew our activity brought learners to see earthworms and the soil as living beings that were not static (Coles, 2023).

We advocate for the future of STEM education to include socioecological curriculum development in which STEM educators, with school educators. Create spaces for exploration in coordination with the living world (Barwell et al., 2022). AI tools could help STEM educators and researchers with ideas of what and how to plan socioecological curricula. The sociological STEM activities had the characteristics to build from the learners' and their community's FoK instead of only using top-down or standardized STEM challenges. In this way, we hope to cast a wider net of learners who feel enfranchised in STEM and empowered to use STEM knowledge and practices to be stewards of their local and global environments.

References

Barwell, R., Boylan, M., & Coles, A. (2022). Mathematics education and the living world: A dialogic response to a global crisis. *The Journal of Mathematical Behavior, 68*, 101013. https://doi.org/10.1016/j.jmathb.2022.101013

Bosio, E., & Waghid, Y. (2023). Cultivating students' critical consciousness through global citizenship education: Six pedagogical priorities. *Prospects*, 1–12. https://doi.org/10.1007/s11125-023-09652-x

Bronfenbrenner, U. (1977). Toward an experimental ecology of human development. *American Psychologist, 32*(7), 513–531. https://doi.org/10.1037/0003-066X.32.7.513

Brown, E. R., Steinberg, M., Lu, Y., & Diekman, A. B. (2018). Is the lone scientist an American dream? Perceived communal opportunities in STEM offer a pathway to closing US–Asia gaps in interest and positivity. *Social Psychological and Personality Science, 9*(1), 11–23. https://doi.org/10.1177/1948550617703173

Bryan, L., & Guzey, S. (2020). K-12 STEM education: An overview of perspectives and considerations. *Hellenic Journal of STEM Education, 1*(1), 5–15. https://doi.org/10.51724/hjstemed.v1i1.5

Calabrese Barton, A., & Tan, E. (2018). A longitudinal study of equity-oriented stem-rich making among youth from historically marginalized communities. *American Educational Research Journal, 55*(4), 761–800. https://doi.org/10.3102/0002831218758668

Civil, M. (2016). STEM learning research through a funds of knowledge lens. *Cultural Studies of Science Education, 11*(1), 41–59. https://doi.org/10.1007/s11422-014-9648-2

Coles, A. (2023). Teaching in the new climate regime: Steps to a socio-ecology of mathematics education. In M. Ayalon, B. Koichu, R. Leikin, L. Ruber, & M. Taback (Eds.), *Proceedings of the 46th conference of the international group for the psychology of mathematics education (Vol. 1)* (pp. 17–33). University of Haifa. https://www.igpme.org/publications/current-proceedings/

Costello, E., Girme, P., McKnight, M., Brown, M., McLoughlin, E., & Kaya, S. (2020). *Government responses to the challenge of STEM education: Case studies from Europe. ATS STEM Report #2.* Dublin City University. https://doi.org/10.5281/zenodo.3673600

Cunningham, C., Lachapell, C., Brennan, R., Kelly, G., Tunis, C., & Gentry, C. (2020). The impact of engineering curriculum design principles on elementary students' engineering and science learning. *Journal of Research in Science Teaching, 57*(3), 423–453. https://doi.org/10.1002/tea.21601

Darling-Hammond, L. (2007). Race, inequality and educational accountability: The irony of "No Child Left Behind.". *Race, Ethnicity and Education, 10*(3), 245–260. https://doi.org/10.1080/13613320701503207

Davis, B., Francis, K., & Friesen, S. (2019). *STEM education by design: Opening horizons of possibility* (1st ed.). Routledge. https://doi.org/10.4324/9780429025143

Díaz, B., & Nussbaum, M. (2024). Artificial intelligence for teaching and learning in schools: The need for pedagogical intelligence. *Computers & Education, 217*, 105071. https://doi.org/10.1016/j.compedu.2024.105071

Duncan, R. G., Krishnamoorthy, R., Tan, E., Austin, T., Reichsman, F., Dabholkar, S., & Worsley, T. (2023). Designing science curricula that disrupt disciplinary boundaries towards sociopolitical change: A middle school life science unit. In *Proceedings of the 17th international conference of the learning sciences-ICLS 2023* (pp. 1138–1141). International Society of the Learning Sciences.

Eddy, S. L., Converse, M., & Wenderoth, M. P. (2015). PORTAAL: A classroom observation tool assessing evidence-based teaching practices for active learning in large science, technology, engineering, and mathematics classes. *CBE Life Sciences Education, 14*(2), ar23. https://doi.org/10.1187/cbe.14-06-0095

Engineering in K-12 Education Understanding the Status and Improving the Prospects National Academies Press Washington D.C.

English, L. (2016). STEM education K-12: Perspectives on integration. *International Journal of STEM Education, 3*(3), 1–8. https://doi.org/10.1186/s40594-016-0036-1

Fortus, D., Dersshimer, C., Krajcik, J., Marx, R., & Mamlok-Naaman, R. (2004). Design-based science and student learning. *Journal of Research in Science Teaching, 41*(10), 1081–1110. https://doi.org/10.1002/tea.20040

Freeman, B., Marginson, S., & Tytler, R. (2014). *Age of STEM: Educational policy and practice across the world in science, technology, engineering and mathematics.* Taylor & Francis. https://doi.org/10.4324/9781315767512

Freire, P. (1970). *Pedagogy of the oppressed* (3rd ed.). Bloomsbury Academic.

Gary, K. (2019). Pragmatic standards versus saturated phenomenon: Cultivating a love of learning. *Journal of Philosophy of Education, 53*(3), 477–490. https://doi.org/10.1111/1467-9752.12377

Grohs, J. R., Kirk, G. R., Soledad, M. M., & Knight, D. B. (2018). Assessing systems thinking: A tool to measure complex reasoning through ill-structured problems. *Thinking Skills and Creativity, 28*, 110–130.

Gülay Ogelman, H. (2012). Teaching preschool children about nature: A project to provide soil education for children in Turkey. *Early Childhood Education Journal, 40*, 177–185. https://doi.org/10.1007/s10643-012-0510-4

Haidt, J. (2024). *The anxious generation: How the great rewiring of childhood is causing an epidemic of mental illness.* Penguin Press.

Harper, S. R. (2010). An anti-deficit achievement framework for research on students of color in STEM. *New Directions for Institutional Research, 148*, 63–74. https://doi.org/10.1002/ir.362

Hogg, L. (2011). Funds of knowledge: An investigation of coherence within the literature. *Teaching and Teacher Education, 27*(3), 666–677. https://doi.org/10.1016/j.tate.2010.11.005

Jang, H. (2016). Identifying 21st century STEM competencies using workplace data. *Journal of Science Education and Technology, 25*(2), 284–301. https://doi.org/10.1007/s10956-015-9593-1

Jones, B. L., & Taylor, J. C. (2022). Within the walls of the classroom: How science teachers' instruction can develop students' sociopolitical consciousness. *Science Education, 106*(5), 1021–1031. https://doi.org/10.1002/sce.21727

Khan, S. (2020). After the M in STEM: Towards multispecies' flourishing. *Canadian Journal of Science, Mathematics and Technology Education, 20*, 230–245. https://doi.org/10.1007/s42330-020-00089-4

Khan, S., & Amstrong, A. (2019). Math-a-POLKA: "Mathematics as a place of loving kindness and.". *Journal of the Canadian Association for Curriculum Studies, 17*(1), 1–12. https://doi.org/10.25071/1916-4467.40441

Ladson-Billings, G. (2006). From the achievement gap to the education debt: Understanding achievement in U.S. schools. *Educational Researcher, 35*(7), 3–12. https://doi.org/10.3102/0013189x035007003

Ladson-Billings, G. (2021). I'm here for the hard re-set: Post pandemic pedagogy to preserve our culture. *Equity & Excellence in Education, 54*(1), 68–78. https://doi.org/10.1080/10665684.2020.1863883

Lee, O., & Grapin, S. E. (2022). The role of phenomena and problems in science and STEM education: Traditional, contemporary, and future approaches. *Journal of Research in Science Teaching, 59*(7), 1301–1309. https://doi.org/10.1002/tea.21776

Learning Science in Informal Environments People Places and Pursuits National Academies Press Washington D.C.

Li, Y. (2018). Promoting the development of interdisciplinary research in STEM education. *Journal for STEM Education Research, 1*, 1–6. https://doi.org/10.1007/s41979-018-0009-z

Louie, N., & Wan-Yue, Z. (2020). A socio-ecological framework for research in mathematics education. *Journal for Research in Mathematics Education, 53*(5), 365–371. https://doi.org/10.5951/jresematheduc-2022-0003

Menon, D., Cheng, D. S., & Ngugi, R. W. (2024). Investigating preservice secondary teachers' understanding and enactment of the standards-based practices during microteaching. *Journal for STEM Education Research, 7*(1), 63–95. https://doi.org/10.1007/s41979-023-00093-x

Moll, Amanti, Neff, & Gonzalez. (1992). Funds of knowledge for teaching: Using a qualitative approach to connect homes and classrooms. *Theory Into Practice, 31*(2), 132–141. https://doi.org/10.1080/00405849209543534

Moll, L., Amanti, C., Neff, D., & Gonzalez, N. (2005). Funds of knowledge for teaching: Using a qualitative approach to connect homes and classrooms. In

Funds of knowledge: Theorizing practices in households, communities, and classrooms (pp. 71–87). Lawrence Erlbaum Associates.

Moore, T. J., Johnston, A. C., & Glancy, A. W. (2020). STEM integration: A synthesis of conceptual frameworks and definitions. In C. C. Johnson, M. J. Mohr-Schroeder, T. J. Moore, & L. D. English (Eds.), *Handbook of research on STEM education* (pp. 3–16). Routledge. https://doi.org/10.4324/9780429021381-2

Muhisn, Z. A. A., Ahmad, M., Omar, M., & Muhisn, S. A. (2019). The impact of socialization on collaborative learning method in e-learning management system (eLMS). *International Journal of Emerging Technologies in Learning, 14*(20), 137–148. https://doi.org/10.3991/ijet.v14i20.10992

Murphy, S., MacDonald, A., Wang, C. A., & Danaia, L. (2019). Towards an understanding of STEM engagement: A review of the literature on motivation and academic emotions. *Canadian Journal of Science, Mathematics and Technology Education, 19*, 304–320. https://doi.org/10.1007/s42330-019-00054-w

National Research Council. (2012). *A framework for K-12 science education: Practices, crosscutting concepts, and core ideas.* The National Academies Press. https://doi.org/10.17226/13165

National Research Council. (2014). *STEM integration in K-12 education: Status, prospects, and an agenda for research.* The National Academies Press. https://doi.org/10.17226/18612

Navy, S. L., & Kaya, F. (2020). PBL as a pedagogical approach for integrated STEM: Evidence from prospective teachers. *School Science and Mathematics, 120*(5), 285–296. https://doi.org/10.1111/ssm.12408

Pramling, N. (2010). Unearthing metaphors: Figurativeness in teacher-child talk about soil and related matters. *Early Childhood Education Journal, 38*, 57–64. https://doi.org/10.1007/s10643-010-0386-0

Purzer, S., Cardella, M., & Strobel, J. (2014). *Engineering in pre-college settings: Synthesizing research, policy, and practices.* Purdue University Press.

Ritz, J. M., & Fan, S. C. (2015). STEM and technology education: International state-of-the-art. *International Journal of Technology and Design Education, 25*, 429–451. https://doi.org/10.1007/s10798-014-9290-z

Roehrig, G. H., Dare, E. A., Ellis, J. A., & Ring-Whalen, E. (2021). Beyond the basics: A detailed conceptual framework of integrated STEM. *Disciplinary and Interdisciplinary Science Education Research, 3*(11). https://doi.org/10.1186/s43031-021-00041-y

Silk, E., Schunn, C., & Cary, M. (2009). The impact of an engineering design curriculum on science reasoning in an urban setting. *Journal of Science Education and Technology, 18*(3), 209–223. https://doi.org/10.1007/s10956-009-9144-8

Sneider, C., & Purzer, S. (2014). The rising profile of STEM literacy through national standards and assessments. In S. Purzer, M. Cardella, & J. Strobel (Eds.), *Engineering in pre-college settings: Synthesizing research, policy, and practices* (pp. 3–19). Purdue University Press.

Stanic, G. M., & Kilpatrick, J. (1992). Mathematics curriculum reform in the United States: A historical perspective. *International Journal of Educational Research, 17*(5), 407–417.

Struyf, A., De Loof, H., Boeve-de Pauw, J., & Van Petegem, P. (2019). Students' engagement in different STEM learning environments: Integrated STEM education as promising practice? *International Journal of Science Education, 41*(10), 1387–1407. https://doi.org/10.1080/09500693.2019.1607983

Suazo-Flores, E., Walker, W. S., Kastberg, S. E., et al. (2024). Mathematics education researchers' practices in interdisciplinary collaborations: Embracing different ways of knowing. *Mathematics Education Research Journal.* https://doi.org/10.1007/s13394-024-00489-x

Suazo-Flores, E., Sherva, D., McGrail, C., & Adgerson, A. (2025). A number talk informed by the socio-ecological. *Mathematics Teacher: Learning and Teaching PreK-12, 118*(9), 702-707. https://doi.org/10.5951/MTLT.2024.0249

Suazo-Flores, E., McGrail, C., & Adgerson, A. (2025b). A socio-ecological STEM task: Re-attaching preschoolers to nature. In le Roux K., Coles, A., Solares-Rojas, A., Bose, A., Vistro-Yu, C., Valero, P., Sinclair, N., Makramalla, M., Gutiérrez, R., Geiger, V., & Borba, M. (Eds.), *Proceedings of the 27th ICMI Study Conference, Mathematics education and the socioecological* (pp. 511–518). Quezon City, Philippines. https://icmistudy27.sciencesconf.org/data/pages/ICMI_Study_27_Proceedings_Feb_25_final.pdf

Sulaeman, N., Putra, P., Mineta, I., Hakamada, H., Takahashi, M., Ide, Y., & Kumano, Y. (2021). Exploring student engagement in STEM education through the engineering design process. *Jurnal Penelitian Dan Pembelajaran IPA, 7*(1), 1–16. https://doi.org/10.30870/jppi.v7i1.10455

Trbusic, H. (2014). Engineering in the community: Critical consciousness and engineering education. *Interdisciplinary Description of Complex Systems, 12*(2), 108–118.

Vossoughi, S., Marin, A., & Bang, M. (2023). Toward just and sustainable futures: Human learning and relationality within socio-ecological systems. *Review of Research in Education, 47*(1), 218–273. https://doi.org/10.3102/0091732X231223505